ANIMAL BIOTECHNOLOGY

ANIMAL BIOTECHNOLOGY

R. Sasidhara

Head, Department of Biotechnology
Dhanalakshmi Srinivasan College of
Arts and Science for Women
Perambalur

MJP PUBLISHERS
Chennai 600 005

MJP PUBLISHERS
A unit of Tamilnadu Book House
47, Nallathambi Street
Triplicane, Chennai 600 005

Preface

Biotechnology is the third wave in biological sciences and represents such an interface between basic and applied sciences, where gradual and subtle transformation of science into technology can be witnessed. Our understanding of living organisms has been immensely deepened by the contributions that recombinant DNA methods have made to basic scientific research. We can now isolate, analyse, and manipulate genes and ultimately alter the genetic make-up of living organisms.

Animal biotechnology, the art and science of producing genetically engineered animals, has advanced in the past few years. It is now possible to generate animals with useful novel properties for dairy, meat or fibre production, for environmental control of waste production; that produce useful products of biomedical purposes or other human consumption or that are nearly identical copies of animals chosen for useful traits, such as milk or meat production, fertility and the like.

I have been teaching college biotechnology course since 2000, and have long been aware that students need a single book that offers a reasonably comprehensive introduction to Animal Biotechnology. That is what motivated me to write this book. However, the nature of modern biotechnology presents the author of an introductory text with several challenges: the field is broad, diverse, and fast moving. Its breadth means that many areas must be addressed; its diversity means that no two teachers of what are nominally similar courses will necessarily address the same combination of topics; and its fast pace means that a book is never up to date even in its proof stage. The last challenge can be addressed by a timely schedule of revisions and website updates. The first two challenges are addressed by integrating information from many areas of animal biotechnology to give readers basic information on essential concepts and methods, and an

understanding of how the field is evolving and what developments are on the horizon.

The first chapter presents an overview of the developments leading to animal biotechnology. Chapter 2 throws light on animal cell culture, which has exponentially increasing applications and is an essential tool in almost all biological and medical research laboratories. Chapter 3 and 4 present the review of outline of genetic engineering and gene transfer methods, respectively.

Each of chapters 5 through 13 focuses on a single broad area of research and development to which modern methods of biotechnology are being applied with significant results and intriguing potential for the future. Genetically engineered poultry, swine, goats, cattle and other livestock are also beginning to be used as generators of pharmaceutical and other products, potential sources of replacement organs for humans and models for human disease.

Gene therapy is "A myth turned into reality". The excitement and fascination that gene therapy has generated in the recent past has broadened the horizons beyond limits. But it has concomitantly overburdened the undergraduates of medical and veterinary colleges. Therefore, only the most essential concepts have been included in chapter 8. Aquaculture has the potential for increasing worldwide fish production, and innovative strategies are needed to improve efficiency.

Significance of monoclonal antibodies and vaccines has been explained precisely. Brief account on blood products and regulatory proteins throw light on the medical applications of these biological components.

Chapter 12 describes the purposes and applications of animal feed additives, chapter 13 presents the current achievements of stem cell technology. Finally, Chapter 14 focuses on moral and ethical issues in animal biotechnology. At relevant places throughout the text the societal, legal, and environmental implications of development are discussed.

In addition to breadth of coverage, other features enhance the book's usefulness.

Appendices supplement essential explanations.

- Readers are referred to numerous World Wide Websites for the most up-to-date information in rapidly moving areas.

- The generous use of cross-references to other chapters allows teachers and students to integrate information from different areas of biotechnology and different stages in its development to achieve a better understanding of this complex and broad field.

- No book can completely replace primary research papers, in a rapidly advancing field. However, it can serve as a springboard for further study in specific areas. Therefore, I have suggested several books and many articles, both general reviews and research papers, for further reading.

- A glossary gives definition of key terms.

- A list of abbreviations and a chronological listing of the historical development of animal biotechnology have been included.

To sum up, this book provides information about Animal Biotechnology and is a valuable tool for students as well as teachers.

This book could not have been written without the help of many people. I am indebted to Mr. C. Sajeesh Kumar, Managing Editor, MJP Publishers for his encouragement, support, and guidance during the development of the book. I also thank him for approaching me with the idea of writing a book on Animal Biotechnology and his patient bearing with all the delay on my part. I also acknowledge Ms. P. Parvath Radha, Project Coordinator for her valuable suggestions, careful editing and constructive criticisms of the manuscript. I extend my gratefulness to Mr. J.C. Pillai, Director, MJP Publishers, for bringing out a book on Animal Biotechnology authored by me.

I wish to express my gratitude to the members of the management, Dhanalakshmi Srinivasan Educational Institutions and the Principal Dr. R.M. Meenakshi, for their constant encouragement and moral support. I express my sincere gratitude to all my most revered teachers who sculptured me. Thanks are also due to all my colleagues and dear students who made valuable suggestions enabling me to update this book in its present form. I wish to acknowledge

Ms. L.Rama and Ms.P.Seethalaksmi for their support in document collection and typing.

The unstinting support of my well-wisher and family members is beyond thanks.

R. Sasidhara

Contents

9. Biotechnology of Aquaculture — 257

14. Ethics, Morality and Animal Biotechnology   **443**

Abbreviations

$\alpha 1AT$	-	Alpha, 1-antitrypsin
AAV	-	Adeno-associated virus
Ab	-	Antibody
ADA	-	Adenosine deaminase
Ag	-	Antigen
AI	-	Artificial insemination
ATCC	-	American type culture collection
ATP	-	Adenosine triphosphate
BCG	-	Bacillus of Calmette–Guerian
BFGF	-	Basic fibroblast growth factor
BHK	-	Baby hamster kidney
BSS	-	Balanced salt solution
BST	-	Bovine somatotropin
cDNA	-	Complementary DNA
CFTR	-	Cystic fibrosis transmembrane regulating gene
C_H	-	Constant region of heavy chains
CHO	-	Chinese hamster ovary
C_L	-	Constant region of light chains
CNS	-	Central nervous system
CO_2	-	Carbon dioxide
Cos	-	Cohesive sequence
DEAE	-	Diethyl amino ethyl-
DHFR	-	Dihydrofolate reductase
DMSO	-	Dimethyl sulphoxide
DTT	-	Dithiothreitol
EBSS	-	Eagle's balanced salt solution
EBV	-	Epstein–Barr virus
EDTA	-	Ethylene diamine tetra-acetic acid
EGF	-	Epidermal growth factor
ELISA	-	Enzyme-linked immunosorbent assay
EMS	-	Embryo splitting
EPO	-	Erythropoietin

ES	-	Embryonic stem cell
Fab	-	Fragment with antigen binding
Fc	-	Crystal forming fragment
FDA	-	Food & Drug administration
FIA	-	Fluorescence immunoassays
FISH	-	Fluorescent *in situ* hybridization
FMD	-	Foot-and-mouth disease
FSH	-	Follicle stimulating hormone
GEM	-	Genetically engineered microorganism
GFR	-	Growth factor receptor
GMO	-	Genetically modified organism
HAT	-	Hypoxanthine, aminopterin and thymidine medium
HBsAg	-	Hepatitis B surface antigen
HBSS	-	Hanke's balanced salt solution
HCG	-	Human chorionic gonadotropin
hGH	-	Human growth hormone
HGPRT	-	Hypoxanthine guanine phosphoribosyl transferase
HIV	-	Human immunodeficiency virus
HSC	-	Haematopoietic stem cell
HSV	-	Herpes simplex virus
ICSI	-	Intracytoplasmic injections
Ig	-	Immunoglobulin
INF	-	Interferon
IS	-	Intergenic space
KLH	-	Keyhole limpet haemocyanin
LDLR	-	Low-density lipoprotein receptor
LOH	-	Loss of heterozygosity
LTR	-	Long terminal repeat
mAB	-	Monoclonal antibody
MAS	-	Marker-assisted selection
MCB	-	Master cell bank
MDR	-	Multiple drug resistance
MEM	-	Minimum essential medium
MHC	-	Major histocompatibility complex
NAD	-	Nicotinamide adenine dinucleotide
NHB	-	Normal human brain
NT	-	Nuclear transfer

PCR	-	Polymerase chain reaction
PEG	-	Polyethylene glycol
PC	-	Phosphotidyl choline
PKU	-	Phenyl ketonuria
PS	-	Phosphotidyl serine
PVC	-	Polyvinyl chloride
QTL	-	Quantitative trait loci
RAPD	-	Random amplified polymorphic DNA
RB	-	Retinoblastoma
RFLP	-	Restriction fragment length polymorphism
RIA	-	Radioimmunoassay
RSV	-	Rous sarcoma virus
SCID	-	Severe combined immune deficiency
SIg	-	Surface immunoglobulin
SOP	-	Standard operating procedure
SSR	-	Short sequence repeats
STR	-	Short tandem repeat
STS	-	Sequence tagged sites
TGF	-	Transferring growth factor
TIL	-	Tumour infiltrating lymphocyte
TK	-	Thymidine kinase
TNF	-	Tumour necrosis factor
tPA	-	Tissue plasminogen activator
TPX	-	Metinex and thermonex
VEGF	-	Vascular endothelial growth factor
VNTR	-	Variable number tandem repeats
YAC	-	Yeast artificial chromosome
YEP	-	Yeast episomal plasmid

History of Animal Biotechnology

1891	-	First successful embryo transfer.
1900	-	*In vitro* embryo culture development.
1951	-	FDA approved antibiotics as feed additives for farm animals.
1961	-	Mouse embryo aggregation to produce chimeras.
1966	-	First report of microinjection of mouse embryos.
1970	-	First chimeric mice were produced.
	-	Cytogenetic analysis using chromosome banding.
1973	-	Foreign gene function after cell transfection.
1974	-	Development of teratocarcinoma cell transfer.
1977	-	mRNA and DNA transferred to *Xenopus* eggs.
1980	-	mRNA transferred into mammalian ova.
	-	The cloning technologies of embryo splitting (EMS) and embryonic nuclear transfer (NT) were introduced into dairy cattle breeding.
1980–81	-	Transgenic mice first documented.
1981	-	The term "transgenic" was first used by J.W. Gordon and F.H. Ruddle.
	-	Transfer of ES cells derived from mouse embryos.
1982	-	Transgenic mice and a growth hormone phenotype were produced.
1983	-	Tissue-specific gene expression in transgenic mice.
1985	-	Transgenic pig, sheep, rabbit, fish produced by microinjection.
1987	-	Chimeric "knock-out" mice described.
	-	Retrovirus-mediated transgenic chicken produced.
1989	-	Targeted DNA integration and germ-line transgenic mice.
	-	Microinjection for transgenic cattle.
	-	First sperm-mediated reports in for animals.
1990	-	First clinical trial of gene therapy began.
1991	-	Microinjection for transgenic goats.
1993	-	Germ-line chimeric mice produced using co-culture.
	-	FDA approved the use of bovine somatotropin (BST) to increase milk yield from dairy cows.
1996	-	ES cells used for nuclear transfer.
1997	-	The first cloned sheep "Dolly" and the transgenic sheep "Polly" were revealed.
2000	-	Transgenic sheep carrying human gene for α, 1-antitrypsin was successfully produced.
2003	-	A horse gave birth to Idaho Gem, a healthy mule.

INTRODUCTION

Dogs and Cats: Products of Artificial Selection

The pampered poodle may win in the show ring, but it is a poor specimen in terms of genetics and evolution. Human notion of attractiveness in pets can lead to breeds that might never have evolved naturally. Behind carefully bred traits lurk small gene pools and extensive inbreeding—all of which spell disaster to the health of many highly prized show animals. Pure-bred dogs suffer from more than 300 types of inherited disorders.

The sad eyes of the basset hound make him a favourite in advertisements, but his runny eyes can be quite painful. His short legs make him prone to arthritis, his long abdomen encourages back injuries, and his characteristic floppy ears often hide ear infections. The eyeballs of the Pekingese protrude so much that a mild bump can pop them out of their sockets. The tiny jaws and massive teeth of pugdogs and bulldogs cause dental and breathing problems, as well as sinusitis, bad colds, and their notorious "dog breath". Folds of skin on their abdomens easily become infected. Larger breeds, such as the Saint Bernard, are plagued by bone problems and short lifespan.

We artificially select natural oddities in cats, too. One of every 10 New England cats has six or seven toes on each paw, thanks to the multitoed ancestor in colonial Boston. Elsewhere, these cats are rare. The sizes of the blotched tabby populations in New England, Canada, Australia, and New Zealand correlate with the time that has passed since cat-loving Britons colonized each region. The Vikings brought the orange tabby to the islands off the coast of Scotland, rural Iceland, and the Isle of Man, where these feline favourites flourished.

The American curl cat's origin traces back to a stray female who wandered into the home of a cat-loving family in Lakewood, California, in 1981 and passed her unusual, curled-up ears to future generations. The cause—a dominant gene that leads to formation of extra cartilage lining the outer ear. Cat breeders hope that the gene does not have other less lovable effects. Cats with floppy ears, for example, tend to have large feet, stubbed tails, and lazy nature.

DEVELOPMENT OF BIOTECHNOLOGY

Biotechnology literally is the technology based on biology. It is the application of scientific and engineering principles to the processing or production of materials by biological agents to provide goods and services. The application of biotechnology to animals has a long history, beginning in south-west Asia after the Ice Age, when humans first began to trap wild animal species and to breed them in captivity.

Selective animal breeding has been practised for nearly 10,000 years to produce desirable traits in livestock. Today, breeders are on the move to produce livestock that grow faster and convert animal feed to lean tissue, with less total fat. Breeders have focused not only on manipulating the composition of animal products but also on increasing production, such as by enhancing reproduction rates. Increased milk production, decreased fat content, better wool quality, faster maturation rate, disease resistance and the increase in frequency of egg-laying are a few changes brought about by selective breeding. However, such methods of producing desirable traits take time; many generations of animals must be born and bred before an appropriate trait is finally expressed.

On the other hand, biotechnological approaches are used either to rapidly multiply animals of desired genotypes or to introduce specific alterations in their genotypes to achieve certain useful goals. To achieve the latter more efficiently, as well as to assist in conventional breeding efforts, the entire genomes of animals are being characterized using biotechnology tools. These activities have been arbitrarily grouped under animal biotechnology since they either utilize animal cells to generate products or apply biotechnological tools to enhance the usefulness of animals to human welfare.

Animals have become man's constant companion, ever since their domestication from time immemorial. All domestic animals have undergone drastic changes due to the process of artificial selection by humans, under captive breeding with a restricted gene pool. Consequently, with the passage of time, and continued isolation from the wild gene pool, the tamed animals acquired startling traits, which have been put to use by the man for his survival and sustenance. Some have changed in such a way that they started yielding increased quantity of better milk or wool than others. Some others have displayed rapid growth and gaining of body mass or developed high fecundity, some became more vigorous and some just looked bizarrely different and aroused curiosity. The classical animal breeders have taken into account natural variability of animals and mated them selectively *inter se* to develop innumerable breeds of domestic animals such as horse, dog, sheep, goat and cattle and other livestock animals that we see all around us in the present-day world.

Animal improvement remained no longer a matter of interest of indolent bourgeois society, but it became a matter of necessity for human survival. The incessant population explosion puts a tremendous pressure on biologists to explore the possibility of developing new technologies for animal improvement under the conditions of shrinking resources, at the same time ensuring the quality of food, fibre, drugs and medicines for health care. Even though the production of food from animals is not only expensive but also inefficient from the angle of per capita utilization of energy, nutrients and space to produce similar quantity and quality of biologically assimilable protein (Table 1.1), there is an imperative necessity to continue to do animal farming for food, hide, fibre, various by-products, sources of dietary supplements like necessary amino acids, drugs, vitamins, etc.

The recently developed rDNA technology (recombinant DNA technology) provides us powerful tools that enable us to solve not only the basic food and nutrition problems but also to develop and produce a range of health care products. The coming hundred years would be the era of animal biotechnology and all over the world, the strategies for animal production would change radically as the biologists learn to use these tools.

Table 1.1 Comparative protein yield and use of resources by plant and animal systems

Source of protein	Protein (kg h^{-1}y^{-1})
Plants	
Algae	30×10^3
Potato	0.8×10^3
Rice	0.6×10^3
Peanuts	0.45×10^3
Wheat	0.36×10^3
Animals	
Fish	1.0×10^3
Milk	0.12×10^3
Meat	0.08×10^3

"The use of animals, tissues, organs, cells, subcellular organelles, and/or parts of those structures, as well as the molecules to effect physical or chemical changes needed to generate new products for research and commercialization" constitute the field of animal biotechnology. In spite of such a broad perspective, animal biotechnology is considered to be synonymous to rDNA technology and also to some of the older technologies along with the state-of-the-art cutting-edge technologies and related methods such as cell culture, monoclonal antibodies, bioprocess engineering and manipulation of reproduction. Thus, animal biotechnologists not only manipulate the genomes of the targeted animals, but also the processes that exist in the organism but are out of reach for manipulation. Hitherto, the modification of such traits was either made fortuitously through the microevolutionary forces or by the strategies of artificial selection practised by the animal breeders. Animal biotechnology offers supplementation of selective breeding, helping to effect changes at the organism level by the manipulation of cells and genes within an organism. It creates novel procedures (protocols) for the effective intervention of biological systems and processes with the potential for bringing about modification, which otherwise is not possible.

MAJOR AREAS IN ANIMAL BIOTECHNOLOGY

The cutting-edge state-of-the-art animal biotechnology includes the following major areas of interest:

- Manipulation of the processes of reproduction

- Gene cloning of macroorganisms

- Genetic engineering of microorganisms and molecules including cell engineering (hybridomas) to synthesize desired end products like vaccines, gene probes, monoclonal antibodies and growth products.

CURRENT STATE OF ANIMAL TECHNOLOGY

As a consequence, there is pressure to utilize the potential for biotechnology to improve in productivity in animal agriculture. Recombinant DNA technology now allows us to introduce foreign genes into organisms for the expression of specific new traits. Animals can also be engineered for a variety of purposes, ranging from use as human disease models to introduction of desirable traits into a variety of agronomically important animals.

Genetically engineered poultry, swine, goats, cattle and other livestock are also beginning to be used as generators of pharmaceutical and other products and potential sources for replacement organs for humans. The technology to produce foreign proteins in milk by expressing novel genes in the mammary glands of livestock has already advanced beyond the experimental stage with some products currently under clinical trials.

Transgenic animals not only provide invaluable research tools for studying gene regulation and diseases but they may be genetically modified for the production of pharmaceuticals, vaccines and rare chemicals as well as for food production. This is exciting yet sometimes controversial.

The major goals of biotechnology include the generation of livestock that are more economically produced and products that are more nutritious to the consumer. Livestock research has had two primary focuses: increasing growth rate and muscle development with growth hormone genes from various sources and producing antibodies and recombinant vaccines to increase disease resistance. A more distant goal is to develop animal "bioreactors" for producing rare pharmaceuticals and other medical compounds. Genetically engineered livestock will yield important products in milk or blood for treating a variety of human diseases and health needs. Such products may include the following.

- Human haemoglobin, which could be used during trauma when much blood is lost. Haemoglobin is more desirable than whole blood or red blood cells for transfusions, since it requires no refrigeration and is compatible with all blood types, eliminating the need for blood grouping.

- Human protein C, which helps prevent blood clotting.

- Human tissue plasminogen activator, which is used to treat patients after a heart attack.

- Human α, 1-antitrypsin (hα1AT), which may be useful in treating people who have hα1AT-deficiency, which predisposes them to a life-threatening type of emphysema.

CELL CULTURE PRODUCTS

Animal cell cultures are used to produce virus vaccines, as well as a variety of useful biochemicals which are mainly high molecular weight proteins like enzymes, hormones, cellular biochemicals like interferon, and immunobiological compounds including monoclonal antibodies. Animal cells are also good hosts for the expression of recombinant DNA molecules and a number of commercial products have been/are being developed. Initially, virus vaccines were the dominant commercial products from cell cultures, but at present monoclonal antibody production is the chief commercial activity. It is expected that recombinant proteins would become the prime product from cell cultures in the near future. Transplantable tissues and organs are another very valuable product from cell cultures.

Virus Vaccines

A vaccine is a preparation containing a pathogen either in attenuated or inactivated state. This preparation is introduced into an individual to induce adequate antibody production against the pathogen in question so that the individual becomes protected against infection, at a later date, by that pathogen. The introduction of a vaccine in an individual is called vaccination or immunization as it leads to the development of immunity in the vaccinated individuals to the concerned pathogen. The immunity is induced by the antigen which has its origin from the pathogen and is present in the vaccine. Any molecule that induces production of antibodies specific to itself when introduced in the body of an animal is called antigen. Usually the antigenic function is confined to a rather small portion of the antigenic determinant or epitope.

When an antigen enters the body of an animal, its antigenic determinant ultimately binds to the specific receptors present on the surface of the B lymphocytes which produce antibodies specific to this antigen. This binding stimulates a rapid proliferation of these B lymphocytes so that the proportion of such cells increases drastically; this is called clonal selection (proposed by N. Jerne). As a consequence, the concentration in the blood serum of antibodies specific to the concerned antigen also increases. This phenomenon is the basis of immunization.

Vaccines are one of the earliest examples of biotechnological intervention in human health care. They offer the cheapest and most effective protection against diseases, and for some diseases, e.g. hepatitis B, AIDS, etc., they are the only means of protection.

The procedure of virus vaccine production using cell cultures is essentially and in simple terms as follows. The cells to be used as host are inoculated in culture vessels and, after suitable growth has occurred, the cultures are infected with the virus concerned, and incubated. After an appropriate period of virus multiplication and release, the culture medium containing virus particles is collected and suitably processed. In the case of killed-virus vaccines, the vaccine is usually subjected to a concentration step after virus inactivation. Suitable stabilizers are added to prevent loss of potency and the vaccines are generally stored at low temperatures till use. A rigorous quality control is maintained throughout the entire operation.

Interferons (INF)

Interferons are proteins produced by a cell infected by a virus, and which provide protection from viruses to other healthy cells. The interferon was discovered in 1957 when Isaac and Lindermann observed that virus-free fluid obtained from cultured cells infected with virus protected other cells from virus infection. They called the substance present in these fluids, which interfered with virus infection, as interferon. There are three major types of interferons:

- Interferon-α, produced by leucocytes,

- Interferon-β, produced by fibroblasts and

- Interferon-γ, produced by stimulated T lymphocyte cells, hence also called immune interferon.

The mechanism of protection by interferons appears to be as follows:

When INF reacts with the INF receptors of a cell, the cell enters a state called interferon-induced antiviral state. In this state, a rapid degradation of mRNA occurs if any virus infects the cell. Interferon induces in cells the production of 2,5-adenosine polymerase. When such a cell is infected by a virus, the 2,5-adenosine polymerase is activated to produce 2,5-adenosine (2,5-A), which in turn activates pre-existing but inactive molecules of ribonuclease-L. Activated ribonuclease-L degrades all mRNA present in the cell bringing the protein synthesis to a halt in such cells. This interferes with multiplication of the virus so that virus infection is either stopped or sufficiently slowed down to allow the production of adequate antibodies against the invading virus. The protection due to interferon is non-specific in that interferon induced by any one virus will provide protection against all viruses. It is possible that interferons modify ribosomes so that they no longer translate viral mRNAs, although they are fully capable of translating mRNA of host origin.

Interferons are produced from human leucocytes isolated from donor blood and cultured *in vitro*, and from mouse fibroblast cultures. The production scheme is as follows: Large-scale (1,001 to 10,001) cell cultures are infected with Sendai virus, and incubated for 24 hours after which the supernatant is collected, centrifuged, and used for interferon isolation. The amount of interferon recovered is relatively small, and the normal leucocytes are difficult to culture, preventing scaling up from relatively small inocula. These contribute to the enormous price of the product.

In view of the value of and/or demand for interferons, intensive efforts were made to produce it in genetically engineered organisms, e.g. *E. coli*, yeast, mammalian cell cultures and even in plants. These efforts have drastically reduced the cost (tissue culture technology less than 10% of the initial price) and improved the purity (by several magnitude) of the product.

Recombinant Proteins

Proteins produced by genes transferred into selected host cells by genetic engineering are called recombinant proteins, since they are based on recombinant DNA technology. Recombinant proteins form an important component of biopharmaceuticals, i.e., biotechnology products having pharmaceutical applications. A large number of recombinant proteins are being produced in mammalian cell cultures, some of which, viz. human growth hormone (hGH), tissue plasminogen activator (tPA), erythropoietin and blood clotting factor VIII, are already in therapeutic use. The host cells used for large-scale production of the various recombinant proteins are Chinese hamster ovary line (CHO), baby hamster kidney line (BHK) and mouse mammary cell lines.

The use of animal cell lines has been made possible by the development of culture systems permitting large-scale cultures of animal cells at high densities, and development of media, which minimize or even obviate the use of serum.

The use of animal cells for the production of recombinant proteins is more preferable than the use of microorganisms for the following reasons:

1. The signals for synthesis, processing and secretion of the concerned proteins are better recognized in animal cells than in microbes.

2. The protein products are readily secreted into the medium, which makes their recovery much easier.

3. The patterns of folding and disulphide bridge formation in the recombinant proteins are similar to those of the natural proteins.

4. The multimeric proteins are also assembled correctly.

Hybrid Antibodies

When antibody molecules are modified or designed using recombinant DNA technology to suit specific applications, such antibodies are called recombinant or hybrid antibodies, and the approach itself is termed as antibody engineering. The essential steps (Figure 1.1) in antibody engineering are:

- to develop the antibody design to serve the specific purpose,

- to bring together the DNA sequences that would generate this antibody molecule into a suitable expression vector,

- to introduce this vector/gene construct into a myeloma cell line, where the designed recombinant antibody gene expresses itself,

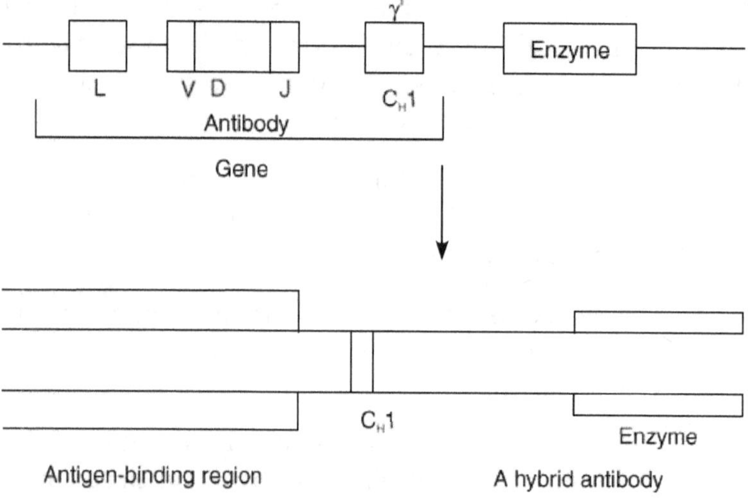

Figure 1.1   A schematic representation of a hybrid antibody and the heavy chain gene producing it. The gene encoding constant region of the heavy chain is fused with the gene encoding the desired enzyme function. The recombinant gene is expressed in a suitable myeloma cell line.

- ❋ to mass-culture the transfected clones of myeloma cell line, and

- ❋ to harvest the recombinant antibodies produced.

It is essential that the myeloma cell line used for transfection is Ig⁻ (Ig –ve), i.e., does not produce any antibody of its own. The hybrid antibodies are monoclonal in nature since each preparation has antibodies of a single specificity.

Monoclonal Antibodies

Antibodies are special types of proteins, ordinarily glycosylated and called immunoglobulins. They are produced in response to antigens; an antibody specifically interacts with the antigen which induced its production. The serum of an animal immunized by a single antigen will contain antibodies of different specificities but reacting to the same antigen. These are called polyclonal antibodies since several different plasma cell clones produce them.

In contrast, a hybridoma clone produces antibodies of a single specificity since the clone is derived from the fusion of a single differentiated (antibody producing) B cell with a myeloma cell, i.e., is a clone of a single B cell. Therefore such antibodies are called monoclonal antibodies.

Monoclonal antibodies are usually produced from hybridoma clones. Each hybridoma clone is derived by the fusion of a myeloma cell and an antibody-producing lymphocyte, and the hybridoma clone producing the desired antibody is identified and isolated. Hybridoma cells are mass-cultured for the production of monoclonal antibodies either *in vivo* in the peritoneal cavity of mice or *in vitro* in large-scale culture vessels.

The chief advantage of monoclonal antibodies is that all the antibody molecules in a single preparation react with a single epitope or antigenic determinant. As a consequence, the results obtained by using monoclonal antibodies are clear-cut, as there is no confusion that arises due to the presence of antibodies of other specificities in the case of conventionally used antisera. The multitude of various applications of monoclonal antibodies may be grouped into the following three categories: diagnostics, therapeutics and purification.

CONCERNS REGARDING EXTANT TECHNOLOGIES

Animal Health

There are well-established guidelines for the application of technologies that maintain animal health, such as standard vaccination against viral and bacterial diseases. Indeed, considerable efforts are being made to expand the range of such technologies in order to prevent epidemic spread of disease in flocks and herds, which are particularly at risk when farmed under intense conditions. Even the therapeutic use of antibiotics to treat animals that have bacterial infections or are in danger of being infected seems not to be controversial, except when antibiotics of medical importance to humans are employed.

Sub-therapeutic Use of Antibiotics

The U.S. Food and Drug Administration (FDA) had approved antibiotics as feed additives for farm animals in 1951. Their use has been extended to fish farming, particularly with the global spread and dramatic increase of aquaculture in tanks and pond-like structures where antibiotics are used for prevention and control of disease rather than to enhance growth.

The treated animals are found to grow more quickly and utilize feed more efficiently than animals on regular feed. At least 19 million pounds of antibiotics are used annually for sub-therapeutic purposes in animal agriculture and generally are added to feed and water (NRC, 1999). Some of these compounds used on livestock, including penicillin, tetracycline and fluoroquinolone are also prescribed to treat human illness and the practice has been shown in a few instances to contribute to antibiotic resistance in human pathogens. It is now generally accepted in the scientific and medical communities that antibiotic resistance can be exacerbated by the widespread improper use of antibiotics.

Assisted Reproductive Procedures

All organisms have intrinsic limitations in their production systems, viz. mating system, genetic variability and maintenance process, which include health as well as nutrition. Animals and plants differ basically from each other from the production point of view. Though the nutritional and environmental stress and diseases are common factors to both of them, however, animal production solely depends upon the health of pasteur/fodder of plants. The pasteur production strategies are interrelated. The plants convert inorganic nutrients from the environment to organic food for animals; animals in turn provide organic manure to plants. The reproductive system is basically different in animals and plants. The plants reproduce both sexually and asexually. When reproduction is through amphimixis (union of gametes) plants produce a large number of gametes, both male and female whereas animals give rise to few. Animals do not reproduce asexually. All livestock animals produce only fertilized eggs each year.

Life cycle in plants is simple, taking only few months to complete whereas in animals, it takes many years from birth to maturity and old age. The low fertility of females in animals has profound impact on genetic manipulation strategies. The existing variability can be compounded and new genetic recombination could be generated very easily in plants, since several thousands of propagules (seed) can be produced from a single hybrid whereas to achieve the same it is very difficult in animals. Even quantitative trait loci (QTL) can be identified through molecular probes very easily in plants, but not so easily in animals. When a desirable genetic variant is isolated, it can be multiplied rapidly in plants, but the same is a very slow process in animals. The important goal of animal biotechnology is to make feasible rapid animal production as that of plants and find out ways and means of genetic manipulations at the cellular, subcellular, tissue or organism level for possible commercialization.

Manipulation of reproductive process is a *sin que non* for the successful exploitation of biotechnological potential of a vast array of gene and genome pools of domestic animals. Historically the manipulation of the reproductive processes started in early 1930s with the development of procedures for artificial insemination (AI). When it was feasible to deep-freeze the semen, the use of AI became widespread in 1960–70s. With the concurrent developments in reproductive research on the female reproductive process soon the procedures were developed for the induction of oestrus cycle and synchronization. Initially this was done using steroid hormones, but subsequently the same was accomplished by using prostaglandins. The embryo transfer technology perfected in 1980s owes its success to the development of methods for super ovulation using gonadotrophic hormones. Since then, the embryo transfer technology has become a routine affair especially because of technological advances in embryo preservation by freezing.

Artificial insemination and embryo transfer technique together offer immense possibilities of enhancing the rate of genetic progress in national breeding programmes. In recent times, with the development of more advanced techniques for manipulating reproduction as well as *in vitro* embryo manipulation and production, the perfection of technologies for gene transfer to a whole animal for the possible production of transgenic livestock animals became a distinct possibility. Basically, the protocol (Figure 1.2) includes the following:

1. removal of oocyte from a genetically superior cow, goat, pig, sheep etc., (or theoretically any animal) during the dioestrous period,

2. *in vitro* maturation and fertilization of the oocytes,

3. culturing of zygote (fertilized oocytes) *in vitro* till blastocyst stage,

4. sexing, splitting and

5. cloning of embryo.

Micromanipulation techniques gave further phillip to the development of protocols for splitting and cloning of sheep and bovine embryos. Marketing of bovine embryos that originate

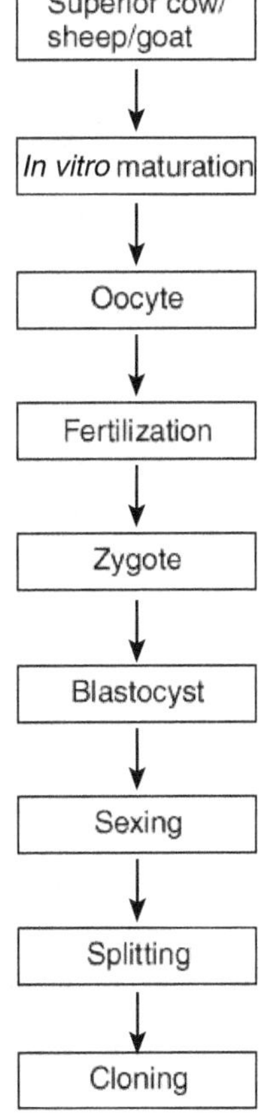

Figure 1.2   Reproductive process

from *in vitro* manipulations, cloned embryos and transgenic animals has become a reality and these procedures can be adapted for the industrial production of animals, with varied applications and utility.

Artificial insemination (AI), associated use of frozen semen, sire testing and sire selection are all part of a combinatorial approach to improve the genetic quality of farmed species. When AI was first introduced into agriculture there was an enormous outcry from the farmers. It gradually has become an accepted practice in agriculture, as well as in human and veterinary medicine. The ability to freeze semen and maintain a high degree of fertilizing ability after thawing extended the power of AI, since a few selected bulls could be used to inseminate many females in different geographical areas. Such bulls could be tested, not only for fertility but also for their ability to sire progeny that produced copious amounts of milk. By maintaining accurate records, breeding value estimations of particular bulls could be calculated. The result was the remarkable increase in the milk production, than noted earlier. On the other hand, the process is leading to potentially destructive inbreeding since many of the select bulls are related. Recently, there has also been a remarkable loss of fertility, with successful progenies resulting from first insemination dropping from 40% to as low as 20% or less in some herds as milk yields have risen.

Embryo recovery and transfer provides the opportunity for a particularly valuable animal to parent more offspring in her lifetime than would be otherwise possible. The embryos also can be frozen and then either stored or transported before they are used to initiate a pregnancy. It is a relatively common technology and has been used to produce an estimated 40,000–50,000 beef calves every year. The approach is to induce, by using hormones, the maturation and release of more than a single egg from the ovaries. Then the animal is inseminated with semen from an equally selected bull and the embryos are collected and transferred individually or in pairs, to the reproductive tract of less valuable cows, which carry the calf to term. A related technique is to use a needle to aspirate immature oocytes from the ovaries (in the case of livestock, oocytes are often taken from slaughtered animals at an abattoir) and to mature the oocytes for about one day in a culture containing hormones. At this stage when the oocytes reach a point midway through the second division of meiosis, they are fertilized with live sperm. In rare instances, fertilization is achieved by a single sperm or sperm head, which is injected through the outer zona pellucida of the oocyte, either beneath the zona or directly into the cytoplasm (intracytoplasmic injections or ICSI). The resulting zygotes are then cultured until the embryo reaches a more advanced stage of development.

These combined techniques form the basis of *in vitro* fertilization. Always IVF is used commercially to preserve the genome of particularly valuable animals that have fertility problems such as blocked oviducts or that respond poorly to superovulation.

In order to manage breeding programmes more intensively, control over the reproductive cycles of the livestock by hormonal intervention has been increased. The technologies involve injecting the animal with hormones either to stop progression through existing oestrous cycle or to mimic the events that lead to selection of one or more mature follicles that will ovulate. Superovulation is a technique designed to mature a cohort of follicles simultaneously.

Splitting or bisecting embryos became an esoteric but well established practice in the 1980s in order to provide zygotic twins. The piece of embryo usually halves, and the two halves are genetically identical in terms of both nuclear and mitochondrial genes. They are placed in an empty zona (the protective coat around early embryos) before being transferred to recipient mothers to carry them to term. Only a small number of the calves are produced in this manner.

Cloning by nuclear transplantation from embryonic blastomeres is an expensive procedure that has its origin in the 1970s. This is distinguished from somatic cell nuclear transfer (the technology that had created Dolly) in the stage of development at which the nuclei are transferred. In the older procedure the cells or blastomeres used were from the morula stage of cell development where the embryo is still an undifferentiated mass and its cells still capable of forming all tissues of a foetus. The cloning technologies of embryo splitting (EMS) and embryonic nuclear transfer (NT) were introduced into dairy cattle breeding in the 1980s.

Modern cloning involves taking an unfertilized egg, removing its chromosomes and introducing the nucleus from a differentiated cell of the animal to be cloned, which is frequently an adult. The introduced nucleus is programmed by the cytoplasm of the egg and directs the development of a new embryo, which is then transferred to a recipient mother to allow it to develop to term. The offspring formed will be identical to their siblings and to the original donor animal in terms of their nuclear DNA, but will differ in their mitochondrial genes and also in the manner in which their nuclear genes are expressed. Cloning from blastomeres has been reported to result occasionally in large calves (and lambs), the so-called large offspring syndrome.

Hormone-treated Cattle

Among the most contentious technologies used in animal agriculture is the use of steroid hormones to increase the rate of weight gain and to reduce accumulation of fat deposits of young heifers and steers as part of the finishing process prior to slaughter. The steroids are administered by slow release from a plastic implant embedded beneath the skin of the ear, which provides "physiologic" circulating levels of the hormone in the bloodstream. The hormones used are mainly Zeranol (a naturally occurring fungal metabolite (Zearalenone) with oestrogenic action), oestradiol, progesterone and testosterone or mixtures of these steroids and trenbolone. The use of diethylstilbestrol was eventually banned from use in the poultry and beef industry because of its adverse effects on humans. Humans are not posed to any risk provided good veterinary practices are employed (e.g. using the correct number of implants and placing implants correctly in the ear cartilage). There are chances for these hormones to pose environmental threat through their leaching into soil and water.

Bovine Somatotropin

The use of Bovine Somatotropin (BST) to increase milk yield from the dairy cows is the subject of trade disputes. The FDA approved BST for use in dairy cattle in the U.S.A in 1993. The Monsanto product, Pasilac, is widely used throughout the U.S. dairy industry, where milk production can be increased as much as 30% without adversely affecting the quality or composition of the milk. The BST is present in low concentrations in milk but has no biological activity in humans. The level of IGF-1, the hormone induced by BST is somewhat elevated but within the "physiologic range" for cows and is probably digested along with other milk proteins in the adult stomach.

The greatest concerns about BST are probably in the area of animal welfare. High-yield milking cows show a greater incidence of mastitis than lower milk producing cows. But mastitis is not exacerbated by BST administration. Another concern is a recent trend to breed heifers only once and then to sustain milk production for as long as 600 days by using BST.

Marker-assisted Selection

Marker-assisted selection involves establishing the linkage between the inheritance of a particular trait, which might be desirable as in the case of milk yield, or undesirable, as in the susceptibility to a disease, with the segregation of particular genetic markers. Thus, even if the gene that controls the trait is unknown, its presence can be inferred from the presence of the marker that segregates with it. This technology has recently become a factor in animal breeding and selection strategies. It is important for studying complex traits.

Initially animals were screened for genes that control simple traits such as horns, which are undesirable in cattle and halothone sensitivity, which segregates with metabolic stress syndrome in pigs. With time, easily identifiable markers will be chosen that accompany the many genes controlling more complex traits such as meat tenderness and taste, growth, calf size and disease resistance.

Chromosome Set Manipulation in Molluscs and Finfish

Altering the chromosome complement of an animal can be a useful way of rendering that animal infertile and is exploited widely in the production of fish and molluscs. Well-timed application of high or low temperatures, certain chemicals or high hydrostatic pressure to newly fertilized groups of eggs can interfere with extension of the second polar body (the last step in meiosis), resulting in "triploid" individuals with three, instead of the two usual chromosome sets. A later treatment can suppress the first cell division of the zygote, resulting in "tetraploid" individuals with four sets of chromosomes. Crossing tetraploids, which are fertile in some species, with normal diploids can produce large number of triploids. Such chromosome set manipulations have been applied to cultured marine molluscs to produce confined stocks of triploids that are unable to reproduce.

The benefit of producing sterile molluscs is maintaining product quality throughout the year. The meat quality of oysters is high just before they spawn, but low after spawning. The product quality of reproductively sterile, triploid oysters remains high year-round.

Triploidy often has been used to reduce the likelihood that introduced finfish species would establish self-sustaining populations. Use of all-female triploid stocks has been suggested as a means of achieving reproductive confinement of transgenic fishes, including Atlantic salmon.

Another technology used on finfish is to farm monosex fish stocks, which are preferred by producers either because one gender grows faster than the other (e.g. males in catfish and tilapia, females in rainbow trout) or because certain species attain sexual maturity before reaching harvest size. Monosex populations have been established through hormone-induced gender reversal. All-male fins can be produced by direct administration of testosterone in feed or all-female by administration of oestrogen.

Animal Cloning

Advances in biotechnology have allowed scientists to make genetically identical copies or clones of animals. Duplication of an organism's genome occurs naturally when identical twins are born or when a plant is grown from a cutting of another plant. When complete animals are obtained from somatic cells of an animal, it is called animal cloning. Cloning is routine in plants but in the case of animals, only limited success has been achieved so far. Earlier, nuclei from a tadpole were transplanted into the cytoplasm of an enucleated fertilized frog egg, and normal frogs were obtained. However, the world really took notice of cloning in 1997 when a group of Scottish researchers announced the birth of Dolly the sheep, which had been cloned using a single cell from an adult sheep. Dolly had only one "parent;" her nuclear genome was exactly like her "mother's" instead of being a combination of two parents. Therefore, Dolly could generally be thought of as her mother's identical twin.

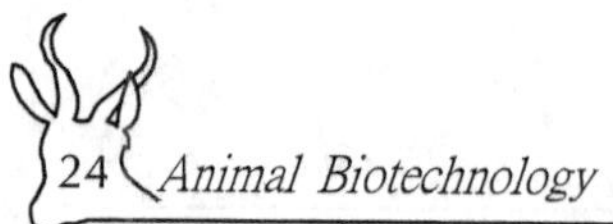

To produce Dolly, scientists took an egg from a sheep and removed its nucleus (which contains the genome or instruction manual), rendering it unable to function or develop. Next, they took a cell with an intact genome from a different adult sheep (Dolly's "mother") and fused it to the sheep egg which lacked a genome. The egg, with its new genome, was stimulated to begin developing into an embryo and was implanted into a surrogate sheep where it grew normally, resulting in the birth of Dolly. Dolly later gave birth to normal lambs.

Udder cells from an adult sheep were first cultured *in vitro*. The cultured cells were arrested in the G_0 phase (quiescent stage) and then fused *in vitro* with enucleated ova of appropriate stage (Figure 1.3). The fusion products were cultured *in vitro* before their transfer into the uterus of surrogate mothers. The rate of success in obtaining normal embryo development is rather low.

Cloning is, in many situations, highly desirable since it allows indefinite multiplication of an elite desirable genotype without the risk of segregation and recombination during meiosis, which must precede sexual reproduction. Obviously, the technique holds a great promise in genetic research, especially in understanding aging and curing genetic diseases.

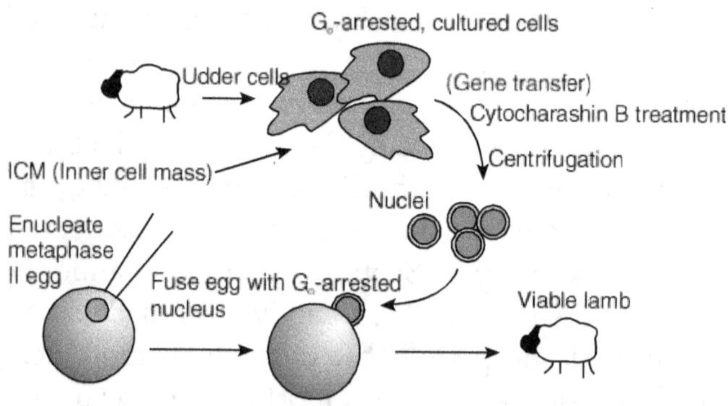

Figure 1.3   Nuclear transfer of somatic cells

Researchers have cloned other mammals including cows, goats, pigs, and mice. However, the overall low rate of successful cloning and frequent occurrence of developmental abnormalities in cloned animals demonstrate the need for further research before cloning will be practical.

It has also been reported that cloned animals may exhibit health problems throughout their life. Cloned animals may age prematurely as Dolly was diagnosed with arthritis at a seemingly young age and cloned mice had a shorter-than-normal lifespan. Additionally, it was demonstrated that cloned mice were both larger in size and heavier than a control group of non-cloned mice.

If advances in animal cloning technology were to overcome the current obstacles, the most obvious benefit would be the ability of a farmer to have a herd of superior performing animals in one generation. Breeding companies could sell cloned embryos in a manner similar to the way in which semen is currently marketed. A potential drawback of this practice would be the loss of genetic diversity in livestock herds, but this could be avoided by limiting the number of cloned embryos of a given animal that were sold.

It has also been proposed that cloning could be used to increase the population of animals in an endangered species. The mouflon sheep, which is a wild Mediterranean sheep with less than 1000 animals remaining, was successfully cloned. Additionally, scientists are attempting to clone an endangered wild Asian ox, called the guar (the first cloned guar died of an intestinal illness shortly after birth) and possibly the giant Panda. Although possible, a recovering population of cloned animals would be hindered by a lack of genetic diversity and would not address the larger issue of how the animal became endangered.

CONCLUSION

Biotechnology has spread its branches to a wide range of fields including animal agriculture. Animals can be engineered for a variety of purposes ranging from use as human disease models to the introduction of desirable traits into a variety of agronomically important animals. An approach that combines breeding with molecular genetics and recombinant DNA technology will lead to greater progress in the future. Major goals of biotechnology include the generation of livestock that are produced more economically and are of high nutritive value. Artificial insemination allows the production of genetically vigorous strains. The coming years will be the era of Animal Biotechnology owing to the current needs of humanity.

Summary

- Biotechnology is the application of scientific and engineering principles to the processing or production of materials by biological agents to provide goods and services.

- Selective breeding, a method of producing desirable traits takes time; many generations of animals must be born and bred before an appropriate trait is finally expressed.

- Recombinant DNA technology helps in engineering animals for a variety of purposes.

- Animal cell cultures are used to produce a variety of useful biochemicals, which are mainly high molecular weight proteins.

- Artificial insemination, associated use of frozen semen, sire testing and sire selection are all part of a combinatorial approach to improve the genetic quality of farmed species.

☑ Marker-assisted selection involves establishing the linkage between the inheritance of a particular trait with the segregation of particular genetic markers.

☑ Chromosome set manipulations have been applied to cultured marine molluscs to produce confined stocks of triploids that are unable to reproduce.

☑ Process of obtaining complete animals from somatic cells of an animal is called animal cloning.

☑ Cloning holds a great promise in genetic research, especially in understanding aging and curing genetic diseases.

REVIEW QUESTIONS

1. Define Biotechnology.
2. What are the products of genetically engineered livestock?
3. What is selective breeding?
4. Explain the role of genetic engineering for livestock improvement.
5. Give a brief note on interferons.
6. State the reasons for using animal cells in recombinant protein production.
7. Brief on artificial insemination.
8. Explain marker-assisted selection.
9. Define cloning. Brief on animal cloning.
10. Give a brief overview of animal biotechnology.

ANIMAL CELL AND TISSUE CULTURE

C ell culture is essentially a novel and exciting technology in the field of modern biology. Currently, this technology is indispensable in several branches of life sciences. Cell culture techniques have laid the foundation for studying the regulation of cell proliferation, differentiation and product formation in carefully controlled conditions.

HISTORICAL BACKGROUND

Harrison (1907) and Carrel (1912) for the first time inaugurated and devised the tissue culture technique in order to study the behaviour of animal cells. In 1907, Harrison preferred to select frog as a source of tissue for his studies. The frog was preferred because it was a cold-blooded animal and further, incubation was not required. In addition, tissue regeneration was more predominant in lower vertebrates than mammals.

The cultivation of tissue *in vitro* by Harrison created a new wave of interest and the medical science extended further interest into warm-blooded animals, where normal and pathological developments are closer to humans. The most favourite choice for tissue culture is embryonated hen's egg. Further, rapid progress in the development of experimental animal husbandry, with genetically pure strains of rodents has made mammals a comprehensive favourite material for tissue culture. The development of tissue culture involves the production of antiviral vaccines and the extensive knowledge on neoplasia.

In addition to virology and cancer research, several other areas of research have come to depend heavily on tissue culture techniques. The introduction of cell fusion technique by Barski *et al.* in 1961 and genetic manipulation by Frederick *et al.* in 1993 has established somatic cell genetics as a major area in the genetic analysis of higher animals. Similarly, monoclonal antibodies and haemopoietic cell lines were developed. The introduction of novel genes into normal human cells help to produce pharmaceutically viable drugs. In addition, cell products such as insulin, interferon, and human growth hormones have been genetically engineered. Other major areas of interest include the study of cell interactions and intracellular control mechanism in cell differentiation and development, besides analysis of nervous functions.

Efficacy of tissue culture technology has been widely used in routine applications in medicine and industry. The chromosomal analysis of cells derived from the womb reveals genetic disorders in the unborn child, and viral infections may be assayed on a monolayer of appropriate host cells. In addition, several other parameters including toxic effect of pharmaceutical compounds and environmental pollution can be measured using *in vitro* assays. Further applications of cell and tissue culture includes homografting and reconstructive surgery using an individual's own cells and it has now become accepted clinical practice in some hospital burn units to biopsy and graft the cultured cells back into the areas of most severe burns.

Tissue culture is the technique in which fragments of excised tissue are grown in culture medium. Some normal functions may be maintained. Original organization of tissue is lost.

Cell culture is defined as the technique in which tissue or outgrowth from an explant is dispersed, most enzymatically, into a cell suspension, which may then be cultured as a monolayer or suspension culture. The characteristic features of cell culture are:

- Development of a cell line over several generations
- Possibility of scale-up
- Loss of some differentiated characteristics in cells

Cell cultures may contain the following three types of cells.

1. Stem cells
2. Precursor cells
3. Differentiated cells

Stem cells are undifferentiated cells, which can differentiate under correct inducing conditions into one of several kinds of cells. Different kinds of stem cells differ markedly in terms of the kinds of cells they will differentiate into. Precursor cells are derived from stem cells and are committed to differentiation, but are not yet differentiated; these cells retain the capacity for proliferation. In contrast, differentiated cells usually do not have the capacity to divide. Some cell cultures like epidermal keratinocyte cultures contain all the three types of cells. In such cell cultures, stem cells constantly provide new cells, which develop into precursors; the precursor cells proliferate and mature into the differentiated cell types. Thus stem cells are necessary for the maintenance of such cultures, which by nature are heterogeneous.

On the other hand fibroblast cultures contain a more or less uniform population of dividing cells at low cell densities ($<10^4$ cells/cm^2), but at high cell densities (10^5 cells/cm^2) are uniformly composed of non-proliferating differentiated cells. The cells begin to proliferate once the cell density is approximately reduced.

Differentiation and cell proliferation are affected not only by cell density but also by factors like serum, Ca^{2+} ions, hormones, cell-to-cell and cell-to-matrix interactions, etc. Generally cell proliferation is promoted by low cell density, low Ca^{2+} (100–600 μm), and high growth factor levels, while differentiation is promoted by the exact opposite conditions and by the presence of differentiation-inducing factors, e.g. cortisone nerve growth factor, etc. The proportion of stem, precursor and differentiated cells are markedly affected by the source tissue used for obtaining the cultures. For example, cultures derived from embryos and those derived from even adult tissues where continuous cell renewal occurs naturally (intestinal epithelium, haemopoietic cells, etc.), stem cells are likely to be more frequent than in other cell cultures. In contrast, cell cultures from tissues where renewal occurs only under stress, e.g. fibroblasts, muscle, etc., may contain only precursor cells.

Cell cultures can be grown as monolayers (adherent cell lines) or as suspension cultures (non-adherent cell lines). Propagation in suspension cultures is limited to haemopoietic cell lines, ascites, tumours and transformed cells (those cells that have become phenotypically modified during *in vitro* culture to become anchorage-independent and are able to grow in layers of several cells thick, as against monolayer growth of non-transformed cells). Therefore cells in culture need a surface or substrate to adhere to so that they are able to proliferate. Cells that are unable to adhere to a substrate are unable to divide, i.e., their growth is anchorage-dependent.

Cells of both adherent and non-adherent types may be i) primary cell lines, ii) immortal cell lines or iii) transformed cell lines. Primary cell cultures are established by inoculating growth medium with cells taken from animal tissue. The excised tissue is fragmented into small pieces with forceps and scissors. Then it is placed in sterile medium in a petri dish. The excised fragment is then treated with proteolytic enzymes such as trypsin to dissegregate the tissue into individual cells. Such a culture may contain variety of differentiated cell types. In the case of connective tissue, fibroblasts start growing within 7 days. Fibroblasts outgrow other cell types. Careful control of medium composition allows selective growth of certain cell types. Subculture of primary culture leads to secondary and tertiary cultures. Primary cell lines have limited lifespan and are also called finite cell cultures. Cells from primary cultures multiply at a constant rate over successive transfers and such cells comprise a cell strain. Human cells generally divide only 50–100 times before dying. The natural aging of cells can be arrested by storage in liquid nitrogen.

Immortal cell lines are capable of unlimited growth in culture and they do not die. They are altered to produce a continuous cell line from cells of primary culture. They are not necessarily malignant cells.

During repeated serial transfer, cell lines can undergo extensive changes in their cultural properties; such cells may grow in clumps rather than in a monolayer, and cells may also be irregularly oriented with respect to each other. Such cells are said to be transformed and are generally neoplastic. Transformed cell lines are either cells derived from tumour cells or cells manipulated by transfection with oncogenes, or obtained by treatment with carcinogens to produce cells with novel phenotype. These cell lines are quite robust. They have low doubling time and low requirement for growth factors. Although transformed cells are much easier to grow in culture, there is some reluctance to use them for large-scale production of biologicals. The concern is due to fear of contamination of products with tumorigenic agents.

FACILITIES FOR ANIMAL CELL CULTURES

Minimal Requirements for Cell Culture

Infrastructure

- Clean and quite sterile area
- Preparation facilities
- Animal house
- Microbiology laboratory
- Storage facilities (for glassware, chemicals, liquids)

Equipment   Laminar-flow, sterilizer, incubator, refrigerator and freezer (–20°C), balance, CO_2 cylinder, centrifuge, inverted microscope, water purifier, haemocytometer, liquid nitrogen freezer, slow cooling device (for freezing cells), pipette washer and deep washing sink are the important equipment required.

Besides the minimal requirements listed above, there are many more facilities that may be beneficial or useful for tissue culture.

These include air-conditioned rooms, containment room for biohazard work, phase-contrast microscope, fluorescence microscope, confocal microscope, osmometer, high-capacity centrifuge and time-lapse video equipment (Table 2.1).

Table 2.1   Tissue culture equipment (Freshney, 1987)

Minimum requirement (essential)	Desirable features (beneficial)	Useful additions
Incubator	Laminar flow hood(s)	–70°C freezer
Sterilizer (autoclave, pressure cooker, oven)	Cell counter	Glassware washing machine
Refrigerator	Vacuum pump	Closed-circuit TV for inverted microscope(s)
Freezer (for –20°C storage)	CO_2 incubator	Colony counter
Inverted microscope	Coarse and fine balance	High-capacity centrifuge (6 × 1 litre)
Soaking bath or sink	pH meter	Cell sizer (e.g. Coulter ZB series)
Deep washing sink	Osmometer	Time-lapse cine-micrographic equipment
Pipette cylinder(s)	Phase-contrast and Fluorescence microscope(s)	Interference–contrast microscope

(Contd.)

Table 2.1   (Continued)

Minimum requirement (essential)	Desirable features (beneficial)	Useful additions
Pipette washer	Portable temperature recorder	Polythene bag scaler
Water purifier	Permanent temperature recorders on sterilizing oven, autoclave	Controlled rate cooler
Bench centrifuge	Roller racks for roller bottle culture	Filing for freezing records and catalogues
Liquid N_2 freezer (~ 35 L, 1,500–3,000 ampoules)	Magnetic stirrer racks for suspension cultures	Centrifugal elutriator centrifuge and rotor
Liquid N_2 storage flask (~25 L)	Pipette drier	Fluorescence-activated cell sorter
	Pipette plugger	Densitometer
	Trolleys for collecting soiled glassware and redistributing fresh supplies	Density meter (for density gradient cell separation)
	Pipette aid	
	Autopipette or other form of automatic dispenser diluter	
	Separate sterilizing oven and drying oven	

CULTURE VESSELS

In tissue culture technology, the cells attach to the surface of a vessel which serves as the substrate and grow. Hence there is a lot of importance attached to the nature of the materials used and the quality of the culture vessel.

Materials Used for Culture Vessels

Glass   Although glass was the original substrate used for culturing, its use is almost discontinued now. This is mainly because of the availability of more suitable and alternative substrates.

Disposable plastics   Synthetic plastic materials with good consistency and optical properties are now in use to provide uniform and reproducible cultures. The most commonly used plastics are polystyrene, polyvinyl chloride (PVC), polycarbonate, metinex and thermonex (TPX).

Types of Culture Vessels

The following are the common types of culture vessels.

- Multi-well plates
- Petri dishes
- Flasks
- Stirrer bottles

Choice of Culture Vessels

The actual choice of a culture vessel depends on several factors.

1. The way cells grow in culture—monolayer or suspension.
2. The quantity of cells required.
3. The frequency of sampling for the desired work.

4. The purpose for which the cells are grown.

5. The cost factor.

In general, for monolayer cultures, the cell yield is almost proportional to the surface area of the culture vessel. Flasks are usually employed for this purpose.

Any type of culture vessel can be used to grow suspension cultures. It is necessary to slowly and continuously agitate the suspended cells in the vessel.

Treatment of Culture Vessel Surface

For improving the attachment of cells to the surfaces and for efficient growth, some devices have been developed. It is a common observation that the growth of the culture cells is better on the surfaces for second seeding. This is attributed to matrix coating of the surfaces due to the accumulation of certain compounds like collagen and fibronectin released by the cells of the previous culture.

There are now commercially available matrices, e.g. matrigel, pronectin, celltak.

Feeder Layers

Some of the tissue cultures require the support of metabolic products from living cells, e.g. mouse embryo fibroblasts. In this case, the growing fibroblasts release certain products which when fed to new cells enhance their growth.

Alternative Substrates for Culture Vessels

In recent years, certain alternatives for culture vessels have been developed. The important alternative artificial substrates are microcarriers and metallic substrates.

Microcarriers   They are in bead form and are made up of collagen, gelating polyacrylamide and polystyrene. Microcarriers are mostly used for the propagation of anchorage-dependent cells in suspension.

Metallic substrates   Certain types of cells could be successfully grown on some metallic surface or even on the stainless steel discs, for instance fibroblasts were grown on palladium.

Use of Non-adhesive Substrates in Tissue Culture

The growth of anchorage-independent cells can be carried out by plating cells on non-adhesive substrates like agar, agarose and methylcellulose. In this situation, as the cell growth occurs, the parent and daughter cells get immobilized and form a colony, although they are non-adhesive.

GROWTH OF ANIMAL CELLS IN CULTURE

The ability to study cells depends largely on how readily they can be grown and manipulated in the laboratory. Although the process is technically far more difficult than the culture of bacteria or yeasts, a wide variety of animal and plant cells can be grown and manipulated in culture. Such *in vitro* cell culture systems have enabled scientists to study cell growth and differentiation, as well as to perform genetic manipulations required to understand gene structure and function.

CULTURE MEDIA

For growing cells *in vitro*, the environment should be as close as that expected *in vivo*. The nutrient media used for animal cell and tissue culture must be able to support their survival as well as growth, i.e., must provide nutritional, hormonal and stromal factors. The medium is an important factor in this. The various types of media used for tissue culture may be grouped into two broad categories:

1. Natural media and
2. Artificial media.

Natural Media

Natural media is the most preferred medium, since it is the cheapest and most convenient. The natural media, which promotes cell growth, falls into the following three categories.

Coagula such as plasma clots   Plasma clots have been in use since long. Plasma clots are now available commercially either as liquid plasma in siliconed ampoules or as lyophilized plasma. This can be reconstituted by the addition of distilled water saturated with carbon dioxide. One can also prepare plasma from a male fowl, where efforts are made to prevent coagulation of blood by adding heparin as an anticoagulant.

Biological fluids such as serum   The most commonly used biological fluid is serum, which is obtained from human adult blood, placental cord blood, horse blood or calf blood. Out of these, human placental cord serum and foetal calf serum seem to be satisfactory. The serum can be obtained as exuded liquid from blood undergoing coagulation and is filtered using millipore filters. The serum should be tested for sterility and toxicity before using and stored at low temperature. The presence of contaminants can be checked by incorporating the serum in a growth medium. The toxicity of sera can be reduced by heat inactivation.

Besides serum, the following are the other biological fluids used as natural media:

- Amniotic fluid
- Ascitic fluid and pleural fluid

- Aqueous humour (from eye)
- Serum ultrafiltrate
- Insect haemolymph
- Coconut milk (usually used for plant tissue culture)

Tissue extracts, of which embryo extract is the most common   Embryo extract is the most commonly used tissue extract. It is a crude homogenate of a 10-day chick embryo clarified by centrifugation. Cohn (1966) fractionated crude extract to give high and low molecular weight fractions. The low molecular weight fraction promoted cell proliferation, while the high molecular weight fraction promoted differentiation of pigment and cartilage cells. These fractions were not fully characterized, but recent evidences suggest that the low molecular weight fraction probably contained peptide growth factors and high molecular weight fraction contained proteoglycans and other matrix constituents. Smith and Schroedl (1992) showed that the embryo extract could be replaced by hemin in the induction of skeletal muscle differentiation.

Artificial Media

Different artificial media have been devised to serve one of the following purposes.

1. Immediate survival
2. Prolonged survival
3. Indefinite growth
4. Specialized functions

The various artificial media developed for cell culture may be grouped into 4 classes:

1. Serum-containing media
2. Serum-free media

3. Chemically defined media

4. Protein-free media

Serum-containing media The various defined media, e.g. eagle's minimum essential media, when supplemented with 5–20% serum are good nutrient media for culture of most types of cells. The serum provides various plasma proteins, peptides, lipids, carbohydrates, minerals, and some enzymes. Serum serves several major functions like providing basic nutrients for cells, hormones, binding proteins and several minerals. It contains several growth factors and protease inhibitors. A major role of serum is to supply proteins and also act as a buffer. However, there are few disadvantages like the following:

- Serum may inhibit growth of some cell types, e.g. epidermal keratinocytes.

- Serum may contain cytotoxic constituents.

- Sometimes serum may interfere with downstream processing.

Serum-free media In view of the disadvantages due to serum, extensive investigations have been made to develop serum-free formulations of culture media. These efforts were mainly based on the following 3 approaches.

1. Analytical approach based on the analysis of serum constituents.

2. Synthetic approach to supplement basal media by various combinations of growth factors.

3. Limiting-factor approach consisting of lowering the serum level in the medium till growth stops and then supplementing the medium with vitamins, amino acids, hormones, etc., till growth resumes.

The advantages of serum-free media are following.

- Improved reproducibility of results over time from different laboratories since variation due to batch change of serum is avoided.

- Easier downstream processing of products from cultured cells.

- Toxic effects of serum avoided.

- Bioassays are free from interference due to serum proteins.

- There is no danger of degradation of sensitive proteins by serum proteases.

- They permit selective culture of differentiated and reproducing cell types from the heterogeneous cultures.

Serum-free media, with all its advantages, faces some disadvantages like

- Most of them are specific to one cell type.

- Reliable serum-free preparations are not available commercially.

- Growth rate and the maximum cell density attained are lower than those with serum-containing media.

Chemically defined media These media contain contamination-free ultra-pure inorganic and organic constituents, and may conain pure protein additives like insulin, epidermal growth factors, etc., that have been produced in bacteria or yeast by genetic engineering.

Protein-free media In contrast to the above, protein-free media do not contain any protein; they only contain non-protein constituents necessary for culture of the cells.

Several assays have been developed for the determination of the best medium for a given cell type, i.e.,

1. Long-term cell multiplication assay in the form of clonal growth assay, cell growth curve analysis, etc.

2. Short-term [³H]-thymidine assay.

It is desirable to use the long-term assays, but often [³H]-thymidine methods are used for screening; these results should be compared with at least the growth curve analysis, or preferrably, clonal growth assay.

INITIATION OF CELL CULTURES

The initiation of cell cultures involves the following conditions:

1. Preparation and sterilization of the substrate (culture vessels)

2. Preparation and sterilization of the medium

3. Isolation of explant

4. Disaggregation of the explant

5. Subculture and cloning

6. Contamination

Preparation and Sterilization of Substrate

Plasticware are supplied sterilized and ready for use, treatments are usually not necessary. Since they are disposable and not reused, they are costlier than glassware which are adequate for most purposes. However, glassware must be carefully washed with non-toxic detergent following, preferably, an overnight soak. They should be thoroughly rinsed in tap water and finally in deionized or distilled water. Glassware are kept in containers or wrapped in aluminium foil and usually sterilized in dry heat (160°C for 1 hour), while screw caps are autoclaved separately.

The quality of glassware washing and sterilization should be checked periodically by examining the glassware with the eye after

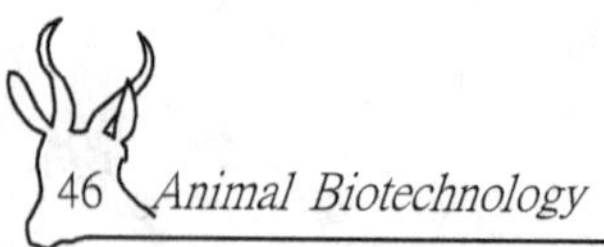

washing, followed by cloning monolayer cells in a low-serum or serum-free medium in the vessels after they have been sterilized.

Preparation and Sterilization of Medium

The various media constituents and other reagents used in cell cultures must be carefully sterilized either by autoclaving or by filtration. Heat-stable constituents like water, salts, supplements like peptone or tryptose, etc. are autoclaved at 121°C for 20 min. But heat-labile constituents like serum, trypsin, proteins, growth factors, etc. must be sterilized by filtration through a 0.2 mm porosity membrane filter. Each filtrate should be tested for sterility to avoid failure due to contamination. Autoclaving is preferred to filtration since it is cheaper, needs less labour and is uniformly effective.

Isolation of Explants

Explants should be dissected out following an appropriate protocol developed to avoid contamination and retain cell survival. The site of opening is sterilized with 70% alcohol; the tissues are removed aseptically, and usually placed in a balanced salt solution, e.g. EBSS or HBSS, supplemented with antibiotics. All subsequent handling should be under strictly aseptic conditions, preferably under laminar-flow cabinets. Care should be taken not to violate ethical or legal boundaries in obtaining the desired explant.

Disaggregation of Explants

Cell cultures are generally started from disaggregated explants. Occasionally, cells do migrate out of the cultured whole tissues; such cells are then separated by trypsin or other enzymatic treatment to establish cell cultures. But in general, the tissues are first disaggregated in one of the following ways:

1. Primary explant technique
2. Mechanical disaggregation

3. Enzymatic disaggregation
4. EDTA treatment

Primary explant technique Harrison (1907) and Corel (1912) were the first persons to develop a primary explant technique. In this technique, a fragment of tissue was embedded in blood plasma or lymph mixed with embryo extract and serum and placed on a slide. The clotted plasma holds the tissue in place. Both the embryo extract and serum will serve as nutrient supplies and stimulate the migration of the explant across the solid substrate. The heterologous serum was used to promote clotting of the plasma. The simplified procedure for primary explant preparation (Figure 2.1) is described below.

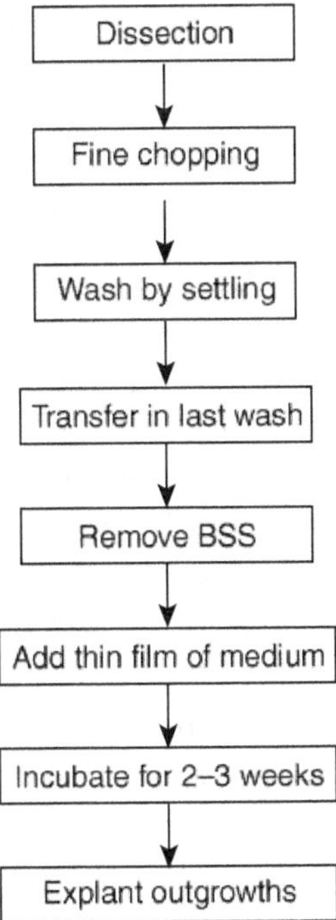

Figure 2.1 Primary explant technique

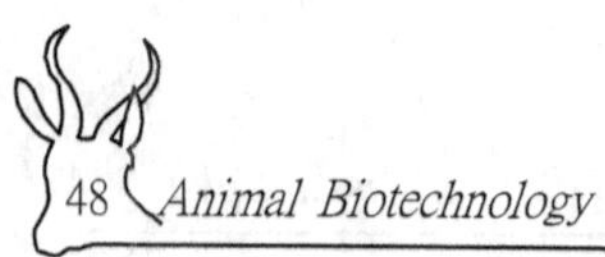

1. Chop the tissue finely, rinse and place on to a medium with a high concentration (40–50%) of serum.

2. Transfer tissue to fresh sterile balanced salt solution (BSS) and rinse, dissect off unwanted tissue like fat or neurotic material and chop finely using crossed scalpels to about 1-mm cubes.

3. Wash the chopped tissue in BSS and allow them to settle; then remove the supernatant fluid two or three times.

4. Transfer the pieces to a culture flask containing 1 ml growth medium per 25 cm² growth surface. Cap the flask and place in an incubator or hot room at 36.5°C for 18–24 hours.

This technique is particularly useful for small amounts of tissue such as skin biopsies where there is a risk of losing cells during mechanical or enzymatic disaggregation.

Mechanical disaggregation   The tissue is carefully sliced and the cells are collected, placed few at a time into a stainless steel sieve of 1 mm mesh in a 9-cm petri dish. The tissue is forced through the mesh into the medium by applying gentle pressure using the piston of a disposable plastic syringe or is repeatedly pipetted. The pieces attach to the substrate due to their own adhesiveness; the dish may be scratched to facilitate attachment, or clotted plasma may be used. Cells grow out from the tissue pieces, which are trypsinized and subcultured. The explants themselves may be retained in the same dish or transferred to a new vessel to obtain further cell growth. The suspension may be diluted and cultured in a suitable medium. Though this method causes mechanical damage to the cells, the suspension is more quickly obtained than by the enzymatic disaggregation. By this method only a good yield of cells in a shorter duration can be obtained but not in efficiency. The steps involved in the mechanical disaggregation method are shown in the Figure 2.2.

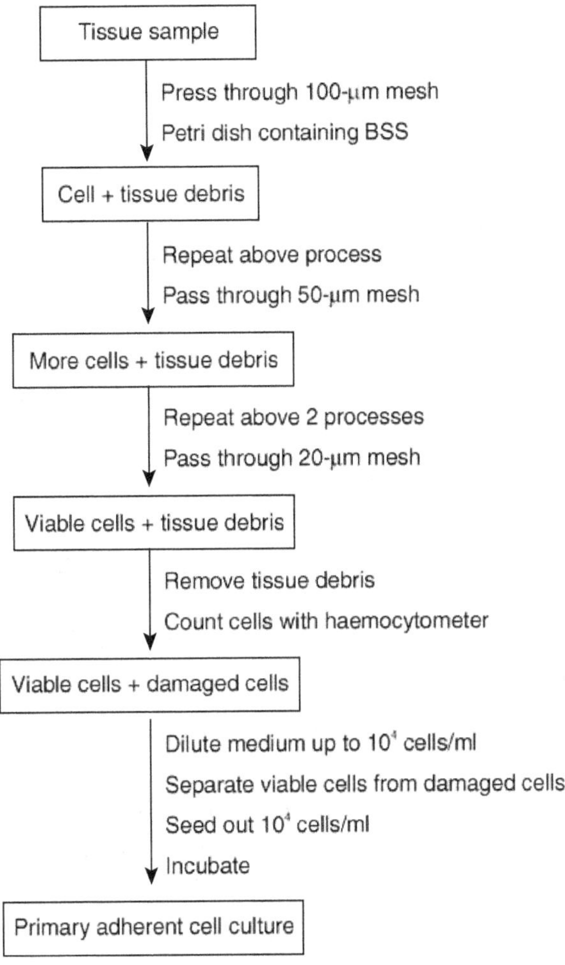

Figure 2.2 Mechanical disaggregation of tissue by sieving (Diagrammatic)

Enzymatic disaggregation Enzymatic or mechanical disaggregation gives a much higher yield of cells in a short duration when compared to primary explant technique. Embryonic tissues disperse more readily and give a higher yield of proliferating cells than new-born or adult tissues. This is because in the latter case, there are fewer undifferentiated cells and more fibrous connective tissue and extracellular matrix. In several tumours, the tissue is mostly destroyed during enzyme treatment and viable cells are obtained with difficulty.

Trypsin (0.25% crude or 0.01–0.05% pure) and collagenase (200–2000 units/ml, crude) are the most commonly used enzymes, but other enzymes like elastase, mucase, papain, etc. have also been used. The steps in disaggregation are as follows:

1. The explant is exposed to the warm enzyme (36.5°C) for a minimum period and the dissociated cells are collected every half an hour.

2. It will take 3–4 hours for the complete disaggregation of tissue. Then trypsin is removed by centrifugation.

3. Alternatively cold trypsinization can be done which involves soaking of tissue in trypsin at 4°C to allow penetration of enzyme, followed by incubation at 36.5°C for a shorter period.

In some cases, trypsin may be either damaging, e.g. epithelial cells, or may be ineffective, e.g. fibrous tissues. Therefore, disaggregation of several normal and malignant tissues is better achieved by collagenase, which readily digests away matrices containing collagen. The method of disaggregation by collagenase is as follows:

1. 0.5 ml of crude collagenase is added to a finely chopped tissue in complete medium.

2. It is then incubated at 36.5°C for 4–48 hours without agitation. Disaggregation is slow in the case of tumour tissues. In that case, it is left in the incubator for up to 5 days.

3. After disaggregation, collagen is removed by centrifugation and cells can be seeded and cultured at a high concentration.

The different steps involved in enzymatic disaggregation of the cells are shown in the Figure 2.3.

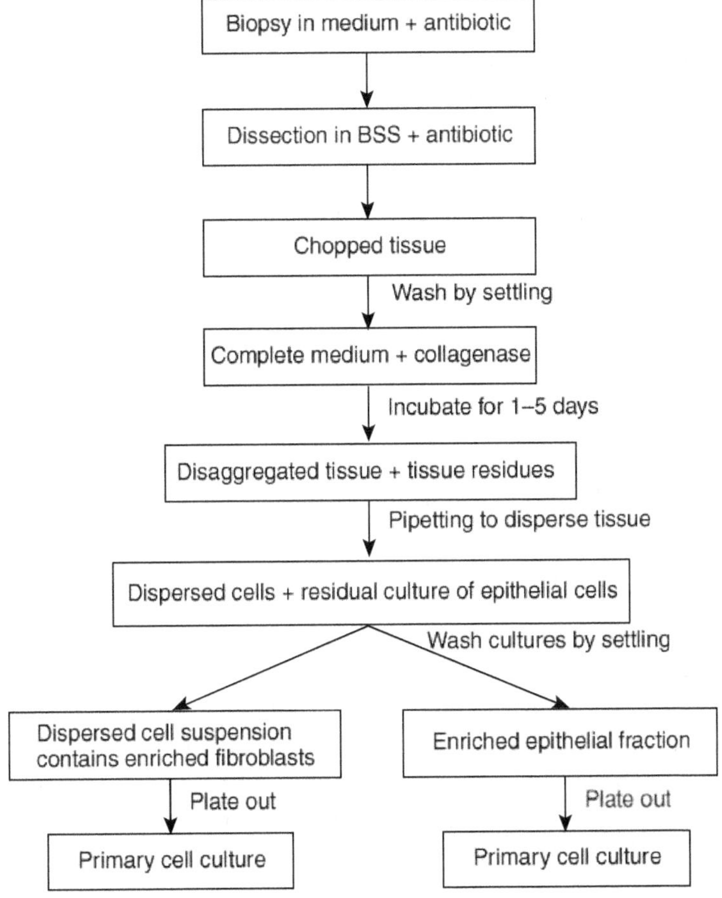

Figure 2.3　Stages in disaggregation of tissue by collagenase

EDTA treatment　Some tissues like epithelium require Ca^{2+} and Mg^{2+} ions for their integrity. Disaggregation of such tissues is readily achieved by a treatment with EDTA (ethylenediamine tetra-acetic acid) prepared in a balanced salt solution. This treatment is generally used for separation of cells from established cultures of epithelial tissues.

Subculture

When mechanically or enzymatically disaggregated tissue pieces or cells are cultured, they attach to the substrate and a monolayer of cells (called primary cultures) develops from them. Primary cultures do not contain many types of cells present in the tissue because they are unable to attach to the substrate and survive; the dead cells are easily eliminated when the medium is poured out during subculture. Further selection will occur during growth of primary cultures and of subsequent cell lines derived from them as faster growing cell types will gradually increase and the slower growing types will go on declining with time.

After a period of time, the available substrate becomes occupied by the monolayer. Cells are then harvested from the primary cultures and inoculated into a fresh vessel. The cultures so obtained are called cell lines. The process of inoculating cultured cells into fresh culture vessels is termed as subculturing.

Contamination

Contamination can occur due to microorganisms, e.g. bacteria, fungi, mycoplasma, viruses and other cell lines. Microbial contamination has been greatly reduced by laminar airflow cabinets and by the use of antibiotics. However, it is important to maintain cell lines without antibiotics to avoid the persistence of some undetected contaminants. Contamination may be checked by looking for:

1. a rapid change in pH

2. cloudiness in the medium

3. granular appearances outside cells (under microscope)

4. unidentified material floating in the medium

Contaminated vessels should be autoclaved prior to opening their mouth for cleaning.

Mycoplasmas are difficult to detect but they do affect cell growth and function. Therefore, cultures should be checked for mycoplasma contamination at least every 3 months using a suitable technique, e.g. fluorescent DNA technique is the most commonly used. Mycoplasma contamination may come from media, sera, trypsin or the worker.

Cross-contamination by other cell lines (from the same or different species) is much more frequent than is realized. This can be checked by using several sensitive and specific techniques. To avoid cross-contamination, bottles of media, reagents, etc. should not be shared among cell lines and even among different workers; pipettes and glassware should not be reused prior to autoclaving once they come in contact with cells.

Bacterial and fungal contaminants can be detected by a series of culture tests using specific test media, e.g. blood agar with defibrinated rabbit blood (5%), yeast extract mannitol broth, nutrient broth with 2% yeast extract, etc.

Testing for the presence of viruses are the most problematic. The various tests employed are

1. cytopatheic effects on cells.
2. adsorption of indicator red blood cells by virus-infected cells in monolayers.
3. sign of degeneration in embryonated chicken eggs following their inoculation with test cells or cell homogenates.
4. cytopathological effects or haemadsorption after co-culturing the test cells with selected substrate cell line.
5. immunological techniques.
6. biochemical assays.

Cross-contamination by cells of a different species can be detected by fluorescent antibody staining, isozyme analyses, and analysis of the chromosome complement (karyotype) with or without a banding pattern. Similarly, cross-contamination by cell lines of the same species are based on i) isozyme patterns which are often characteristic for different cell lines, ii) blood group and histocompatibility antigens, iii) Giemsa banding pattern of chromosomes and iv) DNA fingerprinting (a highly sensitive and reliable approach).

Evolution of cell line The mixture of cells in a single cell suspension may be used as a primary culture or starter culture. The primary culture is subcultured by transferring into culture dishes or flasks containing special growth nutrients at optimal growth conditions. Consequently, some cells attach to the surface and proliferate to yield single cell line, in spite of being damaged in suspension. Therefore, while subculturing, the suspension should be diluted with fresh medium at a certain ratio and transferred into a flask.

The primary culture becomes a cell line only after its first subculture. The subculture is needed when the nutrients present in the medium for cell growth diminish. These may be subcultured several times on the fresh medium and propagated accordingly. By doing so, a continuous growth of cells is maintained. This results in multiple copies of a single type of cell with a negligible amount of non-proliferating cells. During the course of repeated subculture and selection, the cell line gets evolved and properly established, consisting of rapidly proliferating cells (Figure 2.4). Thus the unaltered form of cell line (only for a limited number of generations) is called the continuous cell line, which propagates in logarithmic ways. Any change in the continuous cell line may discontinue the increase in cell number. This may be brought about by chemicals, spontaneous mutation or viruses (e.g. Epstein–Barr virus). The phenomenon of alteration in continuous cell line is called *in vitro* transformation.

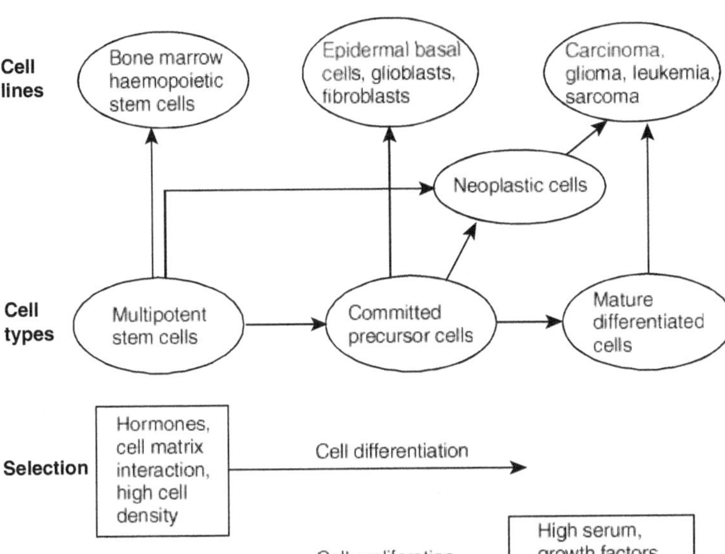

Figure 2.4 Culture conditions affecting cell types and origin of cell lines

MAINTENANCE OF CELL LINES

Thousands of cell lines have been developed from human and other animal tissues; many of them are finite, while others are continuous cell lines. The continuous cell lines themselves may have originated from normal (through transformation) or tumour tissues. American type culture collection (ATCC) maintains a large number of cell lines acquired from their originators; these cell lines are characterized before preservation and/or distribution on request to research workers. A cell line is multiplied and divided into a (i) seed stock and (ii) a working stock. The **seed stock** is characterized and then maintained in frozen state, preferably in liquid nitrogen at −196°C. Seed stock ampoules are used to produce distribution stocks so that the cell line does change with time due to genetic instability, selection, senescence or transformation. On the contrary, **working stock** or **distribution stock** consists of material multiplied from the seed stock and used for further studies or for distribution.

Freeze preservation of animal cells is now routine in all cell line banks. A cryoprotective agent like DMSO or glycerol is generally added to minimize injury to cells during freezing and thawing.

In order to maintain the cell line and proliferation of the primary culture produced on a substrate or in suspension medium, the primary culture needs to be subcultured. Hence cell proliferation has become an important feature. The primary culture may have variable growth fraction, depending on the type of cells present in culture. After the first subculture, the growth fraction is usually high (80% or more).

Many types of cells can be derived from tissue and a more homogeneous cell line can be obtained by a series of subculture of cells derived from the primary culture. The culture now called a cell line can be propagated, characterized and stored. The term "cell line" implies the presence of several cell lineages either similar or distinct. Among these cell lineages, a particular cell lineage with specific properties is identified in the bulk of cells of that culture, which is described as a "cell strain".

Every cell line is given a code or designation in order to differentiate from the several other cell lines used from the same source. For example, normal human brain is designated as NHB, a cell strain is designated by a number such as NHB_1, NHB_2, etc. and if cloned, a clone number NHB_{2-1}, NHB_{2-2}, etc. is given.

For finite cell lines the doubling of the population should be estimated and indicated by a slash number, e.g. $NHB_{2/2}$ and with increase by one for a split ratio of 1 : 2 ($NHB_{2/2}$, $NHB_{2/4}$, etc.) and so on.

Most of the monolayer primary cultures or continuous cell lines need a periodic change of medium (i.e., removing old medium and adding fresh medium), whether or not the cells are proliferating. In cultures where cells are proliferating, the usual practice in subculturing an adherent cell line involves the following steps.

1. Remove the medium and add PBSA along the side of the flask opposite the cells to avoid dislodging of cells. Rinse the cells and discard the PBSA. This is designed to remove traces of serum that would inhibit the action of trypsin.

2. Dissociate cells in the monolayer with trypsin or other enzyme.

3. Incubate until cells round up (~5–15 min.). When the bottle is tilted, the monolayer should slide down the surface.

4. Add medium (0.1–0.2 ml/cm^2) and disperse cells by repeated pipetting over the surface bearing the monolayer. To disperse the cell line, pipette the cell suspension up and down for few minutes to get single cell suspension.

A single cell suspension is desirable at subculture to ensure an accurate cell count and uniform growth.

Intervals between change of medium and subculturing vary from one cell line to another depending upon the rate of growth or metabolism. Rapidly growing cell lines like HeLa are subcultured once per week and the medium changed every four days. Slower growing cell lines are subcultured every 2, 3 or 4 weeks and the medium is changed every week between subcultures.

There are several factors which are used to assess the need for replacement of medium. These include i) drop in pH ii) cell contraction iii) cell type and iv) cell morphology.

The primary cultures or continuous lines are continuously grown in suspension because of two reasons:

1. They are non-adhesive (e.g. many leukemias and murine ascite tumours).

2. They are kept in suspension mechanically and can be subcultured like microorganisms. Trypsin treatment is not required and the whole process is quicker and less traumatic for the cells.

Monolayer cultures are convenient for cytological and immunological observations, cloning mitotic "shake off" (for cell synchronization in chromosome preparation) and are sites of extraction without centrifugation.

LARGE-SCALE CULTURE OF CELL LINES

Cell cultures are used for obtaining useful products like biochemicals (interferon, interleukins, hormones, enzymes, antibodies, etc.) and virus vaccines (for polio, mumps, measles, rabies, foot-and-mouth disease, etc.). For these objectives, large-scale cell cultures are essential; fermenters of up to 10,000 litres are used for this purpose. The scaling up of cell cultures may be done as 1) monolayer cultures, 2) suspension cultures, or 3) immobilized cell systems.

Monolayer Culture

Monolayer cultures are essential for anchorage-dependent cells. Scaling up of such cultures is based on increasing the available area by using plates, spirals, ceramics and microcarriers. The various culture vessels used are briefly described below.

Roux bottle   It is commonly used in the laboratory and is kept stationary so that only a portion of its internal surface is available for cell anchorage (Figure 2.5a). Each bottle provides approximately. 175–200 cm² surface area for cell attachment and occupies 750–1000 cm³ space.

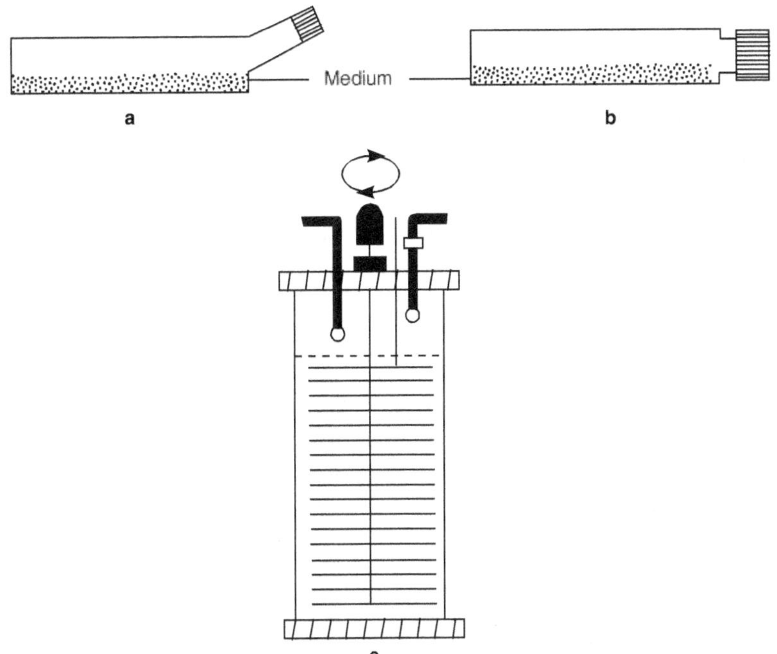

Figure 2.5   Figures showing three types of culture vessels used for large-scale animal cell cultures. (a) Roux flask, (b) Roller bottle (c) Stack plate vessels multitray unit.

Roller bottle   This vessel permits a limited scale up as it is rocked or preferably rolled so that its entire internal surface is available for anchorage. Several modifications of roller bottle further enhance the available surface (e.g. spira-cel and glass tube) (Figure 2.5b).

Multitray unit   A standard unit has 10 chambers stacked on one another, which have interconnecting channels; this enables the various operations to be carried out in one go for all the chambers. Each chamber has a surface area of 600 cm² and the total volume of the unit is 12.5 litres. This polystyrene unit is disposable and gives good results similar to plastic flasks (Figure 2.5c).

Synthetic hollow fibre cartridge   The fibres enclosed in a sealed cartridge provide a large surface of fibres. The capillary fibres are made up of acrylic polymer, are 350 μm in diameter with 75 μm thick walls. The medium is pumped in through the fibre; it perfuses through the fibre walls and becomes available to the cells. The surface area available is very high (up to 30 cm²/ml of medium volume). The system is mainly used for suspension cells, but is also suitable for cell anchorage if polysulphone type fibre is used.

Optical culture system   It consists of a cylindrical ceramic cartridge in which 1-mm² channels run through the length of the unit, and a perfusion loop to a reservoir is provided for environmental (medium, gas, etc.) control. It gives about 40 cm² surface area/ml of medium. It is suitable for virus, cell surface antigen and monoclonal antibody production, and for both suspension and monolayer cell cultures.

Plastic film   Teflon (fluoroethylene propylene copolymer) is biologically inert and highly permeable to gas. Teflon bags (5 × 30 cm) filled with cells and medium (2–10 mm deep) serve as good culture vessels; cells attach to the inside surface of the bags. Alternatively, teflon tubes are wrapped around a reel with a space and the medium is pumped through the tube; cells are grown on the inside surface of the tube. (A culture vessel called stericell is available).

Heli-cell vessels   These vessels are packed with polystyrene ribbons (3 mm × 5–10 mm × 100 μm) that are twisted in helical shape. The medium is pumped through the vessel, the helical shape of ribbons ensuring good circulation; the cells adhere to the ribbon surfaces.

Bead bed reactor   These reactors are packed with 3.5-mm glass beads (which provide the surface for cell attachment) and the medium is pumped either up or down the bead column. Use of 5-mm beads gives better cell yields than that of 3-mm beads.

Microcarrier cultures   Here they use particles of 90–300 mm diameter as substrate for cell attachment. Initially, dextran beads (sephadex A-50) were used by Van Wezel in 1967; these were not entirely satisfactory due to the unsuitable charge of beads and possibly due to toxic effects. The microcarriers available for use at present include dextran, polystyrene, glass, polyacrolein, polyacrylamide, silica, DEAE, sephadex, cellulose and gelatin; the specific gravity of microcarriers ranges from 1.02 to 1.05.

Microcarrier cultures are initiated by harvesting cells from 3 litres of logarithmic phase culture and inoculating them in 1 litre of fresh medium to which 2–3 g/l of microcarriers is then added. The culture is stirred at 15–25 rpm for 3–8 hrs. During this period, cells attach to microcarrier beads and later grow as a monolayer. The volume of culture is slowly increased to 3 litres and stirring is enhanced to the normal rates (20–100 rpm). As the cells grow, the beads become heavier and need to be agitated at higher speeds. The medium needs to be changed every 3 days. Samples of beads can be drawn for observations on cell morphology, growth and number. Harvesting of cells from microcarrier beads is rather simple—stirring is stopped, the medium is drained off, the beads are washed in buffer, treated with trypsin or any other suitable enzyme, the culture is shaken at 75–125 rpm for 20–30 min, stirring is stopped for 2 min and the supernatant is poured and collected.

Scaling up of microcarrier cultures can be done either by increasing the concentration of beads or by enlarging the culture vessel. When high microcarrier concentrations are used, medium perfusion becomes necessary, and efficient filters must be added to allow medium withdrawal without cells and microcarriers. The oxygen supply is problematic; it can be based on the surface aeration, increased perfusion rate of fully aerated medium and sparsing into the filter compartment.

Suspension Culture

In suspension culture, the cells are dispersed in liquid medium and grow freely, and are not attached to any substrate. The medium is agitated so that they do not form sediments. Hence, both chemical (O_2, pH, medium constituents and removal of waste) and physical (the configuration of bioreactor and power supplied to the reactor) factors are optimized.

Stirred bioreactors, continuous flow reactors and airlift fermenters are the 3 main types of reactors used for large-scale suspension cultures.

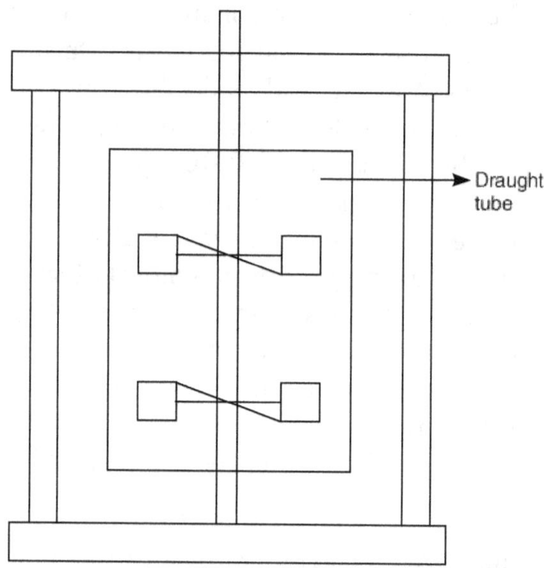

Figure 2.6   A compact loop bioreactor (diagrammatic)

A fully automated bioreactor is used, which maintains the physiochemical and biological factors to optimum level and maintains freely the suspended cells in an agitated liquid medium having low viscosity. The most suitable bioreactor used is a compact-loop bioreactor that consists of marine impellers (Figure 2.6). It also relies on the integrated monitoring and control of physical, chemical, biological and biochemical parameters.

Immobilized Cultures

Cultures based on immobilized cells offer several advantages:

1. Higher cell densities (50–200×10^{-6} cells/ml).
2. Stability and longevity of cultures.
3. Applicablity to both suspension and monolayer cultures.
4. Protection of the cells by many systems from shear forces due to medium flow.
5. Less dependence of cells at higher densities on external supply of growth factors, which saves culture cost.

There are two basic approaches to cell immobilization:

- Immurement
- Entrapment

Immurement cultures In such cultures, cells are confined within a medium-permeable barrier. Hollow fibres packed in a cartridge is one such system. The medium is circulated through the fibre while cells in suspension are present in the cartridge outside the fibre. This is extremely effective for scales up to 1 litre and gives cell densities of 1–2×10^8 cells/ml; sophisticated units can yield up to 40 g monoclonal antibodies/month. Membranes permitting medium and gas diffusion are also used to develop bioreactors of this type.

The cells may be encapsulated in a polymeric matrix by adsorption, covalent bonding, cross-linking or entrapment; the materials used as matrix are gelatin, polylysine, alginate and agarose. This approach has the following advantages.

- It effectively protects cells from mechanical damage in large fermenters.

- It allows production of hormones, antibodies, immunochemicals and enzymes over much longer periods than is possible in suspension cultures.

- The medium diffuses freely into the matrix and into the cells, while cell products move out into the medium. For production of larger molecules like monoclonal antibodies, agarose in a suspension of paraffin oil is preferable to alginate since the latter does not allow diffusion of such products out of the alginate beads.

Entrapment cultures   In this approach, cells are held within an open matrix through which the medium flows freely. An example is the opticell in which the cells are entrapped within the porous ceramic walls of the unit. The cells can also be enmeshed in cellulose fibres, e.g. DEAE, TLC, etc. These fibres are autoclaved and washed as prescribed, and added in a spinner/ stirred bioreactor at a concentration of 3 g/l. Porous microcarriers are small (170 μm–6000 μm) beads of gelatin, collagen, glass or cellulose, which have a network of interconnecting pores. These provide a tremendous enhancement in surface area/volume ratio, permit efficient diffusion of medium and product, are suitable for scaling up, and are equally useful for suspension and monolayer cultures. These can be arranged as fixed bed or fluidized-bed reactors or used in stirred bioreactors. It is expected that future developments will make the immobilized cell systems the most dominant production systems.

CHARACTERIZATION OF CELL LINE

The characterization of various cell lines is required for the following reasons:

1. Correlation with the tissue of origin. (i.e., to identify the lineage to which the cell belongs, to understand the position of cells within a lineage and to check whether the cells have been transformed or not).

2. To maintain the genetic instability and phenotypic variation.

3. To check cross-contamination and for confirmation of species of origin.

4. To identify specific cell lines, which requires demonstration of features unique to that cell line or cell strain.

Following are some of the techniques to characterize cell lines.

Chromosome Banding (Karyotyping)

This technique helps in the identification of individual chromosome pairs, when there is little morphological difference between them. Karyotyping (chromosome banding) is a test to identify chromosome abnormalities as the cause of malformation or disease. This test can count the number of chromosomes and look for structural changes in chromosomes. The results may indicate genetic changes linked to a disease, e.g. Giemsa banding also called G-banding, Q-banding and C-banding.

KARYOTYPING

How the Test is Performed

The test can be performed on a sample of blood, bone marrow, amniotic fluid, or placental tissue. Chromosomes contain thousands of genes that are stored in DNA, the basic genetic material. The specimen is grown in tissue

(Contd.)

culture in the laboratory. Then, the cells are harvested, and the chromosomes are stained and viewed under the microscope. They are photographed to provide a karyotype, which shows the arrangement of the chromosomes. Certain abnormalities can be identified through the number or arrangement of the chromosomes.

How to Prepare for the Test

There is no special preparation for the blood test. To test amniotic fluid, an amniocentesis is performed. Testing on placental tissue is done after a chorionic villus sampling or after a miscarriage. A bone marrow specimen requires a bone marrow biopsy.

Why the Test is Performed

The blood test is usually performed to evaluate a couple with a history of miscarriages or to evaluate an abnormal appearance of the body that suggests a genetic abnormality. The bone marrow or blood test can be done to identify the Philadelphia chromosome that is present in 85% of those with chronic myelogenous leukemia (CML). The amniotic fluid test is done to evaluate a developing foetus for chromosome abnormalities.

Normal Values

- Females: 44 autosomes and 2 sex chromosomes (XX), denoted 46 (X,X)
- Males: 44 autosomes and 2 sex chromosomes (XY), denoted 46 (X,Y)

Abnormal results may indicate Down's syndrome (Trisomy 21 = three copies of chromosome 21 instead of the normal two copies), Trisomy 18, Philadelphia

(Contd.)

chromosome, Klinefelter's syndrome, Turner's syndrome, or other abnormalities.

Additional conditions under which the test may be performed:

- ❋ Multiple birth defects
- ❋ A baby born with genitals that is neither completely male nor female

Risks Involved in this Technique

The risks are related to the procedure used to obtain the specimen. There is a specialized kind of risk in that an abnormal result may have occurred during growth of the cells after they left the body. For this reason it is often prudent to repeat the karyotype test to confirm that the abnormal chromosome constitution is in the body of the patient.

There is a rare difference between the apparent sex of the patient and their chromosomes. For example, a baby may look like it has a penis and be called a boy but turn out to have the chromosomes of a girl. This can raise issues of what gender to raise the child.

Chromosome Painting

It has become possible to locate genes with fluorescently labelled probes that will bind to specific genes on the chromosome. It helps in identifying translocation and determines the species of origin of chromosome. This technique employs *in situ* hybridization technology, which can also be used for extrachromosomal and cytoplasmic localization of specific nucleic acid and can be used to localize specific mRNA species.

- ❋ Chromosome painting has improved the efficiency of screening cells for chromosome abnormalities, in testing chemicals for mutagenicity and for rearrangements associated with tumours.

- ✹ Chromosome painting is also called M-FISH, multicolour FISH.
- ✹ Painting probes detect chromosome rearrangements.
- ✹ Use of the same chromosome paints for chromosomes of different species reveals the extent of chromosome rearrangements since divergence of the species. Such studies reveal extensive synteny between fairly divergent species.

DNA Content Analysis

The amount of DNA per cell is relatively stable in normal cell lines, such as human fibroblast, chick and hamster fibroblast, but varies in cell lines from the mouse and from many neoplasms. The content of DNA can be analysed using flow-cytometer, microfluorometry, ethidium bromide fluorescence, etc. DNA content analysis is useful in characterization of transformed cells that are often aneuploid and heteroploid.

DNA Hybridization

The extracted DNA is cleaved with restriction enzymes, electrophoresed, blotted on nitrocellulose membrane and hybridized with labelled probes. This provides information about species-specific regions or amplified regions of the DNA that are characteristic to that cell line. Thus strain-specific gene amplifications, such as amplification of the dihydrofolate reductase (DHFR) gene, may be detected in cell lines selected for resistance to methotrexate.

DNA Fingerprinting

Since DNA contains a region known as satellite DNA, these are apparently not transcribed and will give rise to regions of hypervariability. When the DNA is cut with specific endonucleases, it can be probed with cDNAs that hybridize to these hypervariable regions. Electrophoresis reveals the variations in fragment length that are specific to the individual from whom the DNA was derived. When analysed by PAGE, each individual's DNA gives a specific hybridization pattern. With this technique, the origin of a cell line and any cross-contamination can be confirmed.

Isoenzymes

The multiple forms of an enzyme catalysing the same reaction are referred to as isoenzymes or isozymes. Isoenzymes differ in many physical and chemical properties, structure, electrophoretic and immunological properties, K_m and V_{max} values.

The isoenzymes can be separated by analytical techniques such as electrophoresis and chromatography. Most frequently, electrophoresis by employing agarose, cellulose acetate, starch and polyacrylamide is used. The crude enzyme is applied at one point on the electrophoretic medium. As the isoenzymes migrate, they distribute in different bands, which can be detected by staining with suitable chromogenic substrates.

The isoenzymes can be used to characterize cell lines, due to polymorphisms among species and sometimes among races or individuals within the species. The isoenzymes can be separated chromatographically or electrophoretically and distribution patterns (zymograms) of species or tissue can be found.

Chromosome Painting, a Powerful Tool for Chromosomal Analysis

Fluorescent *in situ* hybridization (FISH) has been used to detect the location of specific genomic targets using probes that are labelled with specific fluorochromes. The

(Contd.)

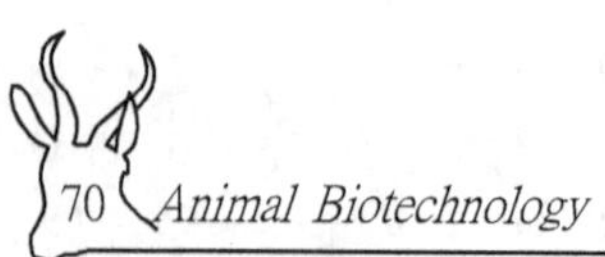

limited number of available, spectrally non-overlapping fluorochromes, has prevented the simultaneous identification of multiple targets including different chromosomes.

Spiecher *et al.* reported a method that is called multiplex-fluorescence *in situ* hybridization (M-FISH). They developed epifluorescence filter sets and computer software that allowed detection and discrimination of 27 different DNA probes hybridized simultaneously to human chromosomes. The technique allowed detection of simple and complex chromosomal rearrangements. In addition, complex chromosomal abnormalities could be identified that could not be detected by the conventional cytogenetic banding techniques. In 1996 Schrock *et al.* reported the development of a similar technique that allows the multi-colour detection of human chromosomes. The technique has been accomplished by allowing 24 combinatorially labelled chromosome-painting probes to hybridize with the human chromosomes. Then, the emitted spectrally overlapping chromosome-specific DNA probes were resolved by using computer separation (classification) of the spectra. This technique can be used for detection of chromosomal abnormalities. On the basis of the location of the probes used, the size of the alteration can be estimated. In addition, the developed technique provides information that complements the conventional banding analysis. With the use of this technique, the authors easily identified presence of numerous chromosomal translocations and unambiguously identified structural alterations including a giant marker chromosome (mar1) in the aneuploid breast cancer cell line, SKBR3.

(Contd.)

Facts

- Collections of nucleic acid sequences, each known to hybridize to only one human chromosome have been assembled.

- Sequences specific for one chromosome are converted to probes labelled with a fluorescent dye. Probe sets for other chromosomes can be made similarly, each with a different fluorescent dye, or specific mixture of dyes.

- Hybridization, under FISH conditions, with one chromosome's paint probe results in specific labelling of that chromosome.

Simultaneous hybridization with all probe sets results in a chromosome spread in which each of the chromosomes in a human haploid chromosome set appears a different colour when viewed with a fluorescence microscope.

Interpretations

- To test whether a specific gene is located on a particular chromosome, FISH is done with the appropriate chromosome paint probe and a gene-specific probe labelled with a colour dye different from the paint probe. The test will also yield an approximate location of the specific gene on that chromosome.

- A chromosome that is the result of a translocation will consist of two segments, each of a different colour. The colours identify which two original chromosomes participated in the translocation. The positions of the colour boundaries on the chromosomes identify the breakpoints of translocation.

Use of Markers

Markers are used to characterize and identify a cell line. These are unique and specific to cell lines, e.g. cell surface antigens, immediate filament proteins, different products such as haemoglobin, melanin, myosin, etc.

Marker gene is also called reporter gene. These genes produce a phenotype which is easily and specifically detected or which allows differential multiplication of the cells. Some examples are thymidine kinase, DHFR, FDA protein, HGPRT, neomycin phosphotransferase.

Thymidine kinase (TK)   Nucleotide synthesis through salvage pathway requires the activity of thymidine kinase. It is necessary in the synthesis of TTP; it phosphorylates thymidine. TK-deficient (TK^-) cells are killed on hypoxanthine aminopterin and thymidine (HAT) medium, which contains DNA nucleosides, hypoxanthine, thymidine and aminopterin.

Aminopterin blocks the de novo (endogenous) pathway of nucleotide synthesis. The cells are forced to follow salvage pathway for their survival. TK^- cells are unable to use thymidine present in the medium for nucleotide starvation on HAT medium, which acts as an extremely efficient selection agent for TK^+ cells.

The *tk* genes can be used as selectable markers only when TK^- host cells are used for transfection. The transfected cells are cultured on the HAT medium on which only TK^+ cells can survive and multiply.

Dihydrofolate reductase (DHFR) (methotrexate resistance) Normal cells are sensitive to methotrexate, even at a very low concentration (0.1 mg/ml). Methotrexate is an inhibitor of DHFR enzyme, involved in the biosynthesis of nucleotides by de novo pathway. Cell lines resistant to methotrexate have been developed. This can be exploited as marker gene for cell line selection.

Neomycin phosphotransferase (G 418 Resistance) G 418 is an aminoglycoside antibiotic, which inhibits protein synthesis. Mammalian cells are sensitive to G 418. Neomycin phosphotransferase confers resistance to G 418. Various constructs (plasmids/cosmids) of neomycin phosphotransferase gene have been prepared and used for transfection of a variety of mammalian cell lines. Transfected cells are efficiently selected in a medium containing G 418.

CDA proteins (PALA resistance) Normal mammalian cells are sensitive to N-phosphoacetyl-L-aspartate (PALA), an inhibitor of CDA protein required for uridine biosynthesis. CDA gene is isolated from Syrian hamster. Transfected mammalian cells are highly resistant to PALA. This can be exploited as a marker gene for selection of cell lines.

VALUABLE PRODUCTS FROM CELL CULTURE

From cultured animal cells, several valuable products such as human monoclonal antibodies, and biochemicals can be produced on a large scale. Several million dollars have been earned from this industry in Europe, America, Africa, Japan and India. This industry has better future interestingly; the genetically engineered cells have revolutionized the cell culture industry. Several specific promoters of human origin are utilized for high expression of foreign genes.

For large-scale production of certain biochemicals, the genetically engineered baculovirus-infected animal cells are also in use in a bioreactor. To fulfil the process, several "perfusion systems" have been developed that retain the cell density and in turn cell productivity. For commercial production of products, a large-scale cell culture system and scaling up of process are required. Therefore, "Master cell banks" (MCBs) are established to meet out the demand.

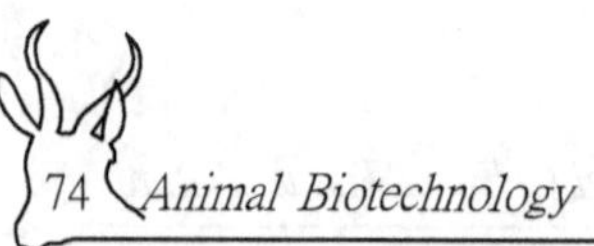

Some of the important products which are produced from animal cell cultures are:

1. Enzymes (asparaginase, collagenase, urokinase, pepsin, renin)

2. Hormones (luteinizing hormone, follicle stimulating hormone, erythropoietin)

3. Vaccines (foot-and-mouth disease vaccine, vaccines for influenza, measles and mumps, rubella, rabies)

4. Monoclonal antibodies

5. Interferon, etc.

Some of the products of medical use derived from animal cell cultures, and their applications are listed in Table 2.2.

Table 2.2 Some products of medical use derived from animal cell cultures

Products	Application
1. **Erythropoietins (EPO)**	
EPO-α	Anaemia resulting from cancer and chemotherapy
EPO-β	Anaemia secondary to kidney disease
2. **Human growth hormones**	
hGH	Human growth deficiency in children, renal cell carcinoma
Somatotropin	Chronic renal insufficiency
3. **Monoclonal antibodies (therapeutic)**	
Anti-lipopolysaccharide	Treatment of sepsis
Murine anti-idiotype /human B cell lymphoma	B cell lymphoma

(Contd.)

Table 2.2 (Continued)

Products	Application
4. **Monoclonal antibodies (diagnostics)**	
Anti-fibrin 99	Blood clot
Tem-Fab (breast)	Blood cancer
PR-356, CYT-356-in-111	Prostate adenocarcinoma
5. **Plasminogen activator**	
Urokinase type plasminogen activator	Acute myocardial infarction, acute stroke
Tissue type plasminogen activator	Pulmonary embolism
Recombinant plasminogen activator	Deep-vein thrombosis
6. **Vaccines**	
HIV vaccines (gp 120)	AIDS prophylaxis and treatment
Malaria vaccines	Malaria prophylaxis
Polio vaccines	Poliomyelitis prophylaxis

MERITS AND DEMERITS OF TISSUE CULTURE

Although tissue culture offers multiple advantages, there are several demerits associated. Hence, extreme caution is to be taken to avoid indiscriminate use of tissue culture. The advantages and disadvantages are listed in Table 2.3.

Table 2.3   Advantages and disadvantages of the use of animal cell and tissue culture

Advantages	Disadvantages
Controlled physiochemical environment (pH, temperature, oxygen, carbon dioxide, osmotic pressure, etc.)	Expertise is needed
Controlled and defined physiological conditions (constitution of medium, etc.)	10 times more expensive for the same quantity of animal tissue
Homogeneity of cell types (achieved through serial passages)	Unstable aneuploid chromosome constitution
Economical, since smaller quantities of reagents are needed than *in vitro*	Requirement of controlled and defined physiological conditions

APPLICATIONS OF ANIMAL CELL CULTURE

There is a widespread concern that extensive use of animals for laboratory experiments is not morally and ethically justifiable. Animal welfare groups world over are increasingly criticizing the use of animals. Some research workers these days prefer to utilize animal cell cultures, wherever possible for various studies. The major applications of laboratory animal cell cultures are shown in Table 2.4 and listed below.

- Studies on intracellular activity, e.g. cell cycle and differentiation, metabolism.
- Elucidations of intracellular flux, e.g. hormonal receptors, signal transduction.
- Studies related to cell–cell interaction, e.g. cell adhesion and motility, metabolic cooperation.
- Evaluation of environmental interactions, e.g. cytotoxicity, mutagenesis.

Table 2.4   Summary of the applications of animal cell cultures

Category	Application
Intracellular activity	Studies related to cell cycle and differentiation, transcription, translation, energy metabolism, drug metabolism
Intracellular flux	Studies involving hormonal receptors, metabolites, signal transduction, membrane trafficking
Cell–cell interaction	Studies dealing with cell adhesion and motility, matrix interaction, morphogenesis, paracrine control, metabolic cooperation
Environmental interaction	Studies related to drug actions, infections, cytotoxicity, mutagenesis and carcinogenesis
Genetics	Studies dealing with genetic analysis, transfection, transport, immortalization, senescence
Cell products	Wide range of applications of the cellular products formed, e.g. vaccines, hormones, interferons, etc.

- Studies dealing with genetics, e.g. genetic analysis, immortalization, senescence.
- Laboratory production of medical, pharmaceutical compounds for a wide range of applications, e.g. vaccines, interferons, hormones.

There are however, several limitations on the use of animal cell cultures. This is mostly due to the differences that exist between the *in vivo* and *in vitro* systems and the validity of the studies conducted in the laboratory.

SAFETY REGULATIONS

Some of the developed countries have formulated general safety regulations to minimize the risks associated with tissue culture laboratories. For example,

- "Biosafety in microbiological and biomedical laboratories", U.S. Department of Health and Human Sciences (1993).

- "Safe working and the prevention of infection in clinical laboratories" U.K. Health Services Advisory Committee (1991).

Some of the general precautions for the safety of a tissue culture laboratory are listed here.

- Strict adherence to recommendations of regulatory bodies.

- Periodical meetings and discussions of local safety committees.

- Regular monitoring of the laboratories.

- Periodical training of the personnel through seminars and workshops.

- Printing and making the standard operating procedures (SOPs) available to all staff.

- Good record keeping.

- Limited access to the laboratory (only for the trained personnel and selected visitors).

- Appropriate waste disposal system for biohazards, radioactive wastes, toxins and corrosives.

Cell culture systems are very much essential as they enable the scientists to study cell growth and differentiation and also to perform genetic manipulations required to understand gene structure and function. Embryo fibroblasts are one of the most widely studied types of animal cells. The medium used to culture animal cells contains salts, glucose, various amino acids and vitamins, which the cells cannot make for themselves. The media also constitute serum that serves as a source of polypeptide growth factors that are required to stimulate cell division. The initial cell cultures established from a tissue are called primary cultures. Characterization of a cell line is necessary and various techniques such as karyotyping, chromosome painting, DNA fingerprinting, DNA hybridization, etc., are employed. Various marker genes are also used to characterize and identify cell lines. These are unique and specific to cell lines. (e.g. cell surface antigens, immediate filament proteins). Some marker genes are helpful in detection as they produce a phenotype, which is easily detected. (e.g. thymidine kinase, DHFR, CDA protein, HGPRT, etc.).

Summary

- Cell culture is a technique in which tissue or explant is cultured as a monolayer or suspension culture.

- The nutrient media used for animal cell and tissue culture must be able to support their survival as well as growth.

- Mechanical disaggregation, enzymatic disaggregation and EDTA treatment are the techniques employed in disaggregation of explants.

- It is important to maintain cell lines without antibiotics to avoid the persistence of some undetected contaminants.

- The primary culture becomes a cell line only after its first subculture.

- Freeze preservation of animal cells using cryoprotective agents like DMSO or glycerol is now routine in all cell line banks.

- ✔ Every cell line is given a code to differentiate from the several other cell lines used from the same source.

- ✔ Monolayer cultures are essential for anchorage-dependent cells.

- ✔ Stirred bioreactors, continuous flow reactors and airlift fermenters are the 3 main types of reactors used for large-scale suspension cultures.

- ✔ Immurement and entrapment are the two basic approaches to cell immobilization.

REVIEW QUESTIONS

1. Define
   i. cell culture
   ii. tissue culture.
2. Enumerate the characteristic features of cell culture.
3. Brief on stem cells.
4. Brief on primary cell cultures.
5. Give a detailed account of the four classes of artificial media developed for cell cultures.
6. State the conditions involved in the initiation of cell cultures.
7. Elaborate disaggregation of explants.
8. Explain the various tests employed to detect the presence of viruses in cell cultures.
9. Give a brief account of monolayer culture.
10. Explain the techniques followed in characterizing cell lines.

3

GENETIC ENGINEERING

Individual versus Clone

Besides the ethical debate over cloning, biology suggests that premises behind the attempt may be flawed, for a clone is not an exact replica of an individual. Many of the distinctions between an individual and a clone arise from epigenetic phenomena, that is, effects that do not change genes themselves, but alter their expression. There are other distinctions too:

- Telomers of chromosomes in the donor nucleus are shorter than those in the recipient cell.

- In normal development, for some genes, one copy is turned off, depending upon which parent transmits it. That is, some genes must be inherited from either the father or the mother to be active, a phenomenon called genomic imprinting. Genes in a donor nucleus do not pass through a germ-line before they go back to the beginning of development, and thus are not imprinted. Effects of lack of imprinting in clones aren't known.

- DNA from the donor cell had years to accumulate mutations. Such a somatic mutation might not be noticeable if it occurs in one of millions of somatic cells, but it could be devastating if the somatic cell nucleus is used to program development of an entire new individual. Donor DNA introducing mutations from the previous life may contribute to the very low success rates of animal cloning experiments.

- At a certain time in early prenatal development in all female mammals, one X chromosome is inactivated. Whether the inactivated X chromosome is from the mother or the father occurs at random in each cell, creating an overall mosaic pattern of expression for genes on the X chromosome. The pattern of X inactivation of a female clone would most likely not match that of her nucleus donor.

- Mitochondria contain DNA. A clone's mitochondria descend from the recipient cell, not the donor cell.

INTRODUCTION

Recombinant DNA technology is one of the most fascinating fields in biotechnology. It stems from an understanding of functions of DNA and discoveries in molecular biology. The foremost technique in recombinant DNA technology is the transfer of a sequence of DNA (gene) from one organism to another. Hence, it is also called as **Molecular cloning** or **gene cloning** or broadly **genetic engineering** .

Genetic engineering is especially useful to introduce a gene from higher organisms in bacteria or yeast. Since manipulation may result in large-scale production of several pharmaceutically important compounds, the exploitation of microorganisms for the benefit of mankind becomes vital.

Genetic engineering is essentially an insertion of a specific piece of foreign DNA into a cell in such a way that the inserted DNA is able to replicate and pass on to daughter cells during cell division. The inserted DNA is known as **Transgene** and any organism containing the transgene is called **Transgenic** or **Genetically Modified Organism (GMO)**.

Cloning is the use of technology to make an exact genetic copy of a living organism. The term may also be applied to making a copy of simple cells, a gene or a segment of DNA.

The essential steps of gene cloning are as follows:

1. Isolation of the target gene from the DNA of an organism

2. Insertion of the target gene into vector (Plasmids/cosmids)

3. Transformation of the inserted DNA in vector

4. Screening of recombinants and selection of transformed cells

5. Expression of a cloned gene through controlled elements

CLONING TECHNIQUES

Some of the current cloning techniques include somatic cell nuclear transfer embryo splitting and nuclear transplantation.

Somatic Cell Nuclear Transfer

Cells from the original organism, called donor cells, are grown in the laboratory. The female egg (ovum) is prepared by having its genetic material removed. The selected donor cell is placed next to the empty ovum and a small electrical current allows the genetic material from the donor egg to fuse into the ovum. The ovum, with its complete set of genes, is "tricked" by thinking it has been fertilized and develops into an embryo. The embryo is then implanted into a prepared uterus.

Embryo Splitting

The simplest way to create a clone is to prompt a fertilized egg to split into two. Soon after fertilization, the fertilized egg divides

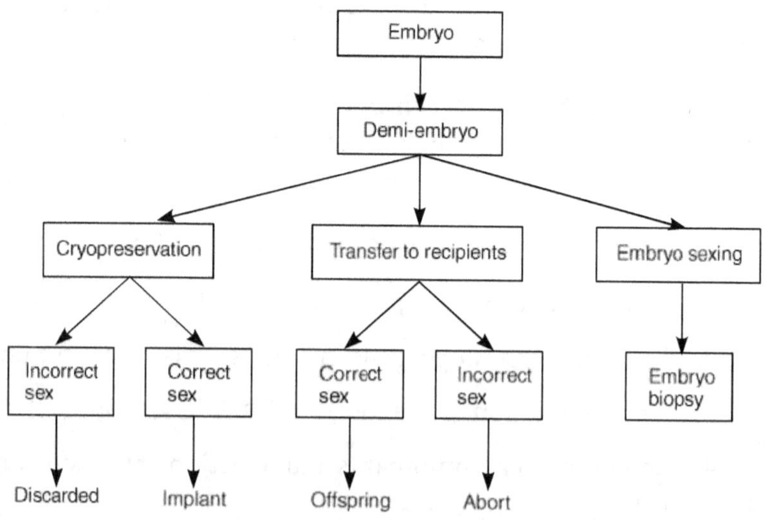

Figure 3.1 An outline of embryo splitting

to form an embryo. The cells of the embryo are called blastomeres. Using special instruments, the coating of the fertilized egg (the zona pellucida) is taken away. The blastomeres are teased apart and each coated with an artificial zona pellucida, which prompts them to start growing as individual embryos. This technique has been employed for stockbreeding. In practice, it is possible to get only a maximum of 4 embryos by using this method (Figure 3.1).

Nuclear Transplantation

It is a process of making multiple copies of the same original cell. The blastomeres form one embryo which is fused with prepared egg cells, using an electric current. The genetic material can be taken from the body (such as the skin or ear) or from a fertilized ovum.

TOOLS IN GENETIC ENGINEERING

The engineering of genetic material was one of the prime goals for biotechnologists in the early 1970s. There was no method available for cutting and pasting of DNA. With the advancement in the new inventions, several tools became available for genetic manipulation of DNA. One of the most remarkable achievements in the emergence of genetic engineering is the discovery of restriction endonucleases, the molecular scissors that contributed significantly towards gene splicing and analysis of cloned genes.

The inventions of different types of vectors and their role in cloning of specific genes of interest have been found to be exemplary. The utility of bacterial plasmids and viruses in accommodating foreign gene of interest and their efficient expression gained significant momentum in recombinant DNA techniques.

Enzymes Involved in Genetic Engineering

Restriction endonucleases   Restriction endonucleases belong to a class of bacterial enzymes that cut DNA molecules internally at specific recognition sites. The discovery of restriction endonucleases was made by Hamilton Smith in 1970, in the bacterium *Haemophilus influenzae.*

Restriction endonucleases cut the DNA backbone at an internal site on the molecule. All restriction endonucleases have a specific recognition sequence within the DNA molecule. These sequences generally fall between 4–6 base pairs in length, and are often known as "palindromic sequences", i.e., the two strands are identical when read both in forward and backward direction, in this region. Each individual enzyme recognizes only one specific sequence. However, the same specific sequence is recognized by two or more different enzymes. Such types of enzymes are called "isoschizomers".

Restriction endonucleases can be classified into types I, II and III. The type I enzyme recognizes the specific sequence, but cuts the DNA 1000 base pairs away from the recognition site. Hence, the use of class I enzyme in genetic engineering is restricted. Class I enzyme requires Mg^{2+} and ATP as cofactors. On the other hand, type II restriction endonucleases recognize the specific sequence and cut the DNA at the same specific site or within the specific sequence. They recognize tetra-, penta- or hexa- and sometimes even hepta-nucleotide sequences. They require Mg^{2+} for restriction digestion. The classical example of class II type restriction is *Eco* RI (Figure 3.2).

Structurally most of the restriction enzymes are made up of 2 equal subunits, that have a molecular weight of 20,000–25,000 Daltons or of a single polypeptide with molecular weight of 30,000–35,000 Daltons.

Restriction endonucleases create staggered end (cohesive/ sticky end or sometimes blunt end (flush end) in the DNA,

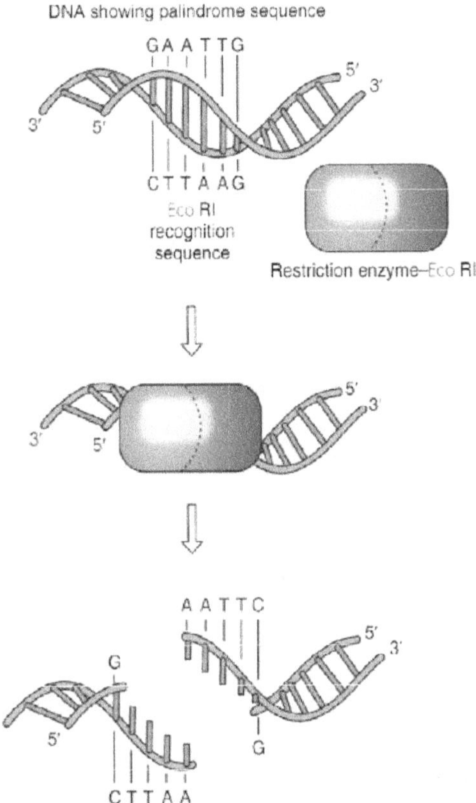

Figure 3.2   Mode of action of *Eco* RI to cleave DNA

depending on the type. (*Eco* RI and Hae III are the examples). Restriction enzymes which produce sticky ends are useful in genetic engineering work (Table 3.1).

Type III restriction enzyme is made up of 2 subunits, one specifies for site recognition and modification and the other for cleavage. In a reaction, it moves along the DNA and requires ATP as source of energy and Mg as cofactor. ATPase activity is lacking in these enzymes. They have symmetrical recognition sites and cleave DNA at specific non-palindromic sequences. One

Table 3.1   Source of type II restriction enzymes, cleavage sites and production of cleavage

Microorganisms	Restriction enzymes	Cleavage sites		Cleavage products	
Bacillus amyloliquefaciens H	*Bam* HI	↓ 5'-GGATCC-3' 3'-CCTAGG-5' ↓ ↑		5-G 3-CCTAG	GATCC-3 G-5
B. globigii	*Bgl* II	5'-AGATCT-3' 3'-TCTAGA-5' ↓ ↑		5-A 3-TCTAG	GATCT-3 A-5
Escherichia coli RY 13	*Eco* RI	5'-GAATTC-3' 3'-CTTAAG-5' ↓ ↑		5-G 3-CTTAA	AATTC-3 G-5
E. coli R₂45	*Eco* RII	5'-GGATCC-3' 3'-CCTAGG-5' ↑		5'-G-3' 3'-CCTAG-5'	5'-GATCC-3' 3'-G-5'
	Hind II	↓ 5'-GTPyPuAC-3' 3'-CAPuPyTG-5' ↓ ↑		5'-GTPy-3' 3'-CAPu-5'	5'-PuAC-3' 3'-PyTG-5'
Haemophilus influenzae Rd	*Hind* III	5-AAGCTT-3 3-TTCGAA-5 ↑		5-A 3-TTCGA	AGCTT-3 A-5

H. parainfluenzae	Hpa I	↓ 5-GTTAAC-3 3-CAATTG-5 ↓↑	5-GTT 3-CAA	AAC-3 TTG-5
	Hpa II	5'-CCGG-3' 3'-GGCC-5' ↑↓	5'-C- 3' 3'-GGC-5'	5'-CGG-3' 3'-C-5'
Klebsiella pneumoniae OK 8	Kpn I	5-GGTACC-5 3-CCATGG-3 ↓↑	5-GGTAC 3-C	C-3 CATGG-5
Nocardia otitidis-caviarus	Not I	5'-GCGGCCGC-3' 3'-CGCCGGCG-5' ↑↓	5'-GC-3' 3'-CGCCGG-5'	5'-GGCCGC-3' 3'-CG-5'
Providencia stuartii	Pst I	5'-CTGCAG-3' 3'-GACGTC-5' ↑↓	5'-CTGCA-3' 3'-G-5'	5'G-3' 3'-ACGTC-5'
Streptomyces albus G	Sal I	5-GTCGAC-3 3-CAGCTG-5 ↓ ↑	5-G 3-CAGCT	TCGAC-3 G-5
Staphylococcus aureus 3AI	Sau 3AI	5-GATC-3 3-CTAG-5 ↑	5- 3-CTAG	GATC-3 5

Arrows indicate the recognition sites

strand of double-stranded DNA is cleaved about 25–27 bp away from the recognition site, which is located in its immediate vicinity.

e.g.   Mbo II   5′ - GAAGA(N)$_8$ - 3′

3′ - CTTCT (N)$_7$ - 5′

Fok I   5′ - GGATGC(N)$_9$ - 3′

3′ - CCTACG(N)$_{13}$ - 5′

DNA ligase   It acts as a key player in recombinant DNA technology as a "molecular glue" or "molecular suture". The main source of this enzyme is the T4 phage. DNA ligase can join two different DNA pieces. The joining process requires ATP, as it needs energy to construct the bond. Normally the process is carried out at 4°C in order to reduce the kinetic energy of the molecule. This enzyme, when joining two DNA pieces, forms a covalent bond between the 5′ phosphoryl of one strand and 3′ hydroxyl of the adjacent strand. Hence it catalyses the end-to-end joining of DNA duplex at the base-paired end. Mertz and Davis (1972) for the first time demonstrated that cohesive termini of cleaved DNA molecules could be covalently sealed with *E. coli* DNA ligase and were able to produce recombinant DNA molecules.

The *E. coli* DNA ligase uses nicotinamide adenine dinucleotide (NAD$^+$) as a cofactor, while T4 DNA ligase requires ATP for the same. Both the enzymes contain an –NH$_2$ group on the lysine residue. In both cases, the cofactor breaks into AMP (adenosine monophosphate), which in turn adenylates the enzyme (E) to form enzyme –AMP complex (EAC). EAC binds to the nick containing 3′OH and 5′PO$_4$ ends on a double-stranded DNA molecule. The 5′ phosphoryl terminus of the nick is adenylated by the EAC with 3′OH terminus resulting in the formation of phosphodiester and liberation of AMP (Lehman, 1974).

After formation of phosphodiester, the nick is sealed. T4 DNA ligase has the ability to join the blunt ends of DNA fragments, whereas *E. coli* DNA ligase joins the cohesive ends

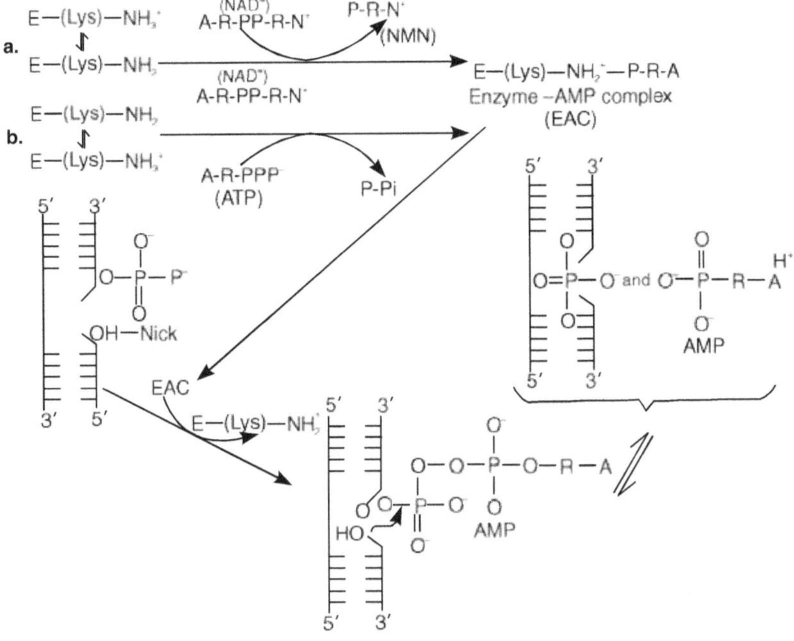

Figure 3.3 Mechanism of DNA ligase enzyme–AMP complex formation and sealing of nick on a double-stranded DNA molecule; **a.** *E. coli* DNA ligase; **b.** T4 DNA ligase; A—adenine; P—PO_4; R—ribose; N—nicotinamide; NMN—nicotinamide mononucleotide

produced by restriction enzymes. Additional advantage with T4 enzyme is that it can quickly join and produce the full base pairs but it would be difficult to retrieve the inserted DNA from vector. However cohesive end ligation proceeds about 100 times faster than the blunt end ligation (Figure 3.3).

Alkaline phosphatase Alkaline phosphatase is a glycoprotein with two identical subunits. The cohesive ends of broken plasmids, instead of joining with foreign DNA, join the cohesive end of the same DNA molecules and get recircularized. To overcome this problem the restricted plasmid is treated with an enzyme, alkaline phosphatase, that digests the terminal phosphoryl group.

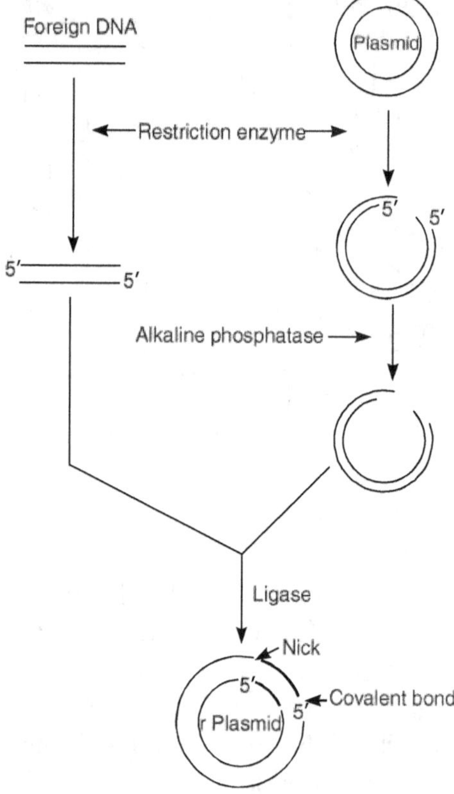

Figure 3.4 Inhibition of recircularization by alkaline phosphatase (to increase recombinant plasmids)

The restriction fragments of the foreign DNA to be cloned are not treated with alkaline phosphatase. Therefore, the 5′ end of foreign DNA fragment can covalently join to 3′ end of the plasmid. The recombinant DNA thus obtained has a nick with 3′ and 5′ P hydroxy ends. Ligase will only join 3′ and 5′ ends of recombinant DNA together if the 5′ end is phosphorylated. Thus, alkaline phosphatase and ligase prevent recircularization of the vector and increase the frequency of production of recombinant

DNA molecules. The nicks between two 3′ ends of DNA fragment and vector DNA are repaired inside the bacterial host cells during the transformation (Figure 3.4).

Reverse transcriptase Reverse transcriptase is used to synthesize the copy DNA or complementary DNA (cDNA) by using mRNA as a template. Reverse transcriptase is very useful in the synthesis of cDNA and construction of cDNA clone bank.

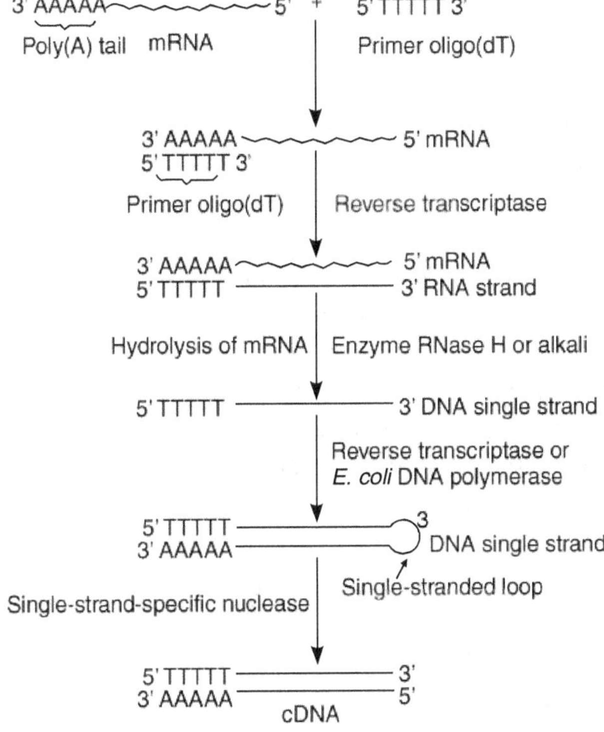

Figure 3.5   Process of formation of cDNA using mRNA as template with the help of enzyme reverse transcriptase

During the 1960s, Temin and his co-workers postulated that in certain cancer-causing animal viruses which contain RNA as genetic material, transcription of cancerous genes takes place most probably by DNA polymerase directed by virus RNA. Then DNA is used as template for synthesis of many copies of viral RNA in a cell. Thus, they found that retroviruses (possessing RNA) contain RNA-dependent DNA polymerase, which is also called as reverse transcriptase. This produces single-stranded DNA, which in turn functions as template for complementary long chain of DNA (Figure 3.5).

DNA polymerase DNA polymerase is a holoenzyme. The larger fragment of this enzyme is known as Klenow fragment. This enzyme polymerizes the DNA synthesis on DNA template and also catalyses a $5' - 3'$ and $3' - 5'$ exonucleolytic degradation of DNA. The DNA polymerase studied by A. Kornberg and his co-workers in *E. coli* in 1956 is now known as DNA polymerase I. The other two enzymes are DNA polymerase II and DNA polymerase III. These have almost similar catalytic activity. DNA pol I has a molecular weight of 109,000 and a single polypeptide chain of about 1,000 amino acid residues. The addition of mononucleotide to the free −OH end of a DNA chain is catalysed by this enzyme. Also it catalyses the other two reactions, i.e., $3' - 5'$ and $5' - 3'$ exonuclease activity. Function of DNA pol II (mol. wt. 1,20,000) is little understood. However, it catalyses $3' - 5'$ exonuclease activity. DNA pol III (mol. wt. about 1,40,000) is about several times more active than the other two. It is a dimer of DNA pol III. It requires an auxiliary protein. It produces a parallel strand of the DNA template in the presence of ATP. It is also used in dideoxy method of DNA sequencing.

RNases RNases cleave the phosphodiester bond between the two adjacent nucleotides. RNases cleave the bond next to uracil and guanine. The main source of this enzyme is *Aspergillus* sp. and bovine pancreas. RNaseH is widely used in cDNA preparation. It removes mRNA from RNA−DNA hybrid. RNase is invariably used in DNA purification where it will eliminate contaminated RNA and set free DNA. This enzyme is also used to remove poly(A) tail.

DNases DNases can digest both the strands of DNA. It can hydrolyse each DNA strand independently in the presence of Mg^{2+} ion. DNase is used in RNA purification where it can eliminate the contaminated DNA and set free RNA, besides its wide utility in footprinting and nick translations.

S1 nuclease It degrades the single-stranded DNA or single strand of double-stranded DNA with cohesive ends. As a result, cohesive ends are converted into blunt ends.

Hence, it is used to remove incompatible ends so that overlapping ends may be developed for annealing of 2 fragments of DNA molecules. The mode of action of S1 nuclease is

$$5' - AG - 3' \qquad \xrightarrow{\text{S1 Nuclease}} \qquad 5' - AG - 3'$$
$$3' - TCTTAA - 5' \qquad\qquad\qquad 3' - TC - 5'$$

Taq polymerase This enzyme is DNA polymerase, which exhibits its activity at a temperature optimum of 72°C. This highly thermostable enzyme is extracted from the thermophilic bacterium *Thermus aquaticus* living in hot springs at a temperature of more than 75°C. *Taq* polymerase consists of a single polypeptide chain with a molecular weight of 90,000 Daltons. The main application of *Taq* polymerase is in polymerase chain reaction (PCR) where it is used to extend primers, and completes the synthesis of the new DNA strand. The specificity and sensitivity of the PCR is also improvized by *Taq* polymerase. Fidelity of DNA synthesis by *Taq* polymerase determines the accuracy of PCR amplification.

MOLECULAR CLONING VECTORS

Vectors are those DNA molecules that can carry a foreign DNA fragment when inserted into it. As a molecular tool, various cloning vectors are currently being used in genetic engineering. These are plasmids, phagemids and cosmids.

Plasmids

Plasmids are the extrachromosomal, self-replicating, and double-stranded, closed and circular DNA molecules present in the bacterial cell. A number of host properties are specified by plasmids—antibiotic and heavy metal resistance, nitrogen fixation, pollutant degradation, bacteriocin and toxin production, colicin factors and phages. The number of plasmids ranges from 10–100 per cell and is known as "high-copy-number plasmids or relaxed plasmids." However, some host cells contain 1–4 copies and are called "low-copy-number plasmids or stringent plasmids." Naturally occurring plasmids can be modified by *in vitro* techniques. Cohen *et al.* (1973) for the first time reported the cloning of DNA by using plasmid as a vector.

A plasmid can be considered a suitable cloning vehicle if it possesses the following features.

- It can be readily isolated from the cells.

- It possesses a single restriction site for one or more restriction enzyme(s).

- Insertion of a linear molecule at one of these sites does not alter its replication properties.

- It can be reintroduced into a bacterial cell, and cells carrying the plasmid with or without the insert can be selected or identified.

- They do not occur free in nature but are found in bacterial cells.

The following are some of the plasmid-based cloning vectors:

pBR 322   It is undoubtedly the ideal cloning vector, commonly employed in genetic engineering. p in pBR 322 signifies plasmid; B is for Boliver and R for Rodriguez, the two scientists who developed pBR 322. It is a small plasmid of 4362 bp. It has the replication module of *E. coli* plasmid Col E1 and two selectable markers (tetracycline, tet^r, and ampicillin, amp^r, resistance gene), and only single or unique recognition sites for 12 different restriction enzymes (Figure 3.6).

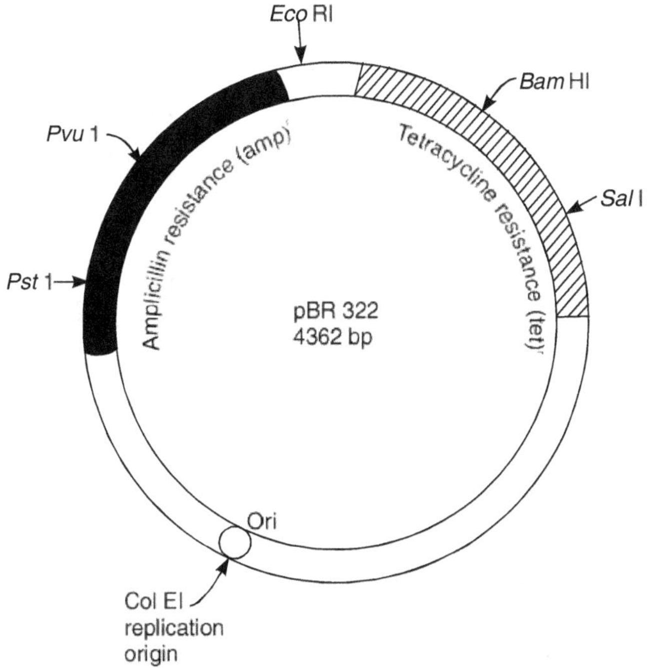

Figure 3.6   Plasmid vector—pBR 322

pUC 19   This plasmid has 2,680 base pairs. It carries gene for ampicillin-resistance, β-galactosidase, *lacZ* and *lacI*. The *lacZ* gene contains short sequence for multiple cloning sites, i.e., *Eco* RI, *Sal* I, *Xma* I, *Bam* HI, *Sma* I, *Pst* I, *Hind* III, etc., and the origin of replication. The main advantages of pUC in genetic engineering are easy selection in the screening process. In pUC 19, p signifies plasmid and the letters U and C stand for the University of California, where it was constructed by J. Messing and J. Vieira. The number 19 distinguishes it from other plasmids. The significant feature of pUC 19 is that it can exist as high copy number, i.e., up to 500 per bacterial cell. The *lacZ* gene is under the control of *lacI* promoter. This mechanism is used in the production of certain foreign fusion proteins (Figure 3.7).

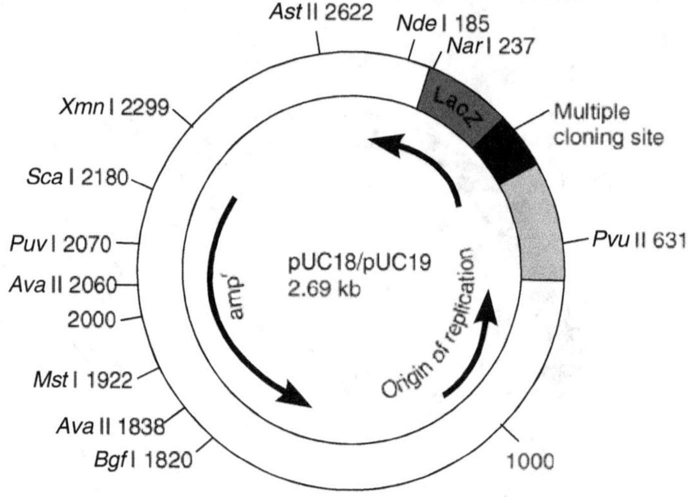

Figure 3.7 Genetic map of pUC 18/pUC19

Virus-based Vectors

Bacteriophage λ-Vectors   In genetic engineering, it is essential to create a vector capable of carrying large pieces of DNA. To meet this demand, the *E. coli* virus, bacteriophage lambda (λ) has been the foremost choice. Lambda virus can accommodate large pieces of inserted DNA.

The bacteriophage λ approaches 2 types of replication inside the host cell. During the lytic cycle it can attach and enter the bacterial cell followed by replication. After a short period of time, newly formed phages are released by cell lysis. In the lysogenic cycle, λ DNA enters bacterial DNA and becomes silent for a period of time. During favourable conditions, λ DNA can become excised and can enter a lytic cycle.

The bacteriophage λ DNA consists of 50 kb, of which 20 kb is responsible for integration and excision during lysogenic

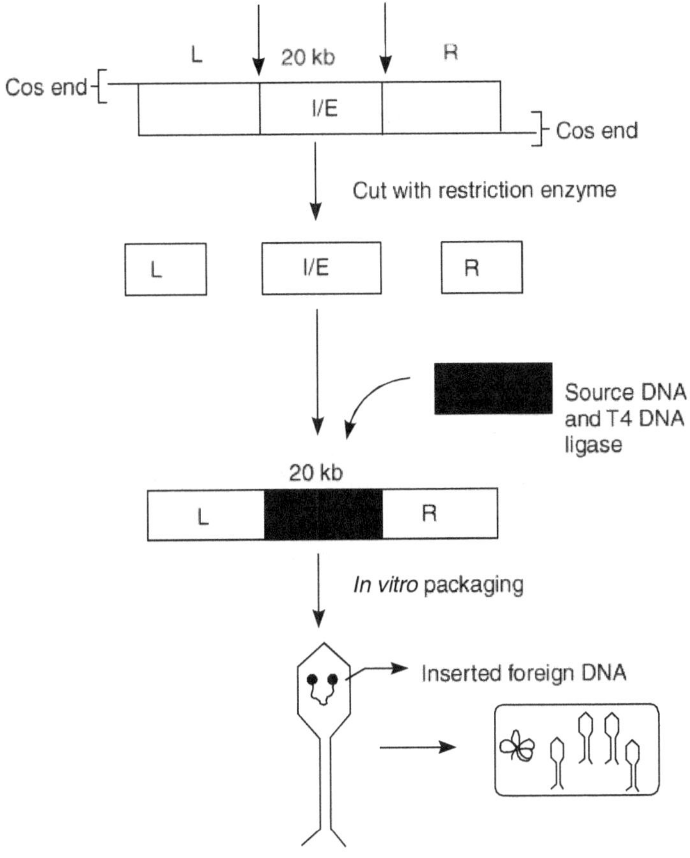

Figure 3.8 Construction of λ phage vector

cycle. This middle region of λ DNA of about 20 kb can be removed and replaced with 20 kb of foreign DNA. The recombinant λ phages are maintained by continual growth on fresh *E. coli* (Figure 3.8).

M13 bacteriophage vector M13 is a single-stranded DNA-containing phage virus. It is filamentous in nature. The protein coat contains three kinds of capsomers. These phages infect cells by adsorbing with the aid of F pili. They can infect only F$^+$ cells. The DNA of wild type M13 comprises approximately 6,402 bases. The presence of 10 genes are essential for replication of the phages.

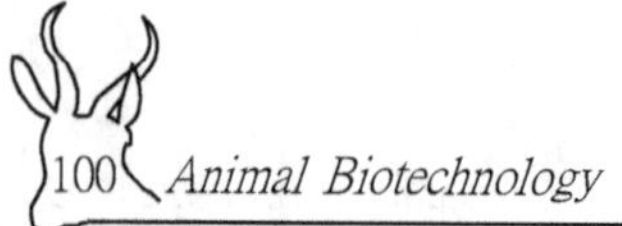

In M13 cloning vector, the 507 nucleotide intergenic sequence (IS) is modified to introduce additional restriction site. Intergenic sequence of wild type is the only region available for modification. The gene *lacZ* is introduced into the IS to get M13 vector. In the screening procedure, the introduced *lacZ* gene produces blue plaques on X-gal agar plates.

Cosmids

Based on the properties of DNA and ColE1 plasmid DNA, a group of Japanese workers showed that the presence of a small segment of phage (DNA containing cohesive end on the plasmid molecules) is a sufficient prerequisite for *in vitro* packaging of this DNA into infectious particles. The cosmids can be defined as the hybrid vectors derived from plasmids which contain *cos* site of λ phage. It was first developed by Collins and Hohn (1978). Cosmids lack genes encoding viral proteins. It has 2 *cos* sites and tetracycline resistance gene. It has a multiple cloning site with 6 restriction sites (Figure 3.9).

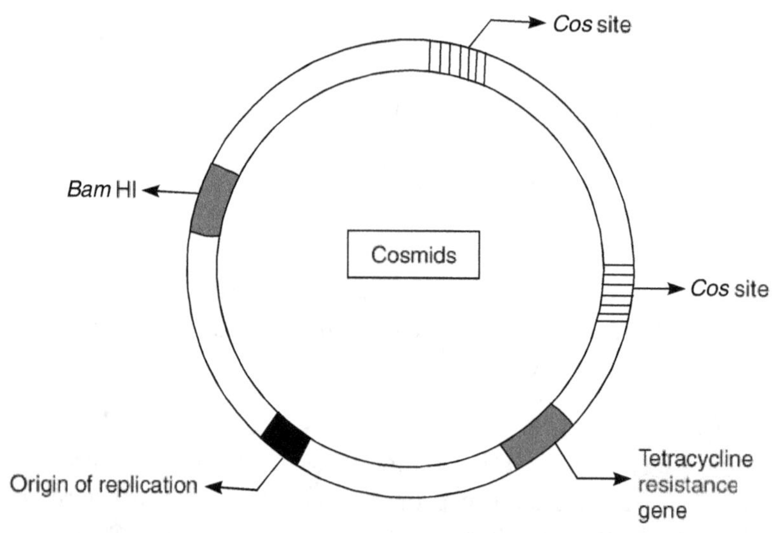

Figure 3.9 Cosmids

VIRAL VECTORS FOR ANIMALS

SV40 or Simian 40 Virus

It contains single double-stranded circular chromosomes. It normally infects monkey cells and induces tumour. The genes of SV40 chromosomes can be divided into two sets. One is known as "early genes", which encodes for tumour antigen and another is "late genes", which encodes for viral coat protein. Simian virus contains small genomes. It cannot accommodate large pieces of foreign DNA. Hence, an essential portion of viral genome is removed. The defective SV40 can no longer cause infection. Therefore with the help of helper virus, the defective virus containing foreign gene can perform infection and participate in gene transfer for mammalian cells.

Retrovirus

Some of the retroviruses have been modified to act as viral vectors to introduce foreign DNA into mammalian cells. The LTR (long terminal repeat) sequences are required for integration of retroviral DNA into the host chromosome. The psi (ϕ) sequence is required to package viral RNA in viral particles. The *gag, pol* genes of the retroviral genome can be replaced with foreign DNA. This results in the formation of recombinant defective retroviral DNA (Figure 3.10). This recombinant DNA lacks the genes required for retroviral replication and assembly of viral particles. To

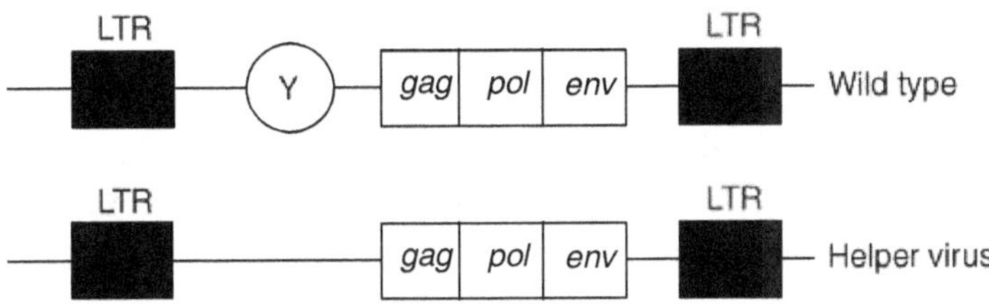

Figure 3.10 Organization of retrovirus vector

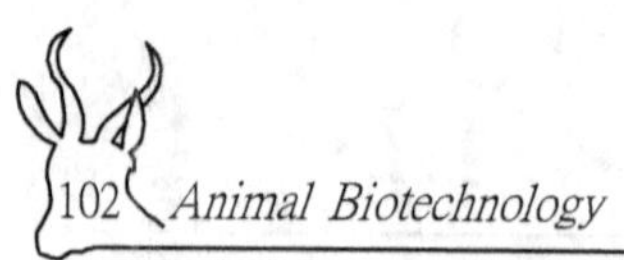

assemble viruses with recombinant DNA, it must be introduced into cells in tissue culture that are infected with helper viruses. Within the cells, recombinant DNA is transcribed and RNA is packaged. These viral particles contain only recombinant viral RNA and can act as vectors to introduce this RNA into target cells.

Retroviruses have single-stranded RNA genomes and these viruses provide the most promising vector system of all. During the process of reverse transcription, a double-stranded proviral DNA molecule is generated, which has all the regulatory functions for viral transcription and translation. Thus any of the normal proviral genes can be replaced with foreign DNA provided the recombinant is propagated in the presence of helper virus. Retroviruses have a number of features that make them particularly useful as gene transfer vectors and these are summarized as follows:

1. The normal replication process involves the stable insertion of a DNA copy of their genome into the host genome.

2. They can infect and transmit their genetic information into a high proportion of recipient cells.

3. Retroviruses have a broad host range, which can be easily modified making it possible to infect cells from a variety of species and cell types.

4. The integrated provirus exists as a stable, predictable structure within the host genome.

5. Infection of mammalian cells with retroviruses does not result in cell death. Rather, infected cells continue to grow and divide while producing large numbers of virus particles.

6. A large amount of foreign gene sequence can be packaged within the virus particle.

Baculovirus

Baculovirus infects invertebrates including many insects. In the baculovirus *Autographa californica*, a multiple nuclear polyhedrosis virus (AcMNPV) which infects insect cells has been developed as a eukaryotic expression vector, with a promising result. This system places a cloned gene into a site in an AcMNPV virus that, in cultured insect cells, produces large quantities of recombinant protein (Figure 3.11). Here, there is exceptionally high gene expression, driven by AcMNPV promoter.

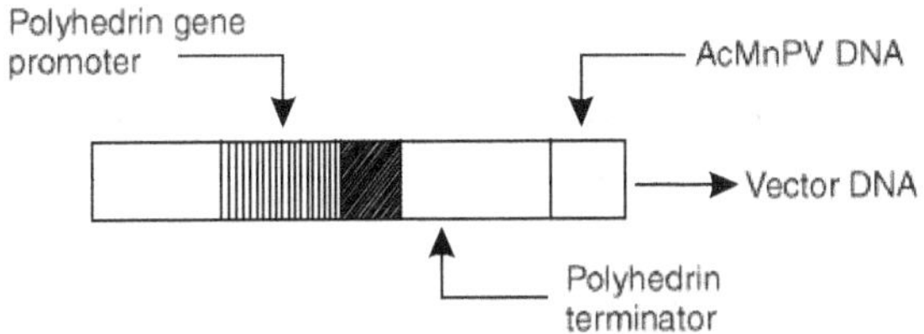

Figure 3.11   Organization of baculovirus vector

SHUTTLE VECTORS

These are the type of vectors capable of replicating in different host systems. Shuttle vectors can be maintained both in *E. coli* and yeast cells. One of the reasons for selecting eukaryotic yeast cell is the presence of well-equipped post-translational modification. Yeast cells are thus designed to be shuttle vectors. Yeast episomal plasmids (YEP) and yeast artificial chromosome (YAC) are shuttle vectors.

CLONING STRATEGIES

Making a cDNA Library

The most important goal of genetic engineering is to clone a particular gene or a genomic fragment of interest. The approach used to clone a specific gene depends to a large extent on the gene of interest. Generally, the procedures start with a sample of DNA such as eukaryotic genomic DNA. The next step is to obtain a large collection of clones made from the original DNA sample. The collection of clones is called a DNA library. This technique is sometimes referred to as "shot gun" cloning because we clone a large sample of fragments so that at least one of the clones will "hit" the desired gene. DNA libraries can be categorized into different types based on the vector used and the source of DNA.

Different cloning vectors carry different amount of DNA, so the choice of vector for library construction depends on the size of the genome being made into the library. Plasmid and phage vectors carry small amounts of DNA, so these vectors are suitable for cloning genes from organisms with small genomes. Cosmids carry larger amounts of DNA, and other vectors such as YACs and BACs carry the largest amounts of all. A phage library is a suspension of phages. A plasmid or a cosmid library is a suspension of bacteria or a set of defined bacterial cultures stored in culture tubes.

The next step is to construct a genomic library or a cDNA library. cDNA or complementary DNA is synthetic DNA made from mRNA with the use of a special enzyme called reverse transcriptase. Since it is made from mRNA, cDNA is devoid of both upstream and downstream regulatory sequences and of introns. Therefore cDNA from eukaryotes can be translated into functional protein in bacteria, an important feature when expressing eukaryotic genes in bacterial hosts.

A cDNA library is based on the regions of the genome transcribed, so it will inevitably be smaller than a complete genomic library, which should contain all of the genome. Although genomic libraries are bigger, they do have the benefit of containing genes in their native form, including introns and regulating sequences. If the purpose of constructing the library is a prelude to cloning an entire genome, then a genomic library is necessary at some stage of screening a cDNA or genomic library.

One of the key elements required to identify a gene during cloning is a probe. A probe is normally a cloned piece of DNA that contains a portion of the sequence for which we are searching. The probes either radioactive or non-radioactive, are added to a solution, and filters containing immobilized clones are then bathed in the solution. The principle behind this step is that the probe will bind to any clone containing sequences similar to those found on the probe. This binding step is called hybridization (Figure 3.12).

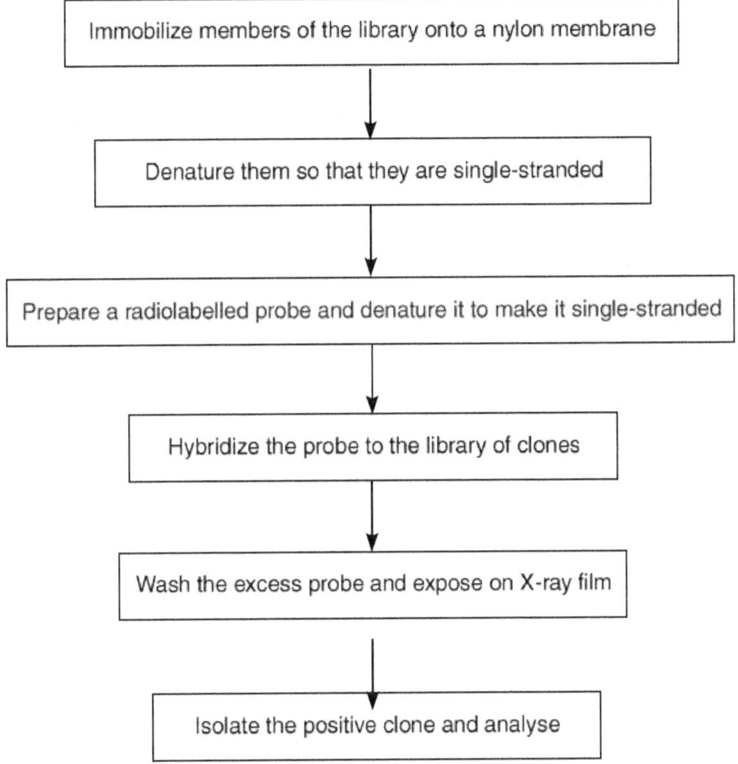

Figure 3.12   Screening a cDNA or genomic library

The hybridization step is performed at a non-stringent temperature that ensures that the probe will bind to any clone containing a similar sequence. At the same time, some non-specific hybridization will occur because some of the clones will contain limited, but not significant similarity to the probe. The washing step is performed at a stringent temperature that is high enough to wash the probe off all clones to which it has bound in a non-specific manner. But it is important that the temperature is not so high that it washes the probe off the clones that contain sequences that are similar or identical to the probe itself.

GENETIC ENGINEERING GUIDELINES

With the success of Boyer–Cohen experiments (in 1973), it was realized that recombinant DNA technology could be used to create organisms with novel genes. This created worldwide commotion (among scientists, public and government officials) about the safety, ethics and unforeseen consequences of genetic manipulations. Some of the phrases quoted in the media during those days are given:

- Manipulation of life

- Playing God

- Manmade evolution

- The most threatening scientific research

It was feared that some new organisms created inadvertently or deliberately for warfare, would cause epidemics and environmental catastrophes. Due to the fears of the dangerous consequences, a cautious approach on recombinant DNA experiments was suggested.

In 1974, a group of ten scientists led by Paul Berg wrote a letter that simultaneously appeared in the prestigious journals—

Nature, Science and *Proceedings of the National Academy of Sciences*. The dangers of DNA technology were printed out in that letter (highlights given below):

"Recent advances in techniques for isolation and rejoining of segments of DNA now permit construction of biologically active recombinant DNA molecules *in vitro*. Although such experiments are likely to facilitate the solution of important theoretical and practical biological problems, they would also result in creation of novel types of DNA elements whose biological properties cannot be completely predicted. There is a serious concern that some of these DNA molecules could prove biologically hazardous". The letter also appealed to molecular biologists worldwide for a moratorium on many kinds of recombinant DNA research, particularly those involving pathogenic organisms.

Asilomar Recommendations

In February 1975, a group of 139 scientists from 17 countries held a conference at Asilomar, a conference centre in California, USA. They assured the uneasy public that the microorganisms used in DNA experiments were specifically bred and could not survive outside the laboratory. These scientists formulated guidelines and recommendations for conducting experiments in genetic engineering.

NIH GUIDELINES

National Institute of Health (NIH), USA, constituted the Recombinant DNA Advisory Committee (RAC) which issued a set of stringent guidelines to conduct research on DNA. RAC was in fact overseeing the research projects involving gene splicing and recombinant DNA.

Some of the important original NIH recommendations on recombinant DNA research relate to the following aspects.

- Physical (laboratory) containment levels for conducting experiments.

- Biological containment—the host into which foreign DNA is inserted should not proliferate outside the laboratory or transfer its DNA into other organisms.

- For research on pathogenic organisms, elaborate, controlled and self-contained rooms were recommended.

- For research on less dangerous organisms, units equipped with high quality filter systems should be used.

- No deliberate release of any organism containing recombinant DNA into the environment.

It may be noted here that although the NIH guidelines did not have the legal status, most institutions, companies and scientists voluntarily complied.

Relaxation of NIH Guidelines

It was in the 1980s that the original NIH guidelines were considerably relaxed by NIH-RAC, based on the experience and experimental data obtained from the NIH-sponsored studies on recombinant DNA research.

It is a fact that the genetic engineering research flourished and progressed rapidly after relaxation of NIH guidelines. It may however be noted that NIH-RAC continues to be a watchdog over the DNA technology experiments.

Pharmaceutical Products of Recombinant DNA

As the recombinant DNA technology progressed, many pharmaceutical compounds of human health care are being produced through genetic manipulations. Most countries consider that the existing regulations for approval of pharmaceuticals of commercial use are adequate to ensure safety since the process by which the product is manufactured is irrelevant. Thus, the recombinant DNA product (protein, vaccine, drug) is evaluated for its safety and efficacy like any other pharmaceutical product.

Genetically Engineered Organisms (GEOs)

Recombinant DNA research has resulted in the creation many genetically engineered organisms. These include microorganisms, animals and plants. The latter two respectively result in transgenic animals and transgenic plants.

THE FUTURE OF GENETIC ENGINEERING

DNA technology has largely helped scientists to understand the structure, function and regulation of genes. The development of new/modern biotechnology is primarily based on the success of DNA technology. Thus, the present biotechnology (more appropriately molecular biotechnology) has its main roots in molecular biology.

Biotechnology is an interdisciplinary approach for applications to human health, agriculture, industry and environment. The major objective of biotechnology is to solve problems associated with human health, food production, energy production and environmental control.

It is an accepted fact that recombinant DNA technology has entered the mainstream of human life and has become one of the most significant applications of scientific research. Biotechnology is regarded as more an art than a science.

After the successful sequencing of human genome, many breakthroughs in biotechnology are expected in future.

Summary

- Transfer of a gene sequence from one organism to another is called as "molecular cloning" or "gene cloning" or "genetic engineering".

- Inserted gene is called as "transgene" and any organism containing the transgene is called "transgenic".

- Restriction endonucleases have contributed significantly towards gene splicing and analysis of cloned genes.

- DNA ligase acts as a key player in recombinant DNA technology.

- Plasmids, phagemids and cosmids are the molecular vectors used in genetic engineering.

REVIEW QUESTIONS

1. What are type II restriction endonucleases and why are they important for recombinant DNA technology?

2. Explain the role of molecular scissors in genetic engineering.

3. Give a brief account of bacterial vectors used in genetic engineering.

4. Describe the use of pBR 322 as a cloning vector. Add a note on its special feature.

5. How do you modify λ phage virus into a cloning vector?

6. Describe viral based cloning vectors for bacteria, animals and plants.

7. Highlight the distinguishing features of plasmids and cosmids as cloning vectors.

8. What are shuttle vectors? Explain their role in genetic engineering?

9. Brief on M13 bacteriophage vector.

10. Give an account of cDNA library.

GENE TRANSFER METHODS IN ANIMALS

INTRODUCTION

The art and science of producing genetically engineered animals have advanced rapidly in the past few years. One challenge in creating transgenic animals is to ensure that the transgene turns on at the right time and in the right tissue. To be functional, the integrated gene must be expressed and regulated appropriately. Thus, the gene to be transferred must be accompanied by the appropriate promoter and regulatory sequences. Some genes require an enhancer that may be located far from the promoter. The engineering of organisms requires fusion of the correct promoter or enhancer and gene-coding sequence. The construct must be incorporated into the chromosome where gene expression is regulated.

There are a number of methods presently employed for genetic engineering of various animal species. Most of these were developed originally in mouse and *Drosophila* models, and have only more recently been extended to other domesticated animals. Access to the germ line of mammals can be obtained by:

1. Direct manipulation of the fertilized egg, followed by its implantation into the uterus.

2. Manipulation of the sperm used to generate the zygote.

3. Manipulation of the early embryonic tissue in place.

4. The use of embryonic stem (ES) cell lines which, after manipulation and selection *ex vivo*, are introduced into early embryos.

5. Phase manipulation of cultured somatic cells, whose nuclei then can be transferred into enucleated oocytes and thereby provide the genetic information required to produce a whole animal.

There are two basic approaches presently in use for inserting DNA into vertebrate germ line cells—transfection and infection with retrovirus vectors. A third approach based on the use of mobile genetic elements, has been commonly used for insects, and is being explored for germ-line modification of vertebrates.

TRANSFECTION

Transfection methods include:

1. Direct microinjection of DNA into the cell nucleus
2. Electroporation
3. Use of polycations
4. Lipofection
5. Sperm-mediated transfection
6. Scrapefection

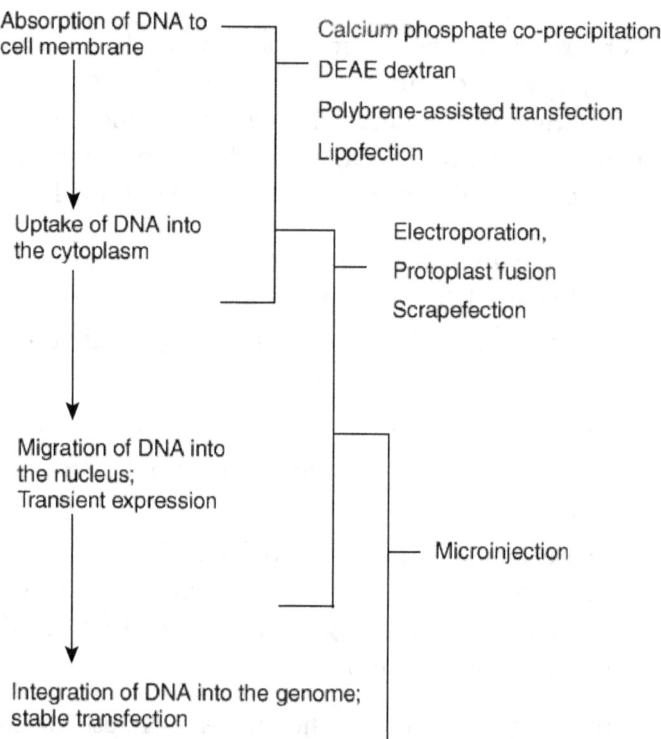

Figure 4.1 The stages of DNA transfection which is involved in a variety of techniques

The manner of introduction of DNA is a technical issue, determined empirically for each system and makes little difference to the final outcome (Figure 4.1).

Microinjection

Microinjection, first developed through use in the mouse, has been the method of choice for most transgenic research.

In this method, the DNA solution is injected directly into the nucleus of a cell or into the male pronucleus of a fertilized one-to-two-cell ovum. Typically, a microinjection assembly consists of a low-power stereoscopic dissecting microscope (to view the ovum and the entire process) and two micromanipulators, one for a glass micropipette to hold the ovum by partial suction and the other for a glass injection needle to introduce the DNA into the male pronucleus. The male pronucleus is much larger than the female pronucleus of fertilized mammalian ova (Figure 4.2).

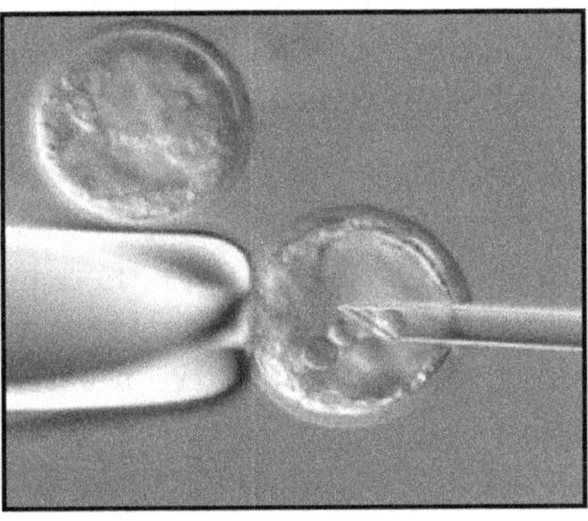

Figure 4.2   Microinjection

The general procedure for microinjection is as follows:

- Donor females are induced to superovulate using appropriate hormone treatments.

- ✿ Female mice are subjected to a regime of pregnant mouse serum gonadotrophin (PMSG), which stimulates growth and development of follicles which contain the developing oocytes.

- ✿ The release of these oocytes (ovulation) is induced by subsequent treatment with human chorionic gonadotrophin (HCG).

- ✿ The superovulated females are then mated with fertile males, and large numbers of fertilized one-to-two-cell ova/embryos are collected surgically. Alternatively, unfertilized ova are collected from superovulated females; the ova are then fertilized *in vitro*.

Transgenics are identified by subjecting the tissue samples to DNA analysis by Southern blot hybridizations with the new gene as a probe or by PCR amplification of the new gene.

In mice, an average of about 3–6% of the progeny derived from microinjected embryos are transgenic; the frequency is much lower in other animals. The transgenic animals contain foreign DNA in their germ cells and, as a consequence, pass it on to their progeny; transgenes show typical Mendelian inheritance.

The transgene integration occurs at random sites in the genome, but in a given cell or embryo usually only a single chromosomal site is involved. However, there is generally a wide variation in the number of copies integrated, ranging from the common one copy to several hundred copies. The multiple copies are integrated at a single site in a head-to-tail arrangement. Consequently, the site of integration of a transgene in different transgenic animals differs greatly and may involve different locations of the same chromosome or different chromosomes. The transgene integration occurs at an early stage of embryo development following microinjection. But often the integration may be delayed, and the transgene remains in the extrachromosomal state during this period.

Microinjection is not without problems. Few injected eggs survive, and not all of those retain the new DNA. The new DNA is targeted at random if it is not targeted to a specific chromosomal locus. Unfortunately, homologous recombination and gene targeting efficiency are extremely low in animals. Microinjected genes often integrate into a single chromosomal locus as head-to-tail concatemers. In addition, many transgenic animals are mosaics because the gene has integrated into only some of the cells. This mixture of transformed and untransformed cells is especially problematic if the gene has not incorporated into the germ cells. All germ cells must be transformed to ensure that the gene is transmitted to the progeny.

Advantages

1. The quantity of DNA delivery can be optimized.
2. DNA delivery is predictable, even into the cell nucleus.
3. Even the smallest cell can be targeted to deliver DNA.

Disadvantages

1. Requires skilled persons.
2. Only one cell receives DNA per injection.
3. Costly instruments.

Electroporation

Electroporation facilitates cells to uptake DNA by reversibly altering the permeability of cell membrane. Electroporation is one of the several standard techniques for efficient transformation. The technique is more suitable for the transformation of plant protoplast and animal cells. In this approach, the transfection mixture containing cells and DNA is exposed for a very brief period (few milliseconds) to a very high voltage gradient (e.g. 4000–8000 V/cm) (Figure 4.3). This induces transient pores in the cell membranes through which DNA seems to enter the cells.

Treatment of cells before they are electroporated increases the frequency of transfection. This is most likely due to the arrest of cells at metaphase and the associated absence of nuclear envelope or to an unusual permeability of the plasma membranes. Linearized DNA is far more efficient in transfection than circular DNA.

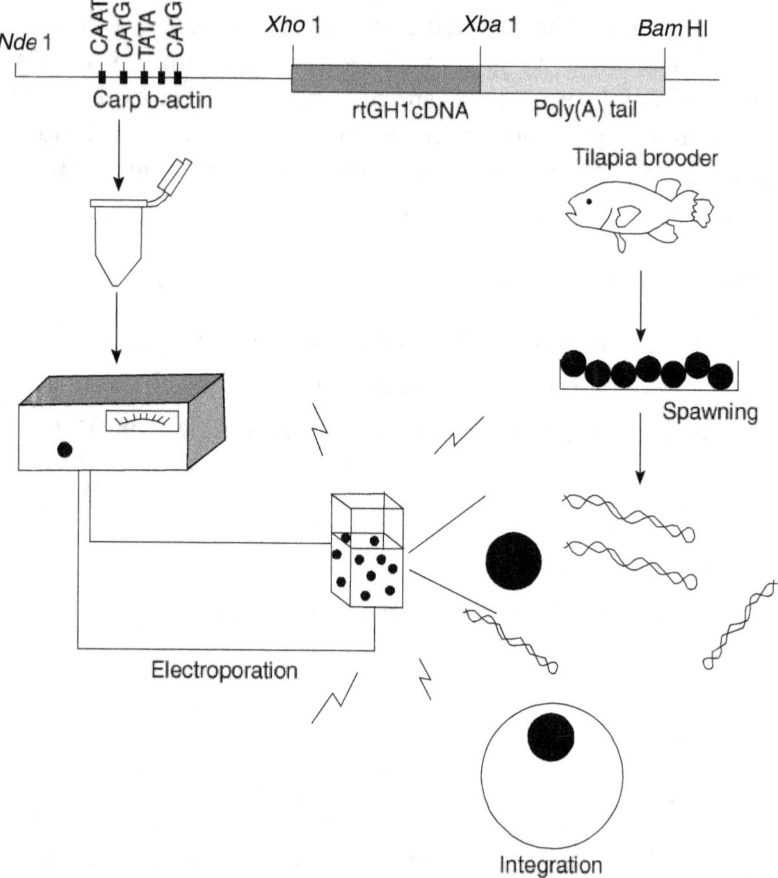

Figure 4.3　Scheme of gene transfer by electroporation

Advantages

1.　Efficient transformation.
2.　Large number of transformed cells obtained.
3.　May not require skilled persons.

Disadvantages

1. Requires protoplast for transformation.

2. Difficulties associated with protoplast regeneration.

3. Risk of obtaining genetic variation in protoplast-regenerated plants.

Use of Polycations

DEAE-dextran (diethyl amino ethyl-dextran) is water-soluble and polycationic, i.e., has a multiple positive charge. It is added to the transfection solution containing the DNA. In some unknown way, DEAE-dextran brings about DNA uptake by the cells through endocytosis. Its possible interaction with the negatively charged DNA molecules and with the components of cell surface plays an important role.

This technique is highly suited for transient transfection used for various molecular biology studies, particularly COS cell lines. However, again for some unknown reason, it is not efficient in producing stable transfection.

Also, direct DNA uptake by protoplasts can be stimulated by using hydrophilic, long-chain polycations such as polyethylene glycol (PEG), poly L-lysine with or without metal ions such as Zn^{2+}, Li^+, Cs^+, Rb^+, K^+, Na^+, Ca^{2+} or Mg^{2+}. The technique is efficient and can be applied virtually to every protoplast system and for transfer of even small quantities of DNA. PEG is an extremely hydrophilic molecule which removes "free water" when placed in a solution. Free water indicates those molecules which interact with charged (usually ionic) molecules soluble in water. Thus, under high PEG concentration (15–25%), ionic macromolecules such as DNA can no longer stay in solution and get precipitated. The protoplast membrane as well as DNA is normally negatively charged, and therefore, charge repulsion is obvious. This repulsion limits the interaction between DNA and protoplasts. At very high concentration, PEG seems to minimize charged repulsion effects thus stimulating DNA uptake by endocytosis without any gross damage to protoplasts.

DNA uptake may be enhanced as a result of the interaction of the polycations at one or more of the following levels:

- Protection of the exogenous DNA against nuclease degradation

- Increase in the permeability of the plasma membrane

DNA transformation by using the polycations "polybrene" (hexamidethrine bromide) is new, although similar techniques have been used for several years in the transformation of animal cells. The major advantages of the polybrene technique are the following:

1. It is less toxic than other polycations.

2. The high transformation efficiency allows very small quantities of plasmid DNA to be used.

3. Since no carrier DNA is used to integrate plasmid DNA into the host genome, the sequences surrounding the site of integration can be analysed.

4. The transformation frequency appears to be independent of the nature of the plasmid.

5. The polybrene method can be used for both stable and transient expression.

Lipofection

The delivery of DNA into cells using liposomes is called lipofection. Liposomes, as defined by Bongham in 1965, are small vesicles in which water is self-contained and the lipid molecules are disposed in bimolecular layers attached by their non-polar interfaces.

Liposomes possess many properties akin to those of biological membranes. They are easy to manipulate, their lipid contents can be varied at will and many substances may be trapped inside the interlamellar spaces. They are found to be ideal carrier systems and are now used in genetic engineering to deliver foreign DNA molecules into a desired cell.

Considerable work has been done on lipofection due to its potential application in targeting gene to specific human tissues for gene therapy. There are 7 approaches for preparing liposomes for DNA delivery.

1. Usually they are prepared by dispersion of phospholipids like phosphatidyl choline (PG) in water by mechanical methods like sonication, which tend to destroy DNA. DNA of up to 1 kb has been incorporated into small sonicated liposomes.

2. However, other techniques allow the entrapment of large DNA sequences into liposomes, e.g. exposure of the anionic lipid phosphatidyl serine (LPS) to Ca^{2+}, and the 2-phase techniques.

    Liposomes prepared by the above approaches are phagocytosed by the cell. The phagocytosis vesicles thus produced ordinarily fuse with lysosomes leading to DNA degradation and low transfection frequencies.

3. The fusion protein of Sendai virus is incorporated into the liposome membrane; this enables the fusion of liposomes with plasmalemma and thereby a direct delivery of DNA into the cytoplasm. A receptor protein is also incorporated which permits a controlled delivery of the DNA into the target cells.

4. The ionizable lipids that undergo phase change in response to the pH of cytoplasm are used to construct liposomes. Such liposomes release DNA into the cytoplasm once they are phagocytosed.

5. Cationic liposomes to which DNA binds on the outside by electrostatic attraction are generally used. These

liposomes cause perturbations in plasma membrane due to which they fuse and the DNA enters into the cytoplasm. Cationic liposomes are available commercially (marketed as Lipofectin by Gibco-BRL).

6. DNA is complexed with a cationic peptide, e.g. gramicidin, which interacts with lipid membranes in a specific manner.

7. The liposomes may be targeted to cell-surface receptors by incorporation of ligand proteins into the liposome membrane. It has been recently proposed to use biotinylated bisanthracycline (which intercalates in double-stranded DNA) for attachment of specific ligand proteins via avidin to the DNA for delivering cationic lipid/peptide-complexed DNA to specific target cells.

Lipofection is the method of choice for transfection of mammalian cells *in vitro*. It has also been used to deliver DNA into live animals by direct injection or intravenous injection. Cationic liposomes have been used in intravenous or intratracheal injection in mice for the expression of marker genes in lungs. Targeted delivery has also been demonstrated by incorporating specific ligand proteins into the liposome membranes. Attempts are being made to deliver the cystic fibrosis gene via nasal or bronchial tissue for stimulating cytotoxic T-lymphocyte response in human patients.

The mechanism of movement of the DNA from cell cytoplasm into the nucleus is unknown. In some cases, DNA movement to the nucleus is greatly facilitated by making DNA constructs that are capable of cytoplasmic translation, and by combining them with RNA polymerase or with a gene that produces RNA polymerase.

Advantages

1. The liposomes protect the DNA from being damaged by the acidic pH and proteases.

2. When liposomes are constructed from material that is compatible with the lipids of the tonoplast, liposome–tonoplast fusion occurs and so the encapsulated DNA should be successfully delivered into the cytoplasm.

The use of liposome for delivery of gene was first described for delivering polio virus RNA. Later liposomes were used in combination with PEG for delivery of plasmid DNA into tobacco protoplast.

Embryonic Stem Cell Gene Transfer

ES cells of mice are pluripotent cultured cells derived from early pre-implantation embryos. ES cells are isolated from the inner cell mass of donor blastocysts of early embryos and can be cultured *in vitro* prior to transfection with a specific gene. The ES cell technology is described as follows (Figure 4.4). The cultured pluripotent ES cells are transfected with the appropriate transgene construct by a suitable transfection technique (preferably by homologous recombination). Transfected ES cells are identified and selected, generally by employing a selectable marker gene, and are cloned. Transformed ES cells are microinjected into animal blastocysts so that they can become established in the somatic and germ-line tissues. They are then passed onto successive generations by breeding founder animals.

Stable transgenic lines are obtained by crossing founder animals that have the gene in their germ cells. The embryos co-cultured and microinjected with transfected ES cells are transferred into surrogate mothers where they complete their development. About 30% of the progeny derived from such embryos contain tissues derived from the ES cells, i.e., they are chimeric. The pluripotent ES cells can give rise to germ cells as well. Therefore, pure transgenic mice can be recovered from the chimeric mice using a suitable breeding scheme.

The production of transgenic animals using transfected ES cell clones may also be achieved by co-culture method. The zona pellucida of embryos 8-cell to morula stage is removed, and the morulae are **co-cultured** with the ES cells; the ES cells are preferentially incorporated into the inner cell mass of the developing embryo. It is also possible to transplant the ES cell nucleus into an enucleated fertilized ovum, but this technique is not commonly employed, as it is quite tedious although all the progeny obtained by this technique are transgenic.

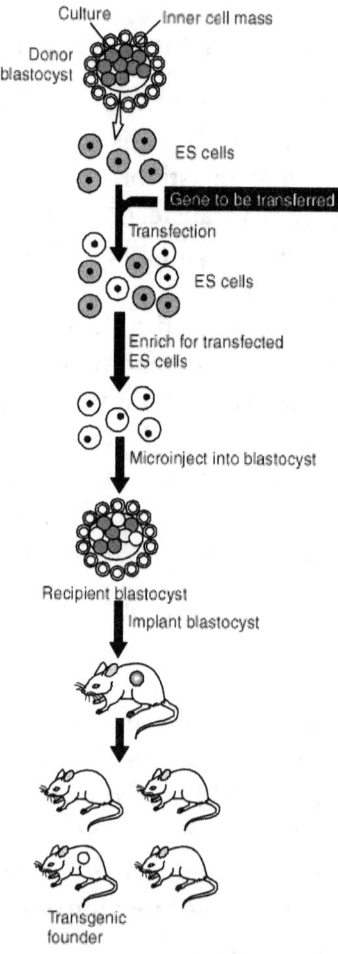

Figure 4.4   Production of transgenic mice by the transfer of embryonic stem cells containing a gene of interest.

Since DNA integration in genome is random, the frequency of targeted gene transfer, i.e., integration of transgene at a specified site, is discouragingly low. Therefore, transfection techniques like microinjection are not suited for targeted gene transfer. ES cell line technology, on the other hand, permits the selection of clone(s) having targeted gene transfers from among several clones showing stable transfection. The desired clone can then be used to produce transgenic animals, having targeted gene transfer.

At present, the ES method is most successful with mice because mouse ES cells are pluripotent and, when integrated into blastocysts, can divide and differentiate in the mouse embryo. However, with current technology, cow, pig, and chicken ES cells do not seem to be pluripotent when placed in blastocyst embryos.

The mouse is used as a model in the study of gene function, and is useful in many areas of medicine. More than 5000 human diseases are caused by genetic defects; thus, by using gene targeting to introduce mutations in experimental animals in order to produce human disease models, researchers may identify effective treatments and gene therapies. Specific genes are inactivated in mouse models to simulate human disease such as Alzheimer's disease, cancer, atherosclerosis, amyotrophia, lateral sclerosis and cystic fibrosis. Methods are used to inactivate gene function and look for changes in the phenotype. Gene targeting also can be used to add genes encoding desirable characteristics. ES cell transformation technology has been used to inactivate specific genes in order to determine their effect on growth and development, or to develop effective treatments.

Sperm-Mediated Transfer

This method of transfection, often practised in fish, relies on sperms, which are the only agents that are guaranteed to get into the egg during fertilization and deliver a package of DNA directly to the female pronucleus. Though this technique is advantageous, there are many limitations. In most organisms it is only a single sperm that enters into egg. So the number of copies of DNA construct that reach the target is quite small when compared to electroporation or microinjection. Moreover, egg has mechanisms to prevent the entry of foreign DNA and its integration into the host genome.

Scrapefection

This technique was initially described as a general method for introducing functional macromolecules into adherent cell lines. It has since been demonstrated to function well in a DNA transfection experiment. The protocol is very simple and involves incubating a washed monolayer of cells with a buffered DNA solution (1–50 (g/ml) prior to scraping with a rubber policeman. The scraped cells are then distributed to fresh culture plates and analysed for transient expression of the transfected DNA 1–5 days later. Up to 80% of scrape-loaded cells were shown to express the DNA and cell viability of 70% was obtained. Stable expression of co-transfected DNAs is also possible. The method is rapid, efficient and is low in cost compared to other transfection procedures. It does not require the formation of a DNA complex or co-precipitate and therefore is a relatively facile operation. As such, this is probably the first choice in determining which transfection technique is adequate for a cell line of choice.

RETROVIRAL INFECTION

Retroviruses can efficiently infect animal cells and integrate into their genomes. They have been used to transfer DNA into a variety of animals during very early development. Retroviruses use RNA as their genome instead of DNA and can infect and replicate in host cells without killing them. During infection, 2 copies of RNA and the enzyme reverse transcriptase are "injected" into the host cell. After infection, reverse transcriptase

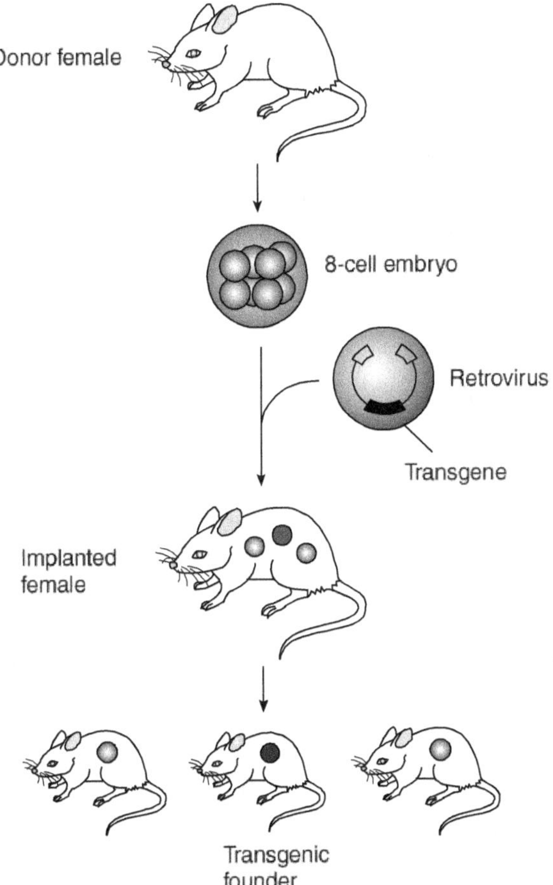

Figure 4.5 Establishing transgenic mice with retroviral vectors

catalyses the synthesis of a double-stranded DNA using RNA as a template. The DNA then inserts into the host genome (provirus). The formation of new virus particles requires viral entry and integration of the DNA copy of the RNA genome and expression of viral genes (Figure 4.5).

Recombinant retroviruses produce virions, which are used to infect animal cells and mice embryos. Generally, early 4–16-celled embryos are used. But when embryos beyond the 4-celled stage are used, often all the cells in an embryo may not be infected by the retrovirus, resulting in chimeric mice. A chimera is an individual which has in its body, cells of 2 or more genotypes. Chimeric individuals produced by transfection arise when some cells of an embryo become stably transfected, while its other cells do not; as a result the individual developing from such an embryo will have cells of 2 genotypes in its body.

Retroviruses can infect a wide variety of cell types, and a large number of each of those cells has a single DNA copy. However, the type of foreign DNA that can be used is severely restricted because only small DNA fragments of approximately 8 kb can be transferred into an organism. To ensure safety during transfection, the virus used as a vector is disabled by the deletion of genes that encode reverse transcription and viral particle packaging. A retrovirus **helper virus** is required for packaging. As an extra safety factor, the helper virus cannot be packaged because of genome modifications. One potential risk however, is that both disabled vectors may recombine to generate a fully functional retrovirus that can integrate into the host genome.

Retroviruses are commonly used to introduce genes of interest into cells and culture or into somatic tissue in experimental animals (Miller, 1997). They also have been used for germ-line modification of fish, molluscs, chicken, mice and more recently, cattle (Chan *et al.*, 1998).

To make a retrovirus vector, a DNA construct containing the gene of interest is flanked by sequences necessary for replication as a virus. The sequences include transcriptional promoters in the long terminal repeats (LTRs), which flanked the integrated DNA, or provirus. Signals necessary for packaging of the transcript in virions, for reverse transcription, and for integration of the resulting DNA also must be included. Introduction of such a DNA construct into cells that express viral proteins, but that are incapable of making infectious virus (i.e., helper, or packaging cells) leads to the creation of infectious

virions containing an RNA copy of the gene of interest. After infection of cells with such virions, the RNA is copied into DNA and integrated at random sites in the cell genome. Again, selectable markers often are included in the construct to select cells containing the desired virus construct.

Summary

- The engineering of organisms requires fusion of the correct promoter or enhancer and gene-coding sequence.

- Transfection and retroviral infections are the two basic approaches in inserting DNA into vertebrate germ-line cells.

- Microinjection is a technique of injecting DNA solution into the nucleus of a cell.

- Electroporation facilitates DNA uptake by reversibly altering the permeability of cell membrane.

- The delivery of DNA into cells using liposomes is called lipofection.

- Embryonic stem cell line technology permits the selection of donors having targeted gene transfers from among several clones showing stable transfection.

REVIEW QUESTIONS

1. Describe various methods of introducing DNA into living cells.

2. Define transfection.

3. Explain different physical methods of gene transfer.

4. What is gene cloning? Give an account of various gene transfer methods.

5. Briefly describe the various transfection techniques for animal cells and ova, and discuss their advantages and limitations.

6. What are embryonic stem cells and how are they derived?

7. Write a note on retroviral infection.

8. Explain scrapefection.

9. Discuss in detail the process of microinjection with a neat-labelled diagram.

10. Write an account of lipofection method of gene transfer.

TRANSGENIC ANIMALS

Pig Parts

In 1902, a German medical journal reported an astonishing experiment. A physician, Emmerich Ullman, had attached blood vessels of a patient dying of kidney failure to a pig's kidney set up by her bedside. The experiment failed when the patient's immune system rejected the attachment almost immediately.

Nearly a century later, in 1997. Robert Pennington, a 19-year-old suffering from acute renal failure and desperately needing a transplant, survived for six and a half hours with his blood circulating outside of his body through a living liver removed from a 15-week-old, 118-pound pig named Sweetie Pie. The pig liver served as a bridge until a human liver became available. But Sweetie Pie had been genetically modified and bred so that her cells displayed a human protein that controlled the complement-mediated hyperacute rejection reaction against tissue transplanted from an animal of another species. Because of this slight but key bit of added humanity, plus immunosuppressant drugs, Pennington's body was able to tolerate the pig liver's help for the few crucial hours.

Successful xenotransplants would help alleviate the organ shortage. A possible danger of xenotransplants is that people may acquire viruses from the organ donors. Viruses can "jump" species, and the outcome in the new host is unpredictable. So far, it is known that a virus called PERV–for "porcine endogenous retrovirus"–can infect human cells in culture. However, a study of several dozen patients who had received implants of pig tissue, for a variety of reasons, revealed that none showed evidence of PERV years later. That study, though, looked only at blood, we still do not know what effect pig viruses can have on human body. Because many viral infections take years to cause symptoms, a new infectious disease in future could be the trade-off for using xenotransplants to solve the current organ shortage.

INTRODUCTION

Gene transfer is an experimental process for modifying the heritable genotypic and phenotypic content of the cell by causing the cell to take up and express gene sequences of purified DNA from donor cells. A successful germ-line transmission of incorporated DNA was also observed. These spectacular results captured the imagination of the scientific community and are the basics for a flourishing research and commercial interest in the area of transgenic animal production.

There have been two basic strategies employed in these efforts. The first involves the production of the transgenic animals by transmitting into the latter a production trait which enhances the animal's value as a producer of conventional animal product. The second strategy in producing transgenic animals is to generate animals that biosynthesize a commercial product other than the conventional animal product.

WHAT ARE TRANSGENIC ANIMALS?

The nuclei of all the cells in every living organism contain genes made up of DNA. These genes store information that regulates how our bodies form and function. Genes can be altered artificially, so that some characteristics of an animal are changed. For example, an embryo can have an extra, functioning gene from another source artificially introduced into it, or a gene can be introduced which can knock out the functioning of another particular gene in the embryo. Animals that have had their DNA manipulated in this way are known as transgenic animals. Transgenic rats, rabbits, pigs, sheep, cows and fish have been produced, although over 95% of all existing transgenic animals are mice.

A transgenic animal is one that carries a foreign gene that has been deliberately inserted into its genome.

WHY ARE THESE ANIMALS BEING PRODUCED?

The two most common reasons are:

1. Some transgenic animals are produced for specific economic traits. For example, transgenic cattle were created to produce milk containing particular human proteins, which may help in the treatment of human emphysema.

2. Other transgenic animals are produced as disease models (animals genetically manipulated to exhibit disease symptoms so that effective treatment can be studied). For example, Harvard scientists made a major scientific breakthrough when they received a U.S. patent for a genetically engineered mouse, called OncoMouse® or the Harvard mouse, carrying a gene that promotes the development of various human cancers.

HOW ARE TRANSGENIC ANIMALS PRODUCED?

Until recently, selective breeding was the only way to enhance the genetic features of domesticated animals. However the combination of the successful transfer of genes into mammalian cells and the possibility of creating genetically identical animals by transplanting nuclei from embryonic tissue into enucleated eggs (nuclear transfer, nuclear cloning) led researchers to consider putting single, functional genes or gene clusters into the chromosomal DNA of higher organisms. Conceptually, the strategy used to achieve this end is simple.

- A cloned gene is injected into the nucleus of a fertilized egg.

- The inoculated fertilized eggs are implanted into a receptive female (because successful completion of mammalian embryonic development is not possible outside of a female).

⚜ Some of the offspring derived from the implanted eggs carry the cloned gene in all of their cells.

⚜ Animals with the cloned gene integrated in their germ-line cells are bred to establish new genetic lines.

The genetic improvement of multicellular organisms by the introduction of relevant transgenes is only slowly being realized. However, transgenesis has become a powerful technique for studying fundamental problems of mammalian gene expression and development, for establishing animal model systems for human diseases, and for using the mammary gland to produce pharmaceutically important proteins in milk. With this application in mind, the term **pharming** was coined to convey the idea that milk from transgenic farm (**Pharm**) animals can be a source of authentic human protein drugs or pharmaceuticals.

The basic methods of producing transgenic animals were described in chapter 5.

The ease with which transgenic animals can be produced and the transfection techniques to be employed depend chiefly on the reproductive biology, husbandry requirements and responses to various experimental procedures of the concerned animal. The following features greatly facilitate gene transfer efforts.

1. Production of larger number of eggs either naturally, e.g. in fish, or in response to hormone regimens used for superovulation, e.g. in mice, pigs, etc.

2. Short breeding cycle, i.e., time taken from birth to reaching the reproductive age, greatly facilitates analysis of the transgenic individuals produced.

3. It is desirable that ovulation occurs throughout the year so that ova are readily available for experimentation.

4. The size and the structure of eggs should be amenable to microinjection, the only transfection technique successful with almost all animal species.

5. In many animal species, e.g. fish, body size may be an important factor to allow sufficient tissue samples to be taken for the detection of transgene integration and function without sacrificing the individual.

6. In the case of mammals where transfected ova/embryos must be transferred into surrogate mothers, the following features are critical:

    ✺ synchronization of females used for superovulation and as surrogate mother by hormone administration.

    ✺ *In vitro* fertilization and culture.

    ✺ Transfer of the embryos into surrogate mothers and completion of the embryo development in them.

    ✺ Production and maintenance of embryonic stem (ES) cell lines capable of giving rise to germ cells, which is essential for the application of ES cell transfer technology.

TRANSGENIC MICE

Mouse is the most preferred mammal for studies on gene transfers due to its many favourable features like short oestrous cycle and gestation period, relatively short generation time, production of several offsprings per pregnancy, convenient IVF, successful culture of embryos *in vitro* at least for a particular period of time, production and maintenance of ES cell lines, availability of a diverse array of genetic stocks, etc.

Transgenic technology has been developed and perfected in the laboratory mouse. Since the early 1980s, hundreds of different genes have been introduced into various mouse strains. These studies have contributed to an understanding of gene regulation, tumour development, immunological specificity, molecular genetics of development, and other biological processes of fundamental interest. Transgenic mice have also played a role in examining the feasibility of the industrial production of human therapeutic drugs by domesticated animals and in the creation of transgenic strains that act as biomedical models for various human genetic diseases. For transgenesis, DNA can be introduced into mice by one of the following methods:

1. retroviral vectors that infect the cells of an early-stage embryo prior to implantation into a receptive female,

2. microinjection into the enlarged sperm nucleus (the male pronucleus) of a fertilized egg, or

3. introduction of genetically engineered embryonic stem cells into an early-stage developing embryo before implantation into a receptive female.

Embryonic stem cells (ES cells) are harvested from the inner cell mass (ICM) of mouse blastocysts. They can be grown in culture and retain their full potential to produce all the cells of the mature animal, including its gametes. The ES cell technology has been described in chapter 4.

The potential for using animals as living factories has been explored in mice, and the technology that has developed is being transferred to other larger animals. Transgenes introduced into the mouse or other animals (e.g. cows and goats) are expressed and their products (e.g. interleukin-2, α-antitrypsin, clotting factor $\bar{x}$) secreted in the milk for isolation. The gene construct must include mammary-specific promoters (for e.g. a promoter from the β-casein gene) and appropriate enhancers. One important protein that is being studied in this way is the cystic fibrosis transmembrane regulator (CFTR). CFTR normally acts as a chloride channel allowing chloride ions to move in and out of cells. When CFTR is mutated, the channel is disrupted. The resulting disruption of ion flow causes mucus to accumulate in organs, especially in the lungs and pancreas. The mucus attracts bacteria that later die and release their DNA, further inhibiting normal organ function. Because the CFTR protein is associated with membranes of transformed cells, it has not been successfully expressed *in vitro.*

Secretion into milk facilitates isolation since the mammary glands produce fat globules in milk, which are encapsulated by plasma membrane. In transgenic animals, a heterologous transmembrane protein could be associated with the plasma membrane of the globule and isolated readily from milk during lactation. The CFTR protein secreted in milk can be used to study protein function so that effective therapies can be developed.

Production of pharmaceutically valuable human proteins from the milk of transgenic animals is called animal pharming. Transgenic mice are developed to have human proteins in the milk, e.g. soluble CD4 protein (for AIDS treatment), urokinase (to dissolve blood clots), etc.

As discussed earlier, mice have been used as a model for human genetic diseases such as Alzheimer disease, arthritis, muscular dystrophy, tumorigenesis, hypertension, neurodegenerative disorders, endocrinological dysfunction, coronary disease, and many others.

Alzheimer disease is a degenerative brain disorder that is characterized by the progressive loss of both abstract thinking and memory and is accompanied by personality change, language disturbances, and a slowing of physical capabilities. Clinical diagnosis of Alzheimer disease is poor, although 1% of the population between 60 and 65 years old and 30% of the population over 80 years old may develop it. Neurofibrillary tangles accumulate within the cell body of the neurons, dense extracellular aggregates called **senile plaques** develop at the ends of neuritis, and brain cells (neurons) are lost in the neocortex and hippocampus of the brain in the patients with Alzheimer disease (Figure 5.1). The core of a senile plaque is composed of a clearly packed, fibrillar structure that traditionally has been called an amyloid body.

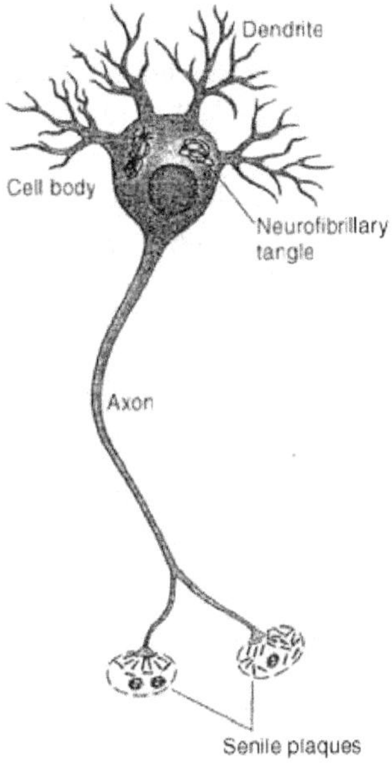

Figure 5.1 Schematic representation of a neuron of the human cerebral cortex, showing some of the histopathological features of Alzheimer disease. Senile plaques containing amyloid deposits and apparent cellular debris accumulate at the synapses of the neurofibrillary tangles contain aggregated cytoskeletal and other proteins. Other changes that occur in the neurons are not depicted here.

The principal component of senile plaques and amyloid bodies is a 4-kilodalton (kDa) protein called $A\beta$ (amyloid β, β-protein, β-amyloid protein, β/A4). There are $A\beta$ proteins with different numbers of amino acid residues, such as $A\beta$ 40 and $A\beta$ 42. All $A\beta$ proteins are products of an internal proteolytic cleavage of the β-amyloid precursor protein (APP). It is not known what causes the accumulation of $A\beta$. Some families that have a high incidence of Alzheimer disease have mutations in the APP gene, a finding that implicates this gene in Alzheimer disease. Unfortunately, it is for the most part impossible to study the onset and pathogenesis of Alzheimer disease in human subjects. Accordingly, an animal model that mimics Alzheimer disease would be an invaluable research tool.

A number of transgenic mice have been generated with full or partial versions of the normal APP gene driven by neuron-specific promoters. In most cases, amyloid plaques, neurofibrillary tangles, neuronal cell death, or disturbed behaviour was not observed. However, neurodegeneration that is similar to Alzheimer disease does occur with a transgene that is made up of the coding region for the terminal 100 amino acids of APP, which contains the $A\beta$ protein sequences.

Gene Targeting in Mice

Gene targeting is the insertion of DNA into a specific chromosomal location; it is often used to inactivate a specific gene in the genome. Gene inactivation is a common means of elucidating how a specific gene functions. In gene targeting, a cloning vector (called a targeting vector), which harbours the gene to be inserted, recombines with a region of the target chromosome that is homologous with a DNA region on the targeting vector. This process is called as homologous recombination, and may be defined as the exchange of identical segments between two DNA molecules that have identical or almost identical sequences. In yeast, gene integration occurs ordinarily by homologous recombination. But in mammals, random DNA integration is far more frequent than homologous recombination. The frequency of integration by homologous recombination appears to be only 0.1 to 1.0% of random integration events; this makes the recovery of transgenics representing targeted gene transfer quite difficult. Recent refinements in the techniques, however, have increased this frequency to 10% or even 50% of random integrations.

The approaches for the identification of homologous recombinations are as follows:

1. When inactivation or activation of the test gene occurs due to homologous recombination, but not by random integration, and produces a selectable phenotype, such cells can be scored or selected for. But there are not many such genes and a great majority of genes of interest do not produce a selectable phenotype.

2. Alternatively, a large number of transfected cells/clones may be screened, often using PCR amplification, to identify those having homologous recombination.

A selectable marker gene (for example, bacterial neomycin phosphotransferase, which confers resistance to the antibiotics neomycin and kanamycin) ensures that the desired recombination event is selected. Antibiotics included in the growth medium of cells transformed with the targeting vector and inserted gene allow only those cells that have the correct insertion into the chromosome to survive.

A gene is first isolated, altered *in vitro* according to the requirements of the experiment, and then targeted to its counterpart on a specific chromosome in cells. A portion of the gene is disrupted by inserting an antibiotic resistance gene such as neomycin into it, with regions of the gene flanking the marker remaining on either side. This marker enables cells that contain targeted DNA to be selected (positive selection). A second marker (a negative marker), located at the end of the targeting vector, is used to select for cells in which the transferred gene is targeted to the specific site rather than to a random site, or for cells in which no insertion occurred.

The marker often is the herpes simplex virus type I thymidine kinase gene (***HSV-tk*** gene). After the gene construct is assembled, the gene and its vector are transferred to embryo host cells, and homologous recombination allows the gene to insert into a specific chromosome region. The transferred DNA lines up next to its chromosomal counterpart in cells so that the identical regions are in alignment. When recombination occurs, there is an exchange of DNA in the identical regions of the gene, but excluding the *tk* marker, which is outside those regions (Figure 5.2).

To select for cells with the foreign gene and its mutation, cells are placed in a medium containing antibiotic (for example, neomycin), which kills cells in which the vector has not inserted into the genome. A drug called ganciclovir selects for targeted insertion by killing any cell that contains the *tk* gene, specifically killing cells harbouring the herpesvirus *tk* gene. Thymidine kinase is produced from ganciclovir, a compound that is toxic to cells. If random insertion occurs, the cells will receive a copy of the *tk* gene and will be killed by ganciclovir. Cells that do not receive the antibiotic resistance gene will be sensitive to the antibiotic in the medium. The surviving cells contain the antibiotic resistance gene and the targeted gene region, but lack the *tk* gene. Therefore, only those cells that have undergone homologous recombination survive treatment with neomycin and ganciclovir.

Screening for homologous recombination often involves using PCR with two primers; one primer is complementary to the transferred DNA that is not present in the chromosomal DNA of host cells, and the other primer is complementary to a region of chromosomal DNA outside the targeted area. The predicted size of the amplified DNA fragment can be confirmed using gel electrophoresis; if a different size is obtained, a random integration occurred. This method is especially important if a selection method, such as the one described for the *tk* gene and ganciclovir, is not used.

ES cells from early mouse embryos are used in gene targeting experiments and can be altered as described for gene targeting experiments. ES cells are placed in mouse embryos, where they become incorporated. These genetically manipulated embryos are placed in surrogate mothers where they develop.

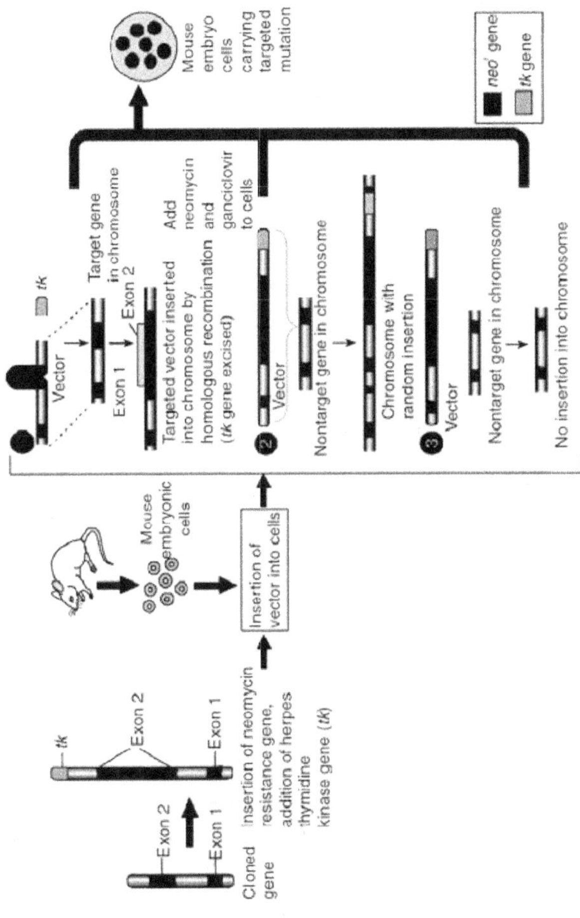

Figure 5.2 Targeted gene replacement in cultured mouse embryonic cells

Coat colours can serve as a marker to track the incorporation of introduced ES cells in mouse embryos. If a mouse has both black and brown coat colour genes, it will be a chimera with a readily detectable black and brown coat. ES cells for example, from a brown mouse would contain a brown coat gene (called agouti gene, expressed as a brown coat even if in single copy) as well as the gene to be targeted. The ES cells to be transferred will contain a gene for a different coat colour than the host embryos (Figure 5.3).

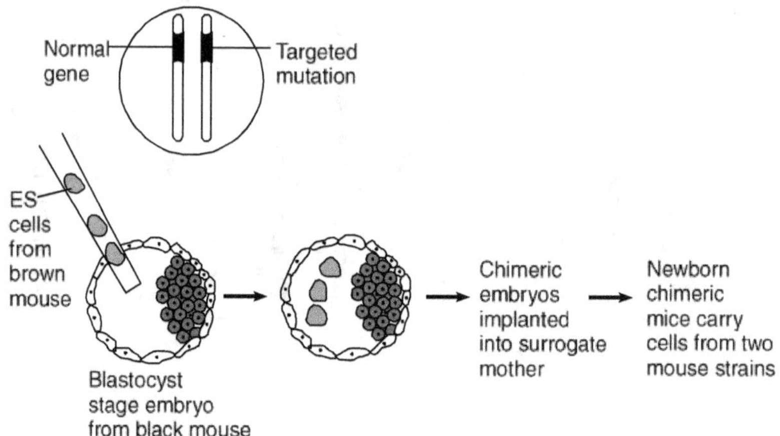

Figure 5.3 Transfer of mouse ES cells into embryos. Coat colour is a marker to determine incorporation of ES cells into embryos

When newborn mice are screened for the presence of a targeted gene, coat colour indicates whether the mice contain transformed ES cells. For example, male chimeric mice with a black and brown coat inherit ES cells from both the host embryo (black coat gene) and the introduced ES cells (brown coat gene). These chimera are selected for mating to a female with a black coat (non-agouti). Only progeny with brown coats are selected for further screening of the DNA to determine which individuals have inherited the targeted gene. Those male and female mice with targeted gene copies are mated to obtain mice with two copies of the targeted gene. Confirmation of mice homozygous for the targeted gene is obtained by DNA testing using either PCR or Southern blot hybridization.

Knock-out and Knock-in Technology

In order to study the relationship between proteins and gene function, scientists now can prevent the manufacturing of the protein by a specific gene. By disabling the gene from a test organism, and then producing descendents that contain the copies of the disabled gene, it is possible to observe the descendents' development in the absence of a particular protein. This practice, referred to as knock-out technology, is an attempt to shut down or turn off a particular gene. Thus far, the mouse has been the mammal in which knock-out technology has been most generally applied. In essence, a **"knock-out"** organism is created when an ES cell is genetically engineered and then inserted into a developing embryo. The embryo is then inserted surgically into the womb of the host (e.g. a female mouse). Once the embryo has matured, a portion of its stem cells will produce egg and sperm with the knocked-out gene. A gene can also be altered in function, in contrast to being deleted. When a gene is altered but not shut down, a **"targeted mutation"** effect is created. This practice is referred to as knock-in technology, whereby a life form has an altered gene **"knocked"** into it. Gene **knock-out/ knock-in** technology is well-established as an experimental tool in mice, due to the availability of ES cell lines. The principle is to take advantage of a rather rare event that occurs after introduction of DNA into cells—homologous recombination between identical sequence in the genome and the transfecting DNA (Bronson and Smithies, 1994). In the most common protocol, a selectable marker such as the neomycin-resistance gene is inserted within a piece of DNA corresponding to the portion of a gene of interest. After transfection of cells by this construct and selection for the marker (by growth in medium

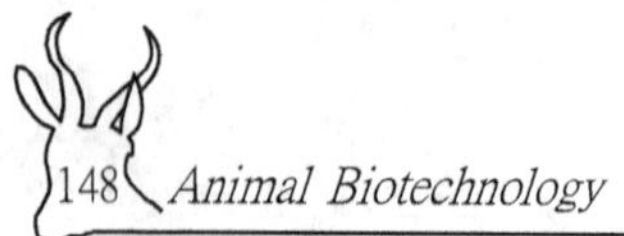

containing the neomycin-related antibiotic G418) the selected cells are screened to identify the small fraction that has one copy of the gene of interest disrupted by the marker. Progeny animals derived from the cells will be heterozygous for the "knocked-out" gene; breeding to obtain homozygous animals is straightforward. Because the process is so inefficient, very large numbers of transfected cells was to be screened, making the use of cultured cells essential, since, it would be impracticable to screen large numbers of progeny from microinjected eggs. The galactosyl transferase-knock-out pigs were generated from cultured foetal fibroblasts manipulated in this way. Nuclei from the cells were then transferred into oocytes.

TRANSGENIC COWS

If the mammary gland is to be used as a bioreactor, dairy cattle, which individually produce approximately 10,000 litres of milk annually, containing about 35 grams of protein per litre, are likely candidates for transgenesis. More specifically, if a recombinant protein was present at 1 gram per litre of milk and it could be purified with 50% efficiency, the yield from 20 transgenic cows would be about 100 kg per annum. Coincidentally, the annual global requirement for protein C, which is used for the prevention of blood clots, is about 100 kg. On the other hand, one transgenic cow would be more than sufficient for the production of the annual world supply of factor IX (plasma thromboplastin component), which is used by haemophiliacs to facilitate blood clotting.

A modified mouse transgenesis DNA microinjection protocol has been developed for cattle (Figure 5.4).

The essential steps are as follows:

1. Collecting oocytes from slaughterhouse-killed cows.
2. *In vitro* maturation of oocytes.

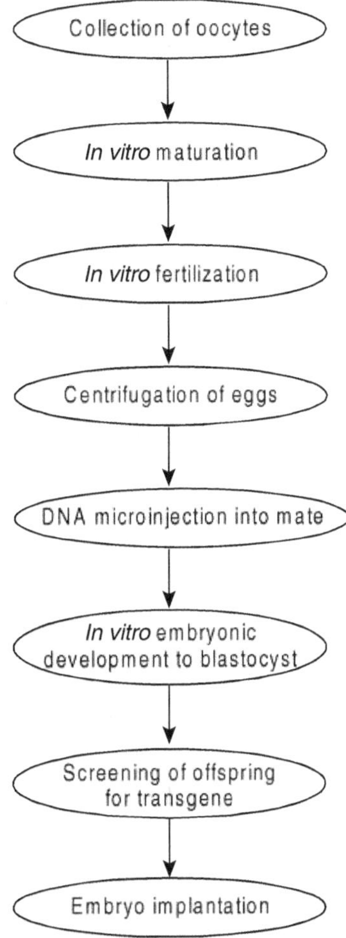

Figure 5.4   Steps in the development of transgenic cattle

3. *In vitro* fertilization with bull semen.

4. Centrifugation of the fertilized eggs to concentrate the yolk, which in a normal egg, prevents the male pronuclei from being readily seen under a dissecting microscope.

5. Microinjection of input DNA into male pronuclei.

6. *In vitro* development of embryo.

7. Non-surgical implantation of one embryo into one recipient foster mother in natural oestrus.

8. DNA screening of the offspring for the presence of the transgene.

When this non-surgical procedure was put to the test, two transgenic calves were obtained from a starting pool of 2,470 oocytes, a result indicating that the methodology is feasible although inefficient. As research in this area continues, the transgenesis procedure will be refined.

Genetic engineering will one day change the composition of milk products. One goal is to increase the protein κ-casein (a phosphoprotein in milk) in cows by over-expressing the κ-casein gene in animals. The end result would be the production of more cheese from milk. Another desirable genetic modification would be to remove lactose from milk. Lactose-free product would be beneficial to the millions of people who are lactose-intolerant and therefore cannot digest milk containing lactose.

Other potential modifications include increased resistance of animals to bacterial, viral and other pathogenic diseases. For example, mastitis, a bacterial infection of the mammary glands, is a common affliction of cows. Resistance in cows would save farmer's money, reduce the use of antibiotics, and prevent much suffering. *In vivo* immunization, transferring the antibody genes for specific antigens to animals, would protect livestock from many diseases and eliminate the need for vaccinations. Genetic engineering could eventually reduce the number of vaccinations, drugs, and veterinarian visits, and ultimately the cost of raising cattle.

The production of recombinant bovine somatotropin (rBST) for use in cows to increase milk production has created a recent controversy. Somatotropins occur naturally in the milk and meat of certain animals, and BST, a protein hormone produced in the cow's pituitary gland, is essential for milk production. When injected into cows, BST stimulates increased milk production and facilitates the efficient conversion of feed into milk rather than body fat. In the past, somatotropin has been obtained from cow pituitaries by extraction, an expensive and time-consuming method. Today, rBST can be produced in *E. coli* after transfer of the *bst* gene. After the injection into cows, rBST increases milk production by approximately 25%. The FDA has deemed products from cows treated with rBST safe for human consumption.

"Annie" was the first cow to be cloned with a gene for an agricultural application. She was born in March 2000, a copy of a pure-breed Jersey cow. Now the scientists have started to test her milk for resistance to mastitis, a bacterial disease. Mastitis is often caused by *Staphylococcus aureus*, bacteria that destroy milk-secreting cells in cow's mammary glands.

Researchers hope that Annie's mammary glands will secrete a protein called lysostaphin that may help her resist attacks from *S. aureus*. They inserted a gene for lysostaphin production obtained from a benign species of *Staphylococcus simulans* that competes with its disease-causing relatives.

Antibiotics cure only about 15% of cows infected with *S. aureus*. It is among the most virulent of mastitis-causing pathogens, causing nearly one-third of all infections in cows. If the lysostaphin works, Annie will have an alternative defence bioengineered right inside her cells.

TRANSGENIC SHEEP

As with cows, experiments with sheep and goats as bioreactors to produce important human proteins or pharmaceuticals seem quite promising.

Although the quantity of milk produced by either a sheep or goat is smaller than that produced by a cow, lactation in sheep and goat yields hundreds of litres of milk per year. With a method very similar to the one used for producing transgenic mice and with transgene constructs that have mammary-gland-specific promoters driving human gene sequences, investigators have created both transgenic sheep and goats that secrete human proteins into milk. The transgene-derived proteins were glycosylated and had biological activities comparable to those extracted from human sources.

The rate of transgenesis in sheep is very low (0.1 to 0.2%). This can be improved only if transgenic viable embryos (after necessary checking) are transferred to surrogate ewes (female sheep). Embryos at 8–16 cell stage can be split into two parts, one for continued culture and the other for detection of integrated genes using polymerase chain reaction. Although microinjection is the most common method for DNA delivery, gene targeting may be increasingly used in future in this approach; embryonic stem cells in culture are transfected with a vector which targets the gene to a particular site by homologous recombination. This technique, though successfully used in mice, has yet to be applied to sheep, where ES cells will have to be isolated first.

The first reports of transgenic sheep were published by J.P. Simons (1986) of Edinburgh. Two transgenic ewes were produced, each carrying about 10 copies of human anti-haemophilic factor IX gene (cDNA) fused with the 10.5 kb BLG (β-lactoglobulin) gene. BLG gene was used because it is necessary for specific expression of gene in mammary glands. Consequently, the gene had a tissue-specific expression and ewes secreted human factor IX (or alpha 1 antitrypsin) into their milk; this human factor IX is active, even though the expression of the transgene is low. The transgenic ewes were born in early summer of 1986 and were successfully mated the same year in December. In 1987, each ewe gave birth to a single lamb. Each lamb inherited BLG-F IX transgene and secreted factor IX in the milk. This programme of the production of transgenic animals by J. P. Simons at Edinburgh was funded by Pharmaceutical Protein Ltd., (Cambridge, UK) due to its commercial appeal.

In another report published in 1991, also from Edinburgh, five transgenic sheep were produced (Alan Colman). In all these cases, transgene involved fusion of the Bovine β-lactoglobulin gene promoter fused to the human α_1 antitrypsin ($h\alpha_1$ AT) gene. Four of these animals were female and one male. In one female, the protein $h\alpha_1$ AT reached a level of 35 g per litre of milk. The protein purified from milk had a biological activity indistinguishable from human plasma-derived antitrypsin. The deficiency of $h\alpha_1$AT leads to a lethal disease emphysema, which is a common hereditary disorder among Caucasian males of European descent. Hence, any strategy giving high yield of protein economically will be most welcome. In view of this, transgenic sheep with $h\alpha_1$ AT gene will prove very useful as a bioreactor.

Until recently, the transgenes introduced into sheep inserted randomly in the genome and often worked poorly. However, in July 2000, success at inserting a transgene into a specific gene locus was reported. The gene was the human gene for **alpha1-antitrypsin**, and two of the animals expressed large quantities of the human protein in their milk. The protocol of generating a transgenic sheep is as follows:

Sheep fibroblasts (connective tissue cells) growing in tissue culture were treated with a vector that contained the following segments of DNA.

1. Two regions homologous to the sheep *COL1A1* gene. This gene encodes type 1 collagen. (Its absence in humans causes the inherited disease osteogenesis imperfecta.)

    This locus was chosen because fibroblasts secrete large amounts of collagen and thus one would expect the gene to be easily accessible in the chromatin.

2. A neomycin-resistance gene aided in isolating those cells that successfully incorporated the vector.

3. The human gene encoding **alpha1-antitrypsin**.

   Some people inherit two non-functioning or poorly-functioning genes for this protein. Its resulting low level or absence produces the disease **alpha1-antitrypsin deficiency** (**A1AD** or **Alpha 1**). The main symptoms are damage to the lungs (and sometimes to the liver).

4. Promoter sites from the beta-lactoglobulin gene. These promote hormone-driven gene expression in milk-producing cells.

5. Binding sites for ribosomes for efficient translation of the mRNAs.

Successfully transformed cells were then fused with enucleated sheep eggs and implanted in the uterus of an ewe (female sheep). Several embryos survived until their birth, and two young lambs have lived over a year. When treated with hormones, these two lambs secreted milk containing large amounts of **alpha1-antitrypsin** (650 μg/ml; 50 times higher than previous results using random insertion of the transgene).

Recombinant DNA technology can also be used to increase the ability of sheep for wool growth. Wool is obtained from Angora, Mohair and Merino sheep. A high class Merino sheep has over 10^8 hair follicles in the skin, but only 25% of hair follicles possess hairs. This is due to low level of keratin in the cells. Keratin is rich in the sulphur-containing amino acid, cystine. High level of cystine favours the growth of wool fibres. The following three technologies have been used to develop transgenic sheep for maximum wool production.

Roger, at the University of Adelaide, Australia, isolated the genes coding for serine acetyl transferase (SAT) and O-acyl serine sulphydrase (OAS) from the bacterium *Salmonella typhimurium*. He then fused the two genes and added a promoter to it. The gene construct was introduced into fertilized eggs. The fertilized eggs were allowed to develop into 8–16-celled embryos. The embryos were implanted into a sheep. The sheep delivered a transgenic lamb, which produces dense, long wool fibres.

In another attempt, Roger introduced genes for isocitrate lyase and malate synthetase of *E. coli* into sheep. These enzymes helped the cells to bypass certain steps in the TCA cycle and helped to synthesize more amino acids. Hence the transgenic sheep produced more wool.

R.A. Leng and his co-workers introduced the genes for isocitrate lyase and malate synthetase into rumen bacteria and the sheep were fed with the recombinant bacteria. The bacteria synthesize more amino acids and provide them to the sheep cells. The sheep skin cells therefore get more amino acids and hence the skin produces dense long wool fibres.

TRANSGENIC GOATS

Transgenic goats were successfully produced recently by groups headed by John Mc Pherson and Karl Ebert in U.S.A. These transgenic goats expressed a heterologous protein (a variant of human tissue-type plasminogen activator = LAtPA) in their milk. This protein is used for dissolving blood cells, i.e., for treatment of coronary thrombosis. A cDNA representing LAtPA was linked with either the murine whey acid promoter (WAP) or a β-casein promoter in an expression vector and used for injecting early embryos obtained surgically from the oviducts of superovulated dairy goats. These injected embryos were either immediately transferred to the oviducts of recipient females (surrogate mothers) or cultured for 72 hours (blocked at 8–16-cell stage) before transfer to the uterus of recipient females. Of the 29 offsprings from 36 recipients, one male and one female contained the transgene. The transgenic female delivered five offsprings, one of which was transgenic showing expression of LAtPA at a low level of few milligrams per litre of milk. In another case of a transgenic goat, few grams of LAtPA per litre of milk could be

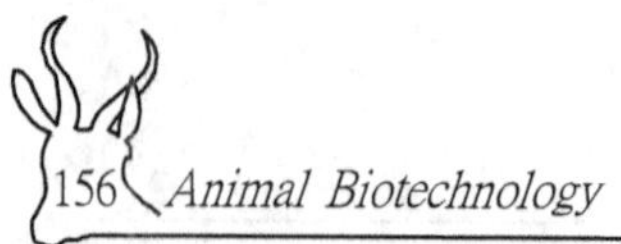

obtained. At this concentration, the dairy goat may become an economically viable bioreactor for human pharmaceuticals.

TRANSGENIC PIGS

The efficiency of the production of transgenic pigs is very low compared to that of the production of transgenic mice. The frequency was as low as 0.6% even when as many as 7000 eggs were injected. Despite this low frequency, transgenic pigs carrying growth hormone (GH) gene from bovine (of "ox" origin) or human, and sheep globin gene have been produced by V.G. Purse at Agriculture Research Service, Beltsville, USA. The pigs carrying hGH gene showed different levels of expression and only 665 of these animals showed detectable levels of hGH and bGH in their plasma. The animals showed detectable levels of hGH and bGH in their plasma. The animals grew a little faster but did not become large; similarly pigs with sheep globin gene did not show any expression of the transgene for unknown reasons. In these transgenic pigs, however, a modest increase of 10–15% in daily weight and 16–18% in feed efficiency was observed. It was also observed that there was a marked reduction in the subcutaneous fat in some of these transgenic pigs suggesting the possibility of producing leaner meat with lower fat content.

It is also reported that a long-term elevation of growth hormone was generally detrimental to health. The pigs had incidence of gastric ulcers, arthritis and several other diseases. Therefore, techniques will have to be developed to manipulate better the transgene expression by a variety of methods (e.g. changing genetic background or modifying the husbandry regimen).

Fertilizing normal eggs with sperm cells that have incorporated foreign DNA has also produced transgenic pigs. This procedure, called sperm-mediated gene transfer (SMGT) may someday be able to produce transgenic pigs that can serve as a source of transplanted organs for humans.

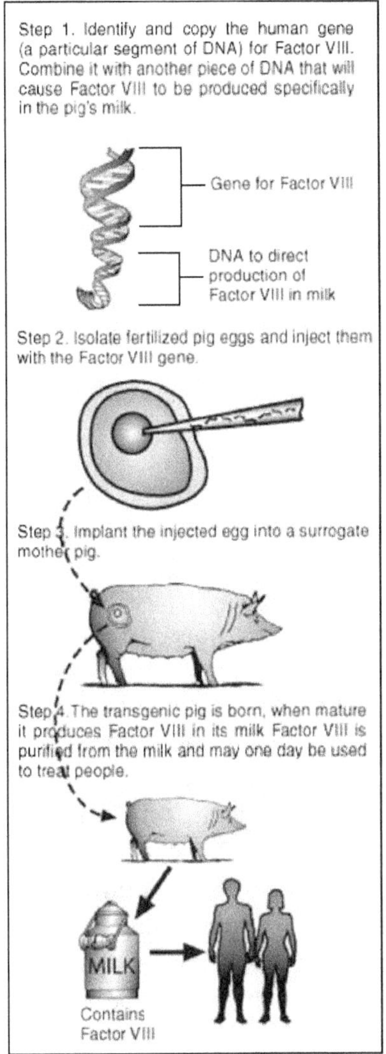

Figure 5.5 Steps in the production of human factor VIII in transgenic pigs.

Factor VIII is a blood clotting factor that is missing in people affected with the bleeding disorder hemophilia A. The steps used by Virginia Tech researchers to produce Human Factor VIII in transgenic pigs are shown in Figure 5.5.

TRANSGENIC BIRDS

Genetic engineering will allow the production of healthier birds used for food in the near future. Chickens, ducks, geese, and small game birds could be made resistant to viral and bacterial diseases. Poultry products could be improved by decreasing the fat content of meat and the cholesterol in chicken eggs. Chicken or duck eggs also could serve as factories for producing valuable proteins. Typically, birds secrete a large amount of ovalbumin in their reproductive tracts that becomes part of the egg. Transgenic birds are engineered to also secrete other types of proteins into the egg, these products could readily be isolated.

Retrovirus vectors have been used to generate transgenic birds. After being disabled to prevent adverse side effects, the virus is used to infect cells at the blastoderm stage of development. Problems associated with retrovirus vectors include the instability of the introduced gene and restrictions on the size of DNA inserts.

Microinjection of DNA into fertilized eggs of birds to produce transgenic strains is not a simple procedure because of several features that are unique to avian reproduction and development. For example, during fertilization in birds, several sperms can penetrate the ovum instead of one, as usually occurs in mammals. Consequently, it is impossible to identify the male pronucleus that will fuse with the female pronucleus. Also, microinjection of DNA into the cytoplasm will not suffice, because this DNA does not become integrated into the genome of the fertilized egg. Finally, even if nuclear DNA microinjection was practicable, the technique would be difficult to implement because the avian ovum, after fertilization, becomes, in rapid succession, enveloped in a tough membrane, surrounded by large quantities of albumin and enclosed in inner and outer shell membranes.

However, it is possible to inject a transgene into the region of the yolk (germinal disc) that contains the female and male pronuclei. The germinal disc is present before the eggshell is formed. After the administration of DNA to a germinal disc, each egg is cultured *in vitro*, and when an embryo forms, it is placed in a surrogate egg to produce a hatchling. One transgenic line of chickens has been established by this strategy. At present, this method is inefficient and technically difficult to carry out on a routine basis.

By the time an avian egg outer shell membrane has hardened, the developing embryo (blastoderm stage) comprises two layers of 40,000 to 80,000 cells. In trial experiments, inoculation of the blastoderm stage with replication-defective retrovirus vectors carrying bacterial marker genes resulted in the production of germ line transgenic chickens and quail. Although some of the transgenic organisms did not produce virus, the use of retrovirus vectors to deliver genes for a product that is to be used as food would ultimately raise questions about its safety, whether real or imagined. Moreover, the size of the transgene that can be introduced into the recipient organism by retrovirus vectors is limited to ~ 8 kb, and occasionally integration at the initial site is not permanent. Consequently, alternative methods of transgenesis have been examined.

A promising technique involves introducing blastoderm cells transformed with liposomes, into the subgerminal space of host embryos (Figure 5.6). A chimeric organism would be produced having some cells from the host and others derived from the transformed donor cells. If the introduced transformed cells become a part of the germ cell line, stable transgenic lines can be produced through matings of founder animals. The proportion of donor cells in chimeras can be increased to enhance the probability of obtaining germ line chimeras if the recipient embryos are irradiated with a dose of 540 to 660 rads for 1 hour before the introduction of the transfected cells. The radiation treatment destroys some but not all the blastoderm cells, thereby increasing the final ratio of transfected cells to recipient cells. It should be possible to produce transgenic chickens in this way, despite the inefficiency of the procedure.

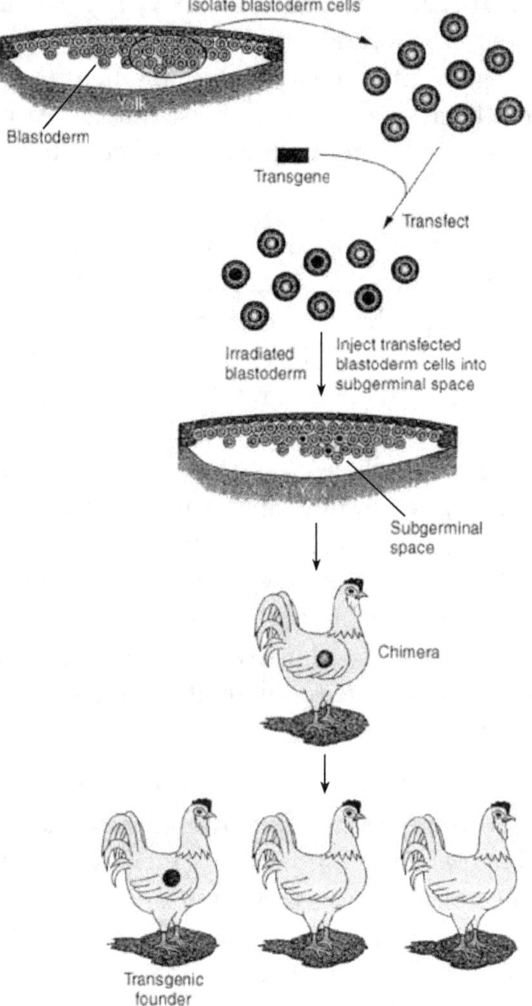

Figure 5.6   Establishing transgenic chickens by transfection of isolated blastoderm cells. Cells from blastoderm donors are removed, transfected by lipofection with a transgene, and inserted into the subgerminal space of an irradiated recipient blastoderm. Some of the resulting chickens may be chimeric, and the chimeras that have the transgene in germ-line cells can be bred to establish transgenic lines.

Chickens

Chickens grow faster than sheep and large numbers can be grown in close quarters. They synthesize several grams of protein in the "white" of their eggs.

Two methods have successfully produced chickens carrying and expressing foreign genes.

- Infecting embryos with a viral vector carrying

    - the human gene for a therapeutic protein.

    - promoter sequences that will respond to the signals for making proteins (e.g. lysozyme) in egg white.

- Transforming rooster sperm with a human gene and the appropriate promoters and checking for any transgenic offspring.

Preliminary results from both methods indicate that it may be possible for chickens to produce as much as 0.1 g of human protein in each egg that they lay.

Not only should this cost less than producing therapeutic proteins in culture vessels, but also chickens will probably add the correct sugars to glycosylated proteins—something that *E. coli* cannot do.

Transgenic chickens could be used to improve the genetic make-up of existing strains with respect to built-in (*in vivo*) resistance to viral and coccidial diseases, better feed efficiency, lower fat and cholesterol levels in eggs, and better meat quality.

TRANSGENIC FISH

As natural fisheries become exhausted, production of this worldwide food resource will depend more heavily on aquaculture. Altering fish genetically by transgenesis is therefore a primary research objective. To date, transgenes have been introduced by microinjection or electroporation of DNA into the fertilized eggs of a number of fish species including carp, catfish, trout, salmon, and tilapia. In fish, the pronuclei are not readily seen under the microscope after fertilization; therefore, linearized transgene DNA is microinjected into the cytoplasm of either fertilized eggs, or embryos that have reached the four-cell stage of development. Unlike mammalian embryogenesis, fish egg development is external; hence, there is no need for an implantation procedure. Development can occur in temperature-regulated holding tanks. The survival of fish embryos after DNA microinjection is high (35–80%) and the production of transgenic fish ranges from 10–70%. The presence of a transgene can be scored by PCR analysis of either nucleated erythrocyte samples or scale DNA. Founder fish can be mated and transgenic lines can be established.

Transgenic fish research was carried out in 1985, and some encouraging results have been obtained. The genes that have been introduced by microinjection in fish include the following:

1. Human or rat gene for **growth hormone**
2. Chicken gene for **δ-crystalline protein**
3. *E. coli* gene for **β-galactosidase**
4. *E. coli* gene for **neomycin resistance**
5. Winter flounder gene for **antifreeze protein**
6. Rainbow trout gene for **growth hormone**

Human growth hormone gene transferred to transgenic fish allowed growth that was twice the size of their corresponding non-transgenic fish (goldfish, rainbow trout, salmon). Similarly antifreeze protein (AFP) gene was transferred in several cases and its expression was studied in transgenic salmon. It was shown that the level of AFP gene expression is still too low to provide protection against freeze. It is contemplated that genes for disease resistance, tolerance to environmental stress, and other biological features will eventually be introduced into both temperate and tropical fish.

TRANSGENIC INSECTS

Research on the genetic engineering of insects is still in the early stages. However, studies have been aimed at the introduction of insecticide resistance into insects. Most insecticides are broad-spectrum chemicals and do not select specifically for harmful pests. If made resistant, beneficial insects, such as ladybugs, honeybees, preying mantids and many species of ants would not be killed by chemical applications in agricultural fields and small gardens. Arachnids (spiders), important insect predators, also are affected by chemical insecticides. Loss of agronomically important insects and other arthropods often leads to outbreaks of insect pests. Engineered insecticide-resistant insect predators could be used in combination with chemicals against insect pests. However, the disadvantage of developing such insects is that they potentially allow increased use of chemical pesticides.

The first step in engineering these insects is to identify and isolate the insecticide-resistance gene. Recently, a fruit fly gene for resistance to cyclodiene insecticides (such as Dieldrin) has been isolated. The most difficult steps will be transforming insects with a gene that is stably maintained and expressed at a high level.

ETHICAL CONCERNS
REGARDING TRANSGENESIS

The scientific outlook of right and wrong opinions about transgenic animals is known as ethics of transgenic animals.

Environmentalists and medical professionals argue about transgenic animals in different ways and have made the following implications.

1. *Creation of new life-forms*

   ✺ The transfer of genes between 2 species may in course of time affect the genome of the species, leading to some damage to the original genome.

   ✺ Transgenic animals may interact with the environment in a way which is not suitable for the sustainability of safe environment, and produce unexpected outcome.

   ✺ As transgenic animals are best models for human diseases, they are used in the study of progress of diseases. But the fact that changing genomes of animals may lead to reduction in breeding ability and lifespan is not taken care of.

2. *Increased use of animals*   Construction of transgenic animals results in altered phenotypes of the host cell. In transgenic disease models, one model has not been used for other diseases or even other workers, as it has the selective target gene alone. So the model animals have to be killed. This is an animal wastage.

3. Gardon (1997) reported that the genetic and physiological characters in most cases of transgenic animals are different from that which were expected. This is mainly due to unexpected interaction of the desired gene with other genes in the genome.

4. Uncontrolled expression of inserted gene may increase the morbidity and mortality, e.g. transgenic animals with inserted growth hormone gene.

PATENTING TRANSGENIC ANIMALS

Winning approval for the patenting of genetically engineered animals is the most recent victory for biotechnology companies in what has been a long struggle to obtain patents on living organisms.

On April 7, 1987, Donald J. Quigg, the Commissioner of Patents and Trademarks, announced that the Patent and Trademark office considers non-naturally occurring non-human multicellular organisms (which of course includes animals) to be patentable.

On April 13, 1988, Harvard University obtained patent number 4,736,866 for a genetically altered mouse used as a model for studying how genes contributed to breast cancer. The "Harvard mouse" set off an emotional debate.

The issue of patenting transgenic animals has also been addressed by lawmakers seeking ways to protect intellectual property, which includes patents, copyrights, trade secrets, and trademarks. The patent debate has been given added heat by dramatic developments in biotechnology in the past 20 years—recombinant DNA technology, cell fusion methods, monoclonal antibody technology, commercial bioprocessing and the like. Much of the debate has focused on patenting animals.

The debate has extended beyond the appropriateness of animal patents to encompass questions about the consequences of commercial uses of patented organisms and about the merits of the technology itself.

Farmers have been especially concerned about the long-term implications of animal patents. Now technology and patented livestock will most likely accelerate the trend towards larger and fewer farms. Small farmers are not equipped to adopt new, patented biotechnologies; nor could most afford them. The high cost of using patented animals would erode the profit margins of many farmers. Unable to compete effectively with larger operations that can adapt to biotechnological developments, small farms will decline still further in number, and fewer, larger farms will remain.

CONCLUSION

Genetic modification of animals by recombinant DNA technology (transgenesis) entails the introduction of a cloned gene(s) into the genome of a cell that might, after proliferation, be present in the germ lines of developed organisms, so that subsequently it may be possible to establish true breeding lineages. Much of the research in this area has centred on developing transgenic mice. Mouse embryonic stem cells are pluripotent (totipotent) and consequently can develop into any type of cell including germ-line tissue. Transgenic mice have been used to develop model systems for human genetic diseases such as Alzheimer disease.

By using similar experimental strategies, transgenic versions of cattle, sheep, goats, birds, pigs, and fish have been generated. It is hoped that transgenesis can be used to enhance existing genetic properties of livestock and provide the genetic basis for novel features. In addition, it is anticipated that the mammary glands especially those of cows, sheep and goats will be used as a biological factory for cloned gene products that can be readily purified in large quantities from milk. The open discussion of ethical, social, and legal implications will be required to ensure that the end products of modern biotechnology will contribute in a positive way towards a better world.

Summary

- Animals with manipulated DNA are known as transgenic animals.

- Transgenesis has become a powerful technique for studying fundamental problems of mammalian gene expression and development.

- ☑ Transgenic technology has been developed and perfected in the laboratory mouse.

- ☑ One transgenic cow would be more than sufficient for the production of the annual world supply of factor IX.

- ☑ Transgenic sheep with $h\alpha_1 AT$ gene will prove very useful as a bioreactor.

- ☑ Transgenic goats express a heterologous protein in their milk, which is used for treating coronary thrombosis.

- ☑ Transgenic birds could be used to improve the genetic make-up of existing strains.

- ☑ The scientific looking of right and wrong opinions about transgenic animals is known as ethics of transgenic animals.

REVIEW QUESTIONS

1. What are transgenic animals?
2. Explain the strategy to achieve transgenic animals.
3. Give a detailed account on the study of Alzheimer disease.
4. Discuss gene targeting in mice.
5. Briefly explain the concept of knock-out and knock-in technology.
6. Briefly explain the steps in the development of transgenic cows.
7. Write short note on transgenic sheep and goats.
8. Discuss the importance of transgenic birds.
9. Explain the ethics of transgenic animals.
10. State the significance of patenting transgenic animals.

IN VITRO FERTILIZATION AND EMBRYO TRANSFER

Intracytoplasmic Sperm Injection

Intracytoplasmic sperm injection (ICSI), available since 1995, has been extremely successful in enabling men with AIDS, paralysis, very low sperm counts, or abnormal sperm to become fathers. But as more ICSI procedures are performed and tests on nonhuman animals continue, potential problems are emerging, based on the fact that ICSI bypasses what one researcher calls "natural sperm selection barriers".

Meanwhile, work on rhesus monkeys is pinpointing the sources of damage to those ICSI embryos that do not continue to develop:

- Injecting sperm at the site of the polar body on the oocyte can disrupt the meiotic spindle, leading to nondisjunction (an extra or missing chromosome).

- Injected sperm DNA does not always condense properly, also leading to nondisjunction.

- Spermatids that have not yet elongated are often unable to fertilize an oocyte. Culturing them in the laboratory until they mature may help improve the odds of success.

- Spermatids may not be completely imprinted, leading to problems in gene expression.

- Mitotic cell cycle checkpoints are altered at the first division of the zygote following ICSI.

- Injected sperm sometimes lacks a protein that normally associates with the sex chromosomes. This may explain the elevation in sex chromosome anomalies.

- Injected sperm can include surface proteins that are normally left outside the oocyte, producing unanticipated effects. They may also include mitochondria from the male.

Says one researcher, " In spite of its potential risks, ICSI still seems to be remarkably safe."

INTRODUCTION

The term *in vitro* means "in glass". In recent years, *in vitro* fertilization (IVF) technology in which the fertilization of egg by sperm takes place at artificially maintained optimum conditions, has revolutionized the field of animal biotechnology, by the production of more and more animals than that of normal course. IVF was first successfully tried in humans over 25 years ago and this fame goes to "Louise Joy Brown", the first test tube baby born in 1978. Since her birth, thousands of children have been born with the aid of this technology.

IVF is the procedure designed to enhance the likelihood of conception in couples for whom other fertility therapies have been unsuccessful or are not possible.

STEPS INVOLVED IN IVF

The steps involved in IVF are as follows:

1. *Obtain eggs from the ovaries of the female donor* For IVF, the collection of eggs is usually performed under transvaginal ultrasound guidance. To accomplish this, a needle is inserted through the vaginal wall into the ovaries using ultrasound to locate each follicle. The follicular fluid is drawn up into a test tube to obtain an egg (oocyte) from the female donor.

2. *In vitro maturation of oocytes* The immature oocytes are inoculated *in vitro*, so that they can mature. However, immature oocytes should be taken out from follicles because they cannot mature in it, but degenerate. The full potential of superovulation as well as all the oocytes can be utilized by IVF technology. Also, the metabolic and hormonal requirement (20%) could be improved.

3. *Collection and preparation of sperm* A semen sample will be obtained from the partner on the day of the oocyte

retrieval. Normally the semen ejaculation should be done 2–5 hours prior and will be prepared for inseminating the collected oocytes. Testicular biopsy can also be performed as a method to extract sperms for IVF.

4. *Insemination of eggs by sperm* Following egg retrieval, the follicular fluid is immediately processed for the identification of eggs, evaluation and preparation for insemination. The eggs are taken in small droplets of culture medium. Each microdroplet comprises of about 10 oocytes.

   The medium should be supplemented with penicillamine, hypotaurin, etc. because they facilitate penetration of sperms into oocytes. Moreover, one dose of sperm consists of about one million sperms per ml of medium.

   After 24 hours, it is evaluated for evidence of fertilization. If no eggs were found to be fertilized, then insemination of eggs is performed by intracytoplasmic sperm injection (ICSI). The fertilized embryo must be maintained at *in vitro* conditions for a few days, before they are placed inside the woman's uterus. This will facilitate the formation of blastocyst. Depending on the couple's wish, the fertilized eggs may be frozen and stored for future use or donated to an anonymous person.

5. *Implantation of embryo into uterus* Embryo in the blastocyst stage is preferable for implantation as it can increase the chances of success while decreasing the likelihood of multiples. Embryos are transferred to the uterus through a fine tube (catheter), which is passed through the cervix at the time of examination. Embryos are placed in a manner so that they reach the top part of the uterus. Typically, 2–4 embryos can be transferred in one treatment cycle.

IVF Without Surgery— Transvaginal Oocyte Retrieval

Due to improvements in ultrasound imaging, surgery is no longer necessary for most *in vitro* fertilization procedures. The technique for recovery of eggs from the ovary is described below. It uses a sonographically guided needle to replace the surgical procedure which previously was used to recover oocytes (eggs). This procedure, called transvaginal oocyte retrieval, requires neither hospitalization nor general anaesthesia.

In order to prepare a proper environment in the woman and to increase the chances of recovering several healthy and mature eggs, the woman will undergo about two weeks of intensive preparation. This will include hormonal therapy with "fertility drugs." Blood tests and ultrasound scans of the ovaries are used to determine the optimal time to retrieve the eggs from the ovary. This optimal time is the moment just before ovulation when the oocytes are almost ready for fertilization.

At the proper time, the woman is put under local anaesthesia and the eggs are visualized by ultrasound and retrieved from the ovary by placing a needle through the vaginal wall. The mild discomfort that the patient feels has been described as similar to a pap smear or endometrial biopsy. After a short rest, the patient will be able to go home and resume normal activities.

The fluid from the follicles is examined under the microscope by the embryologist, who locates the eggs and keeps them in the laboratory under proper physiological conditions. The embryologist will place the sperm with the eggs when they are ready for fertilization. Usually, the eggs will develop into cleaving pre-embryos, whose cells divide 2 or 3 times to become preimplantation embryos (pre-embryos). They are maintained in laboratory dishes, in a nutrient mixture, which acts as a substitute for the environment that would otherwise have been provided by the fallopian tubes.

Using a special catheter, the couple's pre-embryos will be passed through the vagina and into the uterus at the time the pre-embryos would normally have reached the uterus (2+ days after retrieval).

After the pre-embryo placement in the uterus, the patient will lie quietly in bed for about an hour, and then will return home.

Limitations of IVF

Unfortunately, neither conception nor a successful outcome of pregnancy is guaranteed by the IVF procedure. In fact, there are complex and largely unknown factors that limit pregnancy rates following assisted reproductive techniques. Some of the known reasons for failure may include, but are not limited to, the following.

1. There may be a failure in recovering the egg because

    - follicles that contain mature eggs may not develop in the treatment cycle.

    - ovulation has occurred before the time of egg recovery.

    - one or more eggs may not be recovered.

    - pre-existing pelvic scarring and/or technical difficulties prevent safe egg recovery.

2. The eggs that are recovered may not be normal.

3. There may be insufficient semen to attempt fertilization of the recovered eggs because the man is unable to produce vital semen, i.e., the specimen may contain insufficient number of sperms to attempt fertilization.

4. Fertilization of the eggs to form embryos may fail even when the egg and sperm are normal.

5. The embryos may not develop normally or may not develop at all. Embryos that display any abnormal development will not be transferred.

6. Embryo transfer into the uterus may be difficult or not possible, implantation may not occur after transfer or the embryo may not grow or develop normally after implantation.

EMBRYO TRANSFER

The first successful embryo transfer took place in England in the 1890s by a fellow named Walter Heap, his subjects were rabbits. Although that was a success, embryo transfer was not applied commercially until the arrival of the follicle stimulating hormone (FSH), in the 1950s. Initially, the only technique was surgical to both flush and implant the embryos. While these methods were successful, they were very expensive, required a large set-up, and a lot of experience.

Embryo transfer refers to the techniques by which fertilized ova are collected from the reproductive tract of a genetically superior female (donor) and transferred to that of another female (recipient) who is genetically inferior. The main objective of embryo transfer is the improvement of animal population through maximum utilization of superior females. Through artificial insemination, embryo transfer can be achieved at a faster rate than natural mating.

The process of embryo transfer in cattle is as follows:

- The donor cow is inseminated normally 3–4 times at an interval of 12 hours.

- Seven days later, rinsing out the uterus is done to extract the embryos and ova (unfertilized, fertilized or degenerate).

- The good embryos are isolated using a microscope and then transferred into the recipient cows or frozen.

Benefits of Embryo Transfer for Farmers

Basically it multiplies the offspring of the farmers' best animals. Farmers can use their best bulls over their best cow or heifer and get a good calf whereas now the farmer can run an embryo program and possibly get a lifetime's production with one flush.

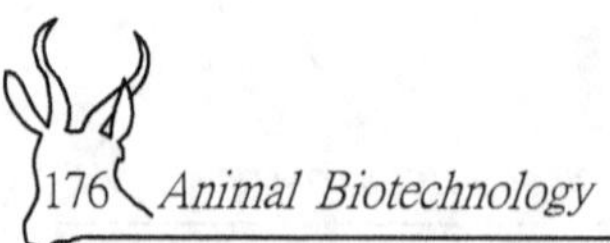

Hormones

There are three hormones namely, follicle stimulating hormone (FSH), prostaglandin (PG) and pregnant mare serum gonadotropin (PMSG). The FSH is taken from a cow's pituitary, which is a gland in the brain that produces growth hormones and sex hormones. The PMSG is taken from a cow's serum, which is from the blood. PMSG and FSH cause the cow to superovulate and the PG causes the cow to cycle.

Preparation

A donor cow chosen for flushing was first inserted with a CIDR (controlled intravaginal drug release). The CIDR stops the cow's menstrual cycle by producing a hormone that makes the cow think that it is pregnant. The cow was then given a series of injections morning and night with a drug such as embryo S, a follicle stimulating hormone that starts off at between 3.2–4.0 ml and works down each day to 0.8 ml. On the third day of injection, prostaglandin, a hormone causing the cow to cycle was also injected. After all the injections have been done and the CIDR has been pulled, the cow was watched to find out the time when it comes on heat. Ten hours after the cow has been on heat, the donor cow was inseminated with two straws of semen. The next three inseminations were done at an interval of 12 hours. Seven days later, the donor cow was flushed and the embryos were isolated and then inserted into the cows or were frozen.

The Flush

After the preparations before the flush, the cow is placed in a head bail. Then the cow is cleaned out, and an instrument called an introducer is put into the cow's vagina and the vet puts his hand up the cow's bottom to help guide the introducer. It is pushed gently through the cervix into the uterus. The centre piece is pulled out and a tube called a catheter is put in through the shell of the introducer and pushed right up into the uterine horn.

Then a small balloon called a cuff at the end of the catheter is blown up. After the cuff has been blown up, a fluid for lubrication and nutrition of the embryos has to be injected. The fluid is a sodium chloride-based phosphate that has a small amount of serum. The serum is an amber-coloured liquid that separates from the clot when the blood hardens which is what we know as a scab. 210–300 ml of this fluid is injected through the tube and into the cow's uterine horn, it is then sent back through the centre tubing of the catheter into an embryo filter. The extra fluid is released while the embryos are kept in the filter by a small grid at the bottom. After one side of the uterus has been flushed, the other side is flushed. When the flush is complete, the embryo filter is washed out into a small container. The vets then search for embryos looking through a microscope. The embryos are graded and put into straws and implanted into the recipient cow or frozen for later use.

Difference between Surgical and Non-surgical Transfers

Besides causing trauma on the recipient, surgical transfer places the embryo further up the uterine horn than non-surgical. There may be a small increase in the conception rate, maybe 5%, however if the recipients are in good shape, it will not be of much benefit. Surgical transfer requires a small cut in the cows flank on the side of the ovulation. The uterus is pulled out and a small hole is made in the uterus. A small catheter is placed in through the hole and the embryo released. Non-surgical is the same as artificial insemination (AI). A special gun is placed up through the cow's vagina and pushed through the cervix. The embryo is placed on the side of the ovulation and released from the gun.

Conception Rate

With embryo transfer, an already fertilized egg is being inserted, which eliminates one of the steps of artificial insemination, so the conception rate should be 5–10% higher than artificial insemination, but a lot depends on the condition of the recipient. On an AI program, the conception rate is 60–65% while on an embryo transfer program it is expected to be roughly the same or a bit higher.

Temperatures for Storage

Fresh embryos can be stored at 37°C for 6–8 hours without much harm but it is recommended that the embryos are inserted into the recipient cow or frozen as soon as possible. The embryos are frozen in liquid nitrogen in which they can be stored forever.

Costs

The embryo transfer program is fairly expensive. Here are some average prices (April 1992):

- The hormones and drugs are approximately Rs. 9600 per donor.
- On a per donor basis you are looking at spending approximately Rs. 4800–14,400 for the flush procedure.
- Transfer fees per embryo vary between Rs. 4500 for non-surgical to Rs. 7400 for surgical.

Cost of providing good quality recipients should not be overlooked. This averages out to around Rs. 14,400–24,000 per recipient.

BOVINE EMBRYO TRANSFER PROCEDURE

The Donor Cow

Any cow that is mature and cycling can be put through an embryo transfer program. The only problem that could stop the cow or heifer from being flushed is abnormalities of the reproductive tract or secondary abnormalities due to disease or injury. Some heifers as young as eight months have been successful in flushing embryos as have cows eighteen years old. Although it is possible for embryos to be flushed from a cow or heifer at these ages, the result obtained from the heifer or cow may not be worth the effort.

A major problem that has to be faced is that the response of an individual cow cannot be predicted. The response of a donor cow can vary from 0–30 or more eggs, the average being around 4–6 depending on the breed, age and maturity. Nutrition is a very important factor, and donors should be on a rising plan of nutrition in the month leading up to the flush. If the donor is not in good condition then it will not produce as many eggs to be fertilized. Also there is the response to the drugs, sometimes the cow will over-respond to the drugs and sometimes under-respond.

The Recipient Cow

Collecting the embryos is only half the job. Then the embryos must be transferred into the recipient cow or frozen for later usage. About 60% of the embryos transferred result in pregnancies. Sometimes it can be as high as 80% and as low as 20%.

It is best for the recipient to be a bit larger than the donor so that it will not have any trouble having the calf. Not all cows can be recipients. Before the cow or heifer can become a recipient, it must be palpated. This is done by the vet putting his hand up the cow's bottom and feeling around to see if everything is in working order. This is done because some cows have a twisted cervix or have had calving problems so it is best to avoid using these cows or heifers for recipients. Just before the flush, the vet has to check the recipient cow or heifer to find out on which side the cow has ovulated so that the embryo can be inserted on that side. The recipient cow should be in good condition for a good conception rate (Figure 6.1).

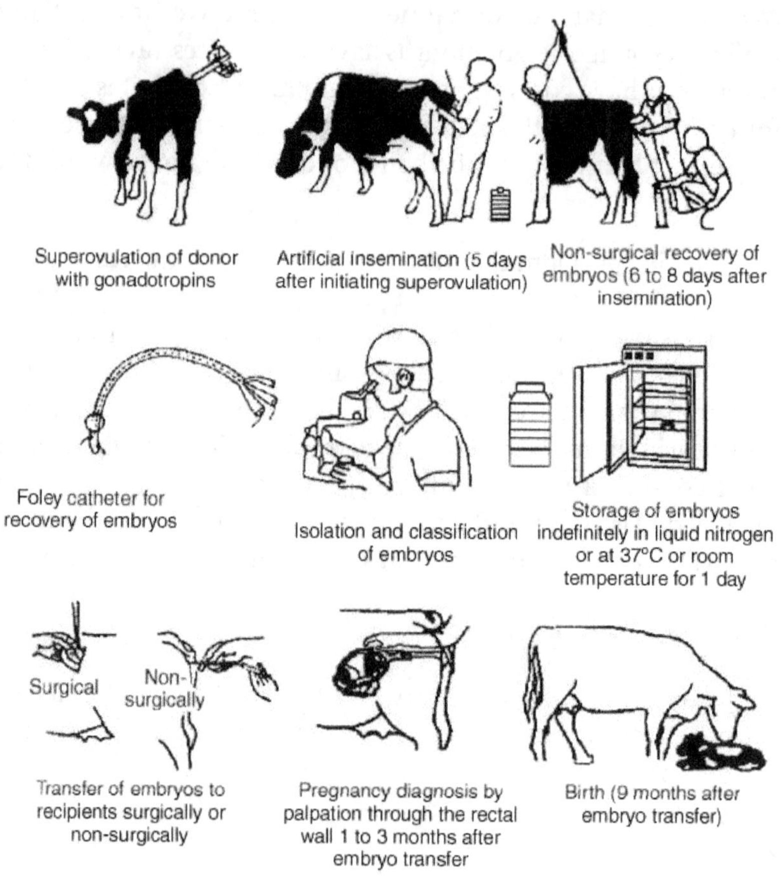

Figure 6.1   Synopsis of bovine embryo transfer procedures

Embryo Transfer as a Management Tool

Embryo Transfer is now accepted as the quickest and most-cost efficient method of increasing the rate of genetic improvement within a herd. For beef and dairy operations, it can be a valuable addition to any breeding program, leading to greater efficiency and profitability.

Embryo transfer offers farmers the chance to

- Produce up to ten or more progeny per year from their best cows.

- Profit from the increased sale of quality genes without losing the bloodlines.

- Extend the productive life of some older cows incapable of carrying another calf by producing further progeny through the use of embryo transfer.

- Conserve the genetics in their herd through the use of embryo freezing for export, domestic sale or future transfers on their own farm.

- Introduce top genes into the farmers' herd, rapidly and economically from other countries.

The top cattle in any herd are rarely available for purchase, however the owners are often willing to sell embryos. The importation of new blood stock can be achieved safely and economically through the purchase of embryos from anywhere in the world, once the quarantine regulations have been fulfilled.

Superovulation

Superovulation or multiple ovulation can be defined as increased ovulatory response by external hormonal therapy above a level that would be expected to occur naturally.

Pregnant mare's serum gonadotrophin (PMSG) which is more recently being called as equine chorionic gonadotrophin (ECG) and a variety of gonadotrophin preparations isolated from animal pituitary tissue are now available from commercial sources and are being used widely for superovulation.

After injecting the gonadotrophin, the females are induced for superovulation. During follicular phase (II phase of oestrous cycle) about 20 ovarian follicles are induced. Eventually these grow and get filled with fluid. The space of the follicle filled with fluid is called antrum and such follicles are called antral follicles. In normal course, only one follicle develops and releases one egg on maturation. However, before ovulation, the follicles laying against the surface of the ovary look large-sized. Therefore immature oocytes from follicles of donor females are recovered surgically.

The females to be superovulated are frequently injected with prostaglandin F2a (PGF2a), so that synchronized oestrous cycle could develop in them. After 10 days of oestrous, they are injected with follicle stimulating hormone (FSH) up to 4 days followed by PGF2a, so that oestrous may be maintained. The FSH treatment induces superovulation. The eggs are artificially inseminated and allowed to get fertilized. After fertilization, embryos undergo developmental stages.

After 6–8 days of fertilization, the embryos are recovered. During this stage, embryos have attained morula/blastula stage and remain in the females' oviduct. A saline solution is drained in the oviduct and embryos get flushed out through the catheter into a storage bottle, where the embryos settle down and the solution is decanted.

Embryos with a small volume of saline are poured into a petri dish. Then they are examined under the microscope. The identified embryos are transferred into synchronized recipient females, when they are in 6–8 days of cycle of embryo development, stored and transported or manipulated as desired. In general, one superovulated female results in 5–6 progenies.

In animals, multiple ovulation with embryo transfer (MOET) allows progeny testing of both male and female. It can increase the genetic gain per year than was possible through artificial insemination alone. MOET also increases the selection intensity through increased number of progenies per cow or buffalo per year.

Embryo Splitting

A very simple method of splitting the embryo into two halves at the late morula or early blastocyst stage of development has recently been devised. The blastocysts are divided in such a way that each half contains inner cell mass (ICM) and trophoectoderm cells. ICM is the mass of cells which develop into foetus. Trophoectoderm is a single layer of trophoblast cells lining the inner side of the zone surrounding the blastocoel.

The split embryos are then transferred to females to produce identical twins. Thus embryosplitting technology has increased the rate of pregnancy.

Embryo splitting can be done as follows:

- Blastocyst stage embryos are transferred for a few minutes into a cell culture medium consisting of hypertonic sucrose and bovine serum albumin (BSA).

- The medium of high osmotic strength enters into the cell membrane of the embryo (zona pellucida) and cells contract due to exosmosis.

- The BSA attaches to the zona pellucida and provides negative charge to the embryo. Then it is transferred into a plastic petri dish containing standard cell culture medium.

- The outer membrane of the embryo is negatively charged and the petri dish is positively charged. The embryo sticks to the surface of the petri dish due to the development of electrostatic interaction of the two charges.

- The petri dish is kept on the stage of an inverted microscope. A micromanipulator equipped with a fine surgical blade is used to cut the embryo roughly into two halves with minimal damage to the cells.

- The hemispherical mass of embryo is referred to as sphere.

- The half or demi embryo is transferred into the recipient's oviduct.

⚛ At this stage, it is essential to study the successful embryo implantation, which is possible with chemical signals from the trophoblast.

Embryo Biopsy

The removal of a small number of cells for genetic analysis is called as "Embryo biopsy". It should be combined with splitting, so that twins will be produced with the identical characters or known genotypes. Biopsy is very essential in breeding so that sex and genetic diseases could be detected. This technology remains in its infancy and can be of profound importance clinically.

Embryo Sexing

In animals, it becomes essential to detect the sex of the embryo before implantation. It is based on the presence of Y or X chromosome. Nowadays, sexed embryos are commercially available but at high cost.

Limitations of Embryo Transfer

Embryo transfer cannot be used widely because of its high cost, technical difficulties and limited supply of embryos from superovulated donors.

FUTURE IDEAS

Australia is quite advanced in embryo transfer, equal to the world. At this point in time scientists are taking embryos and breaking them up into individual cells and making identical animals, this method is called cloning. The cloning aspect is not commercially available yet but should be soon. It really depends on the imagination of the people who are involved with the research. This may lead to farmers taking embryos out of the tank, seven days after examining their own recipients and then transferring the embryo.

Summary

- Fertilization of egg by sperm at artificially maintained optimum condition is called *in vitro* fertilization.

- The full potential of superovulation and all the oocytes can be utilized by IVF technology.

- Embryo in the blastocyst stage is preferable for implantation.

- Embryo splitting technology has increased the rate of pregnancy.

- Embryo biopsy is the removal of small number of cells for genetic analysis.

REVIEW QUESTIONS

1. Define IVF.

2. Explain the steps involved in IVF.

3. What are the limitations of IVF?

4. What is the significance of superovulation?

5. Explain the methods to carry out embryo splitting.

6. Define:
   i. embryo biopsy
   ii. embryo sexing.

7. What are the future prospects of IVF and ET?

8. Discuss in detail the methodology of bovine embryo transfer.

9. State the differences between surgical and non-surgical transfers.

10. What are the limitations of embryo transfer.

7

NEOPLASIA

Vegetables and Cancer Risk

A diet rich in cruciferous vegetables such as broccoli and Brussels sprouts correlates to lowered incidence of colon cancer, according to many population observations. Experiments conducted at the Imperial College of Medicine and the Institute of Food Research in the U.K. cleverly connected the epidemiological evidence with known pieces of the biochemical puzzle:

- Cooking meat induces formation of heterocyclic aromatic amines (HAs). These compounds are absorbed into the lining of the digestive tract and then metabolized by a liver enzyme into mutagens, some of which cause colon cancer.

- Cruciferous vegetables release glucosinolates, which activate "xenobiotic metabolizing enzymes", which metabolize the HAs into less dangerous forms.

In the experiments, 20 healthy men followed three 14-day diet regimens. In the first and third periods, food intake was closely monitored for 12 days to avoid foods that contain enzymes that could interfere with the HA metabolic pathways. On day 13, the men ate a steak in which levels of the two major HAs had been measured. Over the next two days, urine, saliva, and blood were collected at various times. During the middle 14-day period, the men had Brussels sprouts or broccoli soup for breakfast and either vegetable with dinner for the first 12 days, then ate the steak on day 13 and followed up with the body fluid samplings.

The researchers measured effects on metabolism in two ways: they assessed the levels of the two types of HAs in the body fluids, and they also measured the metabolism of a dose of caffeine given 12 hours after the steak meal. Because the enzymes from the vegetables affect liver enzymes that metabolize caffeine much the same way that they affect metabolism of HAs, a change in caffeine metabolism would signal alteration of HA metabolism too. The results of experiments strongly supported the epidemiological evidence; the vegetables lowered the concentration of the HAs in urine by 22 percent. This approach was much faster and safer than conducting dietary experiments over years and waiting to see who developed cancer! The only reported side effects of the dietary regimen were caffeine withdrawal and mild flatulence.

Neoplasia literally means "new growth". A neoplasm, as defined by Willis, is an abnormal mass of tissue, the growth of which exceeds and is uncoordinated with that of the normal tissues and persists in the same excessive manner after the cessation of the stimuli, which evoked the change. In common usage a neoplasm is often referred to as a **tumour** and the study of tumours is called **Oncology**. The tumours may be benign or malignant. Malignant tumours are collectively referred to as **cancers.**

CANCER

Cancer can be defined as a disease involving heritable defects in cellular control mechanism resulting in the formation of malignant and usually invasive tumours. Cancerous cells begin to grow and divide in an unregulated fashion, without regard to the body's need for further cells of its type and has descendants that inherit property to proliferate without responding to regulation. This results in a clone of cells that are able to expand indefinitely. In addition, the cancer cell is not confined to its original tissue but invades other tissues where it proliferates, i.e., metastasis.

Cancer is caused by mutations, but there are two key differences between cancer and other genetic diseases.

1. Cancer is caused mainly by mutations in somatic cells, whereas other genetic diseases are caused solely by mutation in the germ line. Some individuals, however, have inherited genetic mutations that predispose them to develop specific types of cancer.

2. An individual cancer does not reproduce from a single mutation, but rather from accumulation of as few as 3 to perhaps as many as 20 mutations, depending on the type of cancer, in genes, that normally regulate cell multiplication.

Types of Cancer

The most important issue in cancer pathology is the distinction between benign and malignant tumour.

Benign tumour   A common skin wart remains confined to its original location, neither invading, surrounding normal tissue nor spreading to distant body sites. It can be removed surgically.

Malignant tumour   It is capable of invading, surrounding normal tissue and spreading throughout the body, via the circulatory or lymphatic system. The spread of malignant tumour to distant body sites frequently makes them resistant to such localized treatment.

Cancer cells fall into one of the 3 main groups namely carcinomas, sarcomas and leukemias.

1. Carcinomas including approximately 90% of human cancers, are malignancies of epithelial cells.

2. Sarcomas are rare in humans, and are solid tumours of connective tissues such as muscle, bone, cartilage and fibrous tissue.

3. Leukemias accounting for 8% of human malignancies, arise from the blood-forming cells and from cells of the immune system.

There are two phases of carcinogenesis namely initiation of the tumour and proliferation of the tumour.

Initiation of the Tumour

At the cellular level, the development of cancer is viewed as a multi-step process involving mutation and selection for cells with progressively increasing capacity for proliferation, survival, invasion and metastasis.

The first step, the initiation of tumour, is thought to be the result of a genetic alteration, leading to abnormal proliferation of a single cell. Cell proliferation, then leads to the outgrowth of a population of clonally derived tumour cells. Tumour growth requires formation of new blood vessels.

Tumour, whether primary or secondary, requires recruitment of new blood vessels in order to grow to a large mass. In the absence of blood supply, a tumour can grow into a mass of about 10^6 cells, roughly a sphere of 2 mm in diameter. At this point, division of cells on the outside of the tumour mass is balanced by death of those in the centre due to an inadequate supply of nutrients. Such tumours, unless they secrete hormones, cause few problems. However most tumours induce the formation of new blood vessels that invade the tumour and nourish it, a process called angiogenesis.

Many tumours produce growth factors that stimulate angiogenesis whereas others induce surrounding normal cells to synthesize and secrete such factors. Basic fibroblast growth factor (**BFGF**), transferring growth factor α (**TGF-α**) and vascular endothelial growth factor (**VEGF**), which are secreted by many tumours, all have angiogenic properties.

Initiation of the tumour is associated with alkylation of DNA and a transient inhibition of DNA synthesis. At this level no morphological markers are seen, whereas biochemical markers characteristic of altered DNA are demonstrable. Thus, the initiators induce the changes in the nature of cells at molecular level (biochemical level) and they may remain latent for the lifespan of the animal without showing any morphological characteristic to distinguish them from the neighbouring non-neoplastic cells in the tissue.

Promotion of the Tumour

Tumour progression continues as additional mutations occur, within cells of the tumour population. Some of these mutations confer a selective advantage to the cell, such as more rapid growth, and the descendants of a cell bearing such a mutation will consequently become dominant within the tumour population. The process is called **clonal selection**, since a new clone of tumour cells has evolved on the basis of its increased growth rate or other properties (such as survival, invasion or metastasis) that confer a selective advantage. Clonal selection continues throughout tumour development, so the tumour continuously becomes more rapid-growing and increasingly malignant.

Studies of colon carcinomas have provided a clear example of tumour progression during the development of a common human malignancy. The earliest stage in tumour development is increased proliferation of colon epithelial cells which gives rise to small benign neoplasm (an adenoma or polyp). Further rounds of clonal selection lead to the growth of adenomas of increasing size and proliferative potential. Malignant carcinomas then arise from the benign adenoma, indicated by invasion of the tumour cells through the basal lamina into underlying connective tissue. The cancer cells then continue to proliferate and spread through the connective tissue of the colon wall. Eventually, the cancer cells penetrate the wall of the colon and invade other abdominal organs, such as the bladder or small intestines. In addition, cancer cells invade blood and lymphatic vessels, allowing them to metastasize throughout the body.

Promotion of the tumour is associated with the modification in the gene activity, causing a reversal of cell differentiation and after initiation phase, a promoting factor is required to allow expression of the neoplastic phenotype and growth to form a clinical or a frank neoplasm.

Raick (1974) proposed that the ability of the promoter to modulate gene activity, perhaps depressing genes and inducing reversal of the cell differentiation, might be the alteration that is essential for the activation of the gene loci that code for the neoplastic phenotype. The cell proliferation and hyperplasia are the most conspicuous changes induced by the promoters (Figure 7.1).

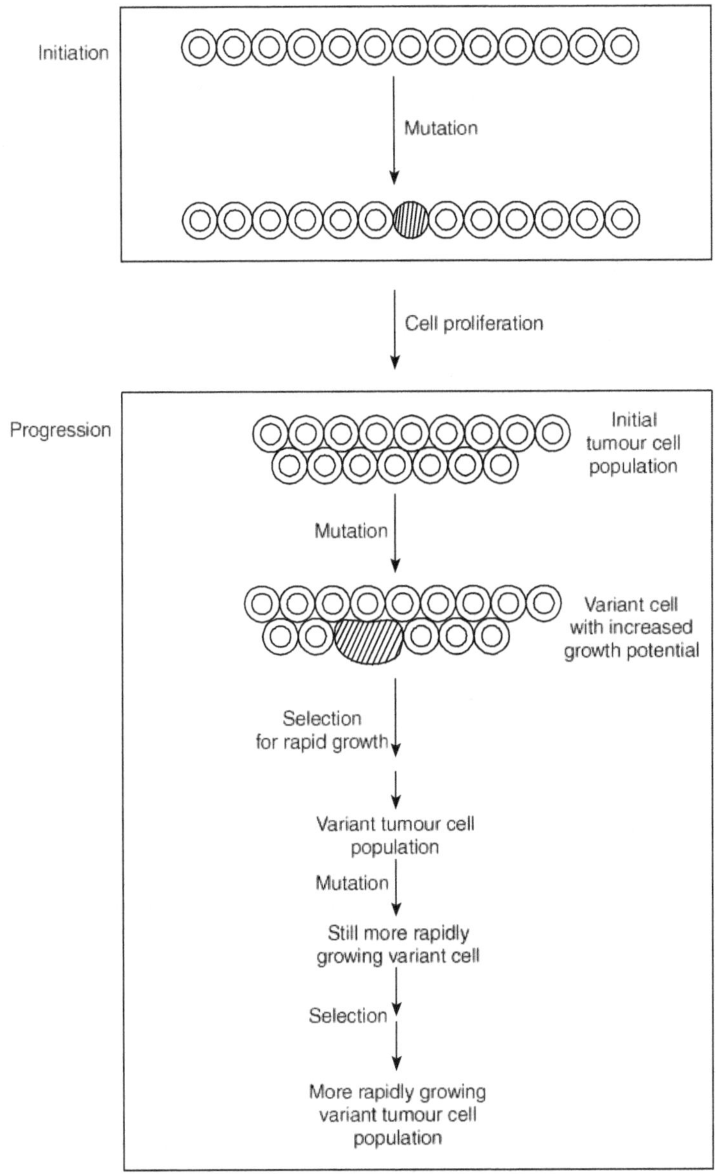

Figure 7.1   Stages of tumour development

MORPHOLOGY OF THE CANCER CELLS

Intermitotic Nucleus

Shape and volume The nucleus of the cancer cell is characterized by the irregularity of its contents and the presence of deep fissures. Altman (1852) described this condition as the **Kernspalten**, which sometimes results in a markedly lobulated appearance. Many neoplastic nuclei appear to be sprouting, deformed by bulbous protrusions, constricted at their base or even completely separated from the central nuclear mass. The presence of deep indentations on the nucleus increases the surface, possibly increasing the nucleo-cytoplasmic interchanges.

The cancer cells are also characterized by the increase in nuclear and nucleolar volume. Nuclear hypertrophy, chromatin lumps or chromocentres and increased DNA have been stressed as features of malignancy but in actual fact none are typical of cancer. Nuclear hypertrophy in malignancy may be the result of multiplication of chromosomes, the number of which is in multiples of the original set. The cancer cells are generally polyploids.

Contents It is confirmed that the nucleus of the cancer cell contains more chromocentres than that of the normal cell. This is the classic conception of the cancer cell nucleus having numerous heterochromatin masses, either dense and voluminous or minute, giving the nucleus a dusty appearance.

In the tumour cells, DNA content is higher than in homologous normal tissue. During the development of tumour, the successive mitoses evidently imply an increase in the total quantity of DNA. Increase in DNA content correlates with the gradual transformation of the normal cells into the cancer cells. Variation in DNA content within the same tumour is correlative with changes in nuclear volume and chromosome count.

RNA synthesis in the nucleus also increases during the tumour formation. Nuclear proteins play a predominant role in the determination of nuclear size. Studies on squamous cell carcinoma has demonstrated that increase in protein shows geometrical progression (two or four times).

Nucleolus

Cancer cells contain a very hypertrophic nucleus, among other signs, a characteristic aid in the diagnosis of cancer. The number of nucleoli in cancer cells is frequently increased compared with normal homologous tissues. The shape may also be irregular.

The nucleolus is of prime importance in protein synthesis, and derangement of its normal activity might be followed by serious chemical changes and consequently by disturbances in the biological behaviour of the cancer cells.

Nuclear Membrane

The nuclear membranes of cancer cells are thickened and irregularly folded.

Mitotic Nucleus

Hansemann (1890) drew attention to the abnormal behaviour of chromosomes in cancer cells. Genetic instability is a remarkable feature of cancer cells. Genetic instability is a remarkable feature of cancer cells, even in human tumours. The chromosomal change is a constant occurrence as a tumour strain evolves. The range of chromosome variation is said to be narrower in tumours derived from single cell clones.

The form of chromosomes is extremely variable in cancer cells. These may be very short square and contracted (contraction chromosomes). Other types of chromosomes are elongated, filamentous and thin. The range between extreme chromosomal sizes is wider in cancer cells than in normal cells. Dicentric and annular chromosomes are also observed.

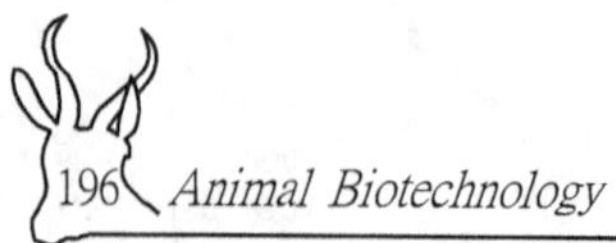

Cytoplasm

The cytoplasm becomes shrunken and the cytoplasmic nuclear ratio decreases. The contact between individual cells in normal cultures is continuous and discrete, and even slight overlapping is rare. Tumour cells rarely show well-defined cell membranes, and the edge-to-edge contact nearly always overlap. Tumour cells show more pseudopodial processes than normal cells. These cells may show considerable, poor or no phagocytic and pinocytic activity.

Endoplasmic reticulum is customarily deficient in malignant tissues as opposed to normal cells, and less differentiated tumour cells show the greatest loss of membranous ergastoplasm. Groups of 3–20 straight and curved membranes were observed in certain tumour cells in electron microscope.

No constant specific feature of Golgi complex has been discovered in tumour cells. It seems to be present in all neoplasms. Greatly hypertrophied in certain benign or malignant tumours, they are only occasionally visible in highly dedifferentiated tumours. It may be inert in anaplastic tumour cells.

The centriole may be hypertrophic in certain tumours. In some it may lose its polarity.

Cancer cells often contain smaller mitochondria than normal cells. In some, these may be larger because they are swollen. The mitochondria become rounded and more or less swollen and progressively lose the majority of their internal crests, become clear, and are finally transformed into apparently empty cases. The cristae also become irregular.

METABOLISM OF CANCER CELLS

The essential feature of the change of normal cell into the cancer cell is that in the neoplasm, the capacity for growth has supplanted that of function. In the cancer cell there is a tendency to converge towards a common enzymatic pattern of activity, whereas the cells of normal organs possess specific chemical characteristics. Certain chemical functions are lost in the transformation of normal to neoplastic cell. The changes are as follows:

Sources of energy   Neoplastic cell has a full complement of oxidative enzymes but has a reduced capacity for oxidation. Anaerobic energy-yielding mechanisms are predominant. No new pathways of metabolism in tumour are distinguishable and no new enzymes or coenzymes are present.

The energy substrate systems   A normal cell is more concerned with function involving energy-releasing catabolism than with growth involving anabolic energy. Tumour cells are concerned primarily with growth, not with function. The specific functional enzyme–substrate systems therefore become superfluous in comparison with those required for the anabolic processes of growth and cell division. Normal cells possess diversified metabolic activities, whereas in tumour cells there is much greater biochemical uniformity.

Competitive struggle   A large neoplasm acquires nitrogenous building blocks from the body stores to satisfy the continual demand for protein synthesis, but the supply of these blocks is not unlimited. Cancer cells appear to exercise priority over the demands of normal tissues for amino acids thus constituting a nitrogen trap. This trap accounts for the wasting and cachexia that is often so characteristic a feature of the later stages of malignancy.

IMMORTALIZATION

Neoplastic cell cultures can grow indefinitely whereas, normal cell cultures die after some generations, e.g. human cell cultures die after 50 generations.

Cancer cells apparently lack proper recognition and communication. Contact inhibition is a process when the cells do not move and grow in a culture; this is because of the contact of plasma membranes of different cells and formation of gap junction. Cancer cells lose the property of contact inhibition, and such cells divide even after forming a monolayer. Loss of contact inhibition enables the cells to dissociate from neighbouring cells and infiltrate other organs. Such cells pass over or under one another, they can grow on top of one another, and they frequently form a gap junction.

The transformed cells have the ability to invade other tissues. Normal cells lack this property. The transformed cells first invade extracellular matrix and then enter blood circulation, these pass through the wall of the vessel and form metastasic tumours. The invasion and metastasis are biological hallmarks of malignancy (Figure 7.2).

Most normal cells must be attached to a rigid substratum in order to grow. However, the transformed cells can grow even when they are not attached to the substratum. The cells that have lost anchorage-dependence generally form tumours with high efficiency when they are injected into animals that cannot immunologically reject the cells.

Tumour cells consume much more glucose than normal cells, because they have to grow and multiply. There is a great increase in the rate of sugar transport across the surface cell membrane after transformation. This increases the intake of glucose by the transformed cells. More transporters (permeases) are available on cell surface proteins.

Cancerous cells secrete a protease called plasminogen activator, which cleaves a peptide bond of plasminogen, a serum protein, converting it to the protease plasmin. It results in loss of actin microfilaments, growth stimulation, etc. Increased plasmin may help the cells penetrate the basal lamina, thus, facilitating invasiveness of transformed cells. The transformed cells secrete proteins called transforming growth factors (TGFs) that can stimulate growth of normal cells.

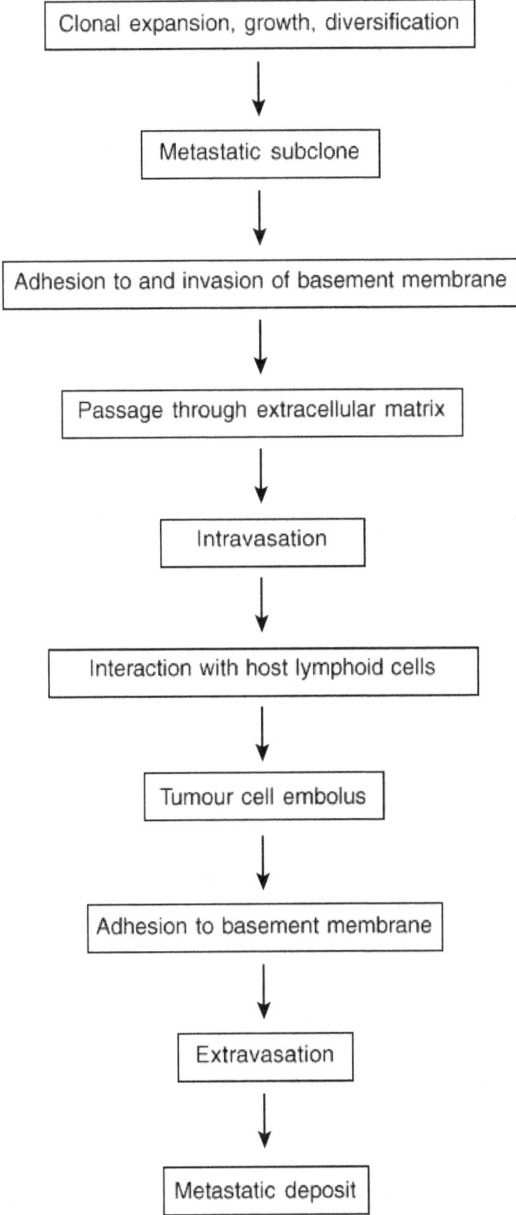

Figure 7.2   The metastatic cascade—a schematic illustration

Normal cells suppress their own growth when the concentrations of any of the many critical nutrients or factors fall below a threshold value. When the concentration of isoleucine, phosphate, epidermal growth factors (EGF) or other substances that regulate growth falls below the necessary level, normal cells go into quiescence. Transformed cells are deficient in their ability to respond by growth arrest to lowered nutrient or factor concentrations.

The surface of transformed cells undergo changes as follows.

- In transformed cells, glycolipids and glycoproteins are modified. Protein-lined N-acetyl neuramine (sialic acid) is decreased.

- Ganglioside content of all lipids decreases.

- Cell surface proteins become more mobile, because of which antibodies can more easily agglutinate surface proteins.

- Links between surface proteins and cytoskeletal elements are modified.

CARCINOGENESIS

Carcinogenesis is a process which induces cancer in normal tissues; carcinogens are agents which are associated with or responsible for **neoplasia**. It has been identified both by studies in experimental animals and by epidemiological analysis of cancer frequencies in human populations.

There are some other factors which predispose to or help in the action of these carcinogens, called co-carcinogens. Radiation and many chemical carcinogens act by damaging DNA and induce mutations. Hence, they are generally referred to as initiating agents.

The carcinogens in tobacco smoke (including bengo or phrene, dimethylnitrosamine and nickel compounds) are the major identified causes of human cancer. Smoking is the undisputed cause of 80–90% of lung cancers.

Other carcinogens contribute to cancer development by stimulating cell proliferation rather than by inducing mutation. Such compounds are referred to as tumour **promoters**, since the increased cell division is required for the outgrowth of a proliferative cell population during early stages of tumour development. **Promotion** refers to that part or period of the process, which covers events from the time the irreversible biochemical subcellular changes have occurred up to the time when a tumour becomes clinically visible.

The pherbol esters that stimulate cell proliferation by activating protein kinase are noted examples. Tumorigenesis can be initiated by a single treatment with a mutagenic carcinogen. In addition to chemicals and radiation, some viruses induce cancer, both in experimental animals and in humans. The common human cancers caused by viruses include liver cancer and cervical carcinoma, which together account for 10 to 20% of worldwide cancer incidence. These viruses are important not only as causative agents of human cancer, but also have a key role in elucidating the molecular events responsible for the development of cancers induced by both viral and non-viral carcinogens. The scheme of cancer carcinogenesis is shown in Figure 7.3 and the fundamental principles of the molecular basis of cancer (carcinogenesis) may be as follows:

1. Nonlethal genetic damage which may be acquired by the action of environmental agents such as chemicals, radiations, viruses or it may be inherited in the germ line.

2. Two classes of normal regulatory genes namely the growth-promoting proto-oncogene and the growth-inhibiting cancer suppressor gene (anti-oncogenes) are the principal targets of genetic damage. Mutant alleles of proto-oncogenes are dominant. In contrast, both normal

alleles of the tumour suppressor genes must be damaged for transformation to occur, so this family of genes is sometimes referred to as recessive oncogenes.

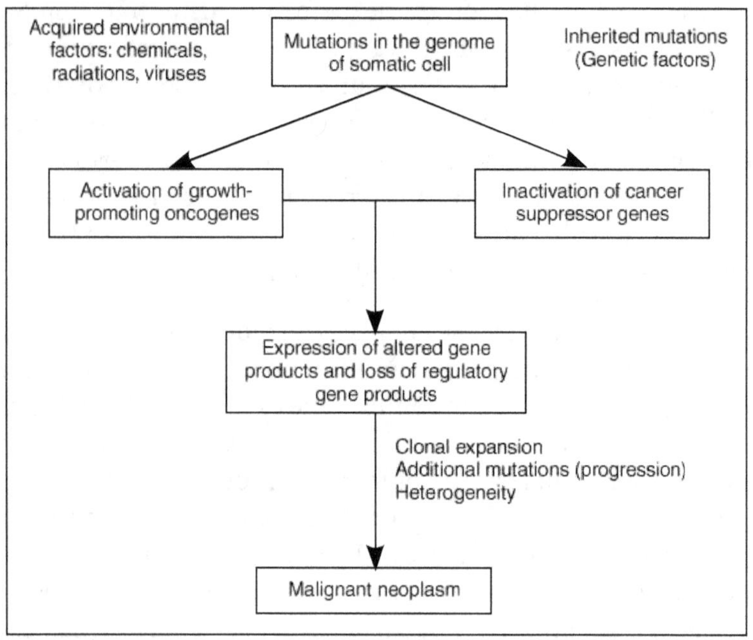

Figure 7.3   Scheme of cancer pathogenesis

3. Carcinogenesis which is a multistep process at both the phenotypic and genetic level. A malignant neoplasm has several phenotypic attributes, such as excessive growth, local invasiveness and the ability to form distant metastases. These characteristics are acquired in a stepwise fashion, a phenomenon called tumour progression.

ONCOGENES AND CANCER

In 1989, J. Michael Bishop and Harold Varmus received the Nobel Prize for their pioneering contribution in the study of **proto-**

oncogenes and their potential to transform into cancer-causing oncogenes. Proto-oncogenes are cellular genes. Molecular hybridization revealed that the viral oncogene sequences are found in the normal cellular DNA.

The studies of tumour viruses revealed that specific genes, namely oncogenes, are capable of inducing cell transformation, thereby providing the first insights into the molecular basis of cancer. The prototype of this highly oncogenic retrovirus is Rous

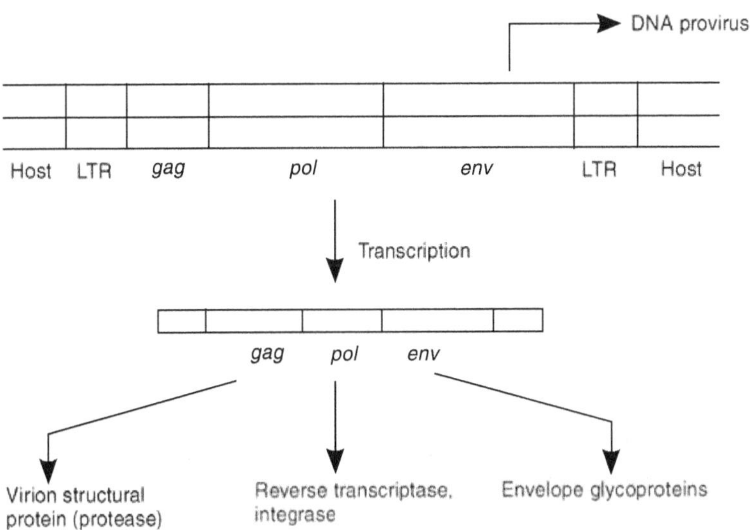

Figure 7.4   A typical retrovirus genome

sarcoma virus (RSV) first isolated from a chicken sarcoma by Peyton Rous in 1911 (Figure 7.4). Viral oncogenes were first defined in RSV, which transforms chicken embryo fibroblasts in culture and induces large sarcomas within 1–2 weeks after inoculation in chickens.

The proteins encoded by oncogenes are called **oncoproteins**. The oncoproteins resemble the normal products of proto-oncogenes with the following exceptions:

- Oncoproteins are devoid of important regulatory elements.

- Their production in the transformed cells is not dependent on growth factors or external signals.

The mechanism by which proto-oncogenes are transformed into oncogenes falls under 2 broad categories:

- Changes in the structure of the gene, resulting in the synthesis of an abnormal gene product (oncoprotein) having an aberrant function.

- Changes in the regulation of gene expression, resulting in enhanced or inappropriate production of the structurally normal growth-promoting protein.

Structural and regulatory changes affecting proto-oncogenes may include point mutations, chromosomal translocations, gene amplification, etc.

The majority of oncogene proteins function as elements of signalling pathways that regulate cell proliferation in response to growth factor stimulation. These oncogene proteins include polypeptide growth factors (GF), growth factor receptors (GFR), elements of intracellular signalling pathways and transcription factors.

Tumour Suppressor Genes

Oncogene drives abnormal cell proliferation as a consequence of genetic alterations that either increase gene expression or lead to uncontrolled activity of the oncogene-encoded protein. Tumour suppressor genes represent the opposite side of cell growth control, normally acting to inhibit cell proliferation and tumour development. In many tumours, these genes are lost or inactivated, thereby moving negative regulation of cell proliferation and contributing to the abnormal proliferation of tumour cells.

Many genes have been identified which, when present in a mutant form, cause cells to become tumorigenic. The normal counterparts of such proto-oncogenes have often found to play a role in the control of cellular proliferation. They may be changed into a functional oncogene by mutations that alter their biochemical properties or that result in a change in their level of expression.

In order to render primary tissue culture cells oncogenic it is often necessary to introduce two different oncogenes that act co-operatively to induce cellular transformation. The *ras* and *myc* oncogenes form such a co-operating pair. The *ras* and *myc* oncogenes were isolated because of their ability to transform cells in culture. Evidence that they actually play a role in naturally occurring cancers was derived from the presence of a potentially oncogenic form of each of these genes in human tumours. Because the point mutation in the *ras* oncogene normally occurs within the same codon, it is possible to use oligonucleotide probes in Southern transfer or in PCR, to detect a change at this position.

The first insight into the activity of tumour suppressor genes came from somatic cell hybridization experiments initiated by Henry Harris *et al.* in 1969. The fusion of tumour cells with normal cells yield hybrids that contain chromosomes from both parents. Such hybrids are usually non-tumorigenic.

ras and *myc* are **dominant** oncogenes; when the oncogenically activated form on the gene is introduced into an immortalized cell line, it becomes tumorigenic, despite the fact that a normal copy of the gene is also present. There are also **recessive** oncogenes or tumour suppressor genes, such as the retinoblastoma *(RB)* gene, that appears to act as a suppressor of cell proliferation. Only if both copies of the gene are inactivated by mutation do cells become tumorigenic. Biochemical studies have shown that in normal cells the RB protein is complexed with several proteins that are potential activators of cellular proliferation. It is assumed that these proteins are rendered inactive when bound to RB and that deletion of the *RB* gene, or mutations that inhibit its binding to other proteins, are oncogenic because of these properties. The *P53* gene encodes another cellular protein that acts as a tumour suppressor, apparently by a similar mechanism. Remarkable point mutations in the *P53* gene have been found in almost half of the human cancers characterized so far, including cancer of the colon, breast, lung and liver. Retinoblastoma, a rare childhood eye tumour if detected early can be successfully treated.

Loss of heterozygosity (LOH) of tumour suppressor genes occurs by mitotic recombination or chromosome mis-segregation. Let us consider a somatic cell that contains one mutant and one

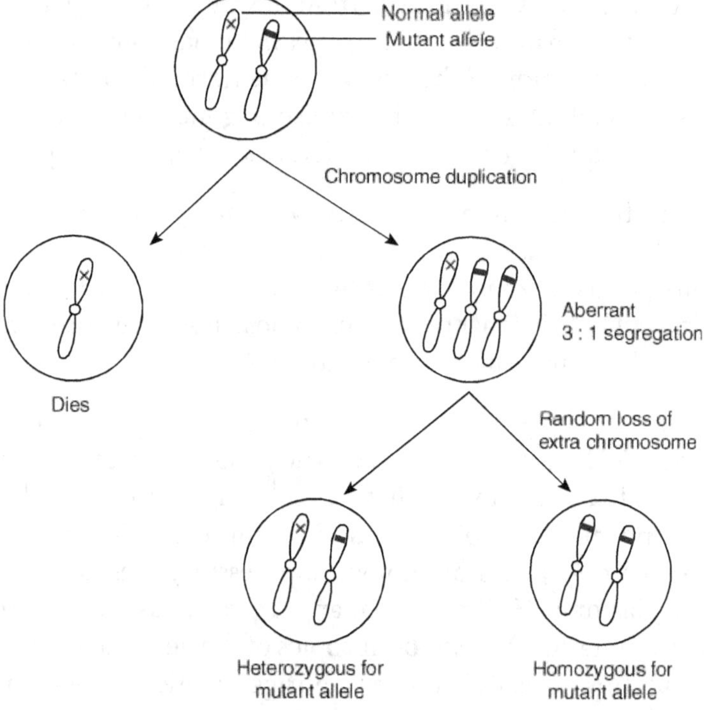

Figure 7.5   Schematic representation of mis-segregation

normal allele of a tumour suppressor gene and discuss the loss or inactivation of normal allele. One common mechanism of LOH involves mis-segregation of the chromosome bearing the heterozygous tumour-suppressor gene during mitosis. In this process, also referred to as non-disjunction, one daughter cell inherits only one normal chromosome, while the other inherits 3 and other normal chromosome as well as two bearing the mutant allele. Such mis-segregation is caused by failure of a mitotic checkpoint, which would normally prevent a metaphase cell, with an abnormal mitotic spindle from completing mitosis. Subsequent loss of one chromosome often occurs, restoring the complement; if the normal chromosome is lost, the resultant cell will contain 2 copies of the mutant chromosome (Figure 7.5).

EXOGENOUS CARCINOGENS

Carcinogen is any substance or agent that produces cancer or increases the risk of development of cancer. Exogenous carcinogens induce cancer by acting as external agents.

Exogenous carcinogens are basically classified into 3 types namely

1. Chemical carcinogens
2. Physical carcinogens
3. Biological (viral) carcinogens

Chemical carcinogens are of extremely diverse structure and include both natural and synthetic products (Table 7.1). Some are direct-reacting and require no chemical transformation to induce carcinogenicity, but others are indirect-reacting and become active only after metabolic conversion. Such agents are referred to as **procarcinogens** and their active end products are called **ultimate carcinogens**. All chemical carcinogens are highly reactive electrophiles that react with the electron-rich atoms in RNA, cellular proteins and mainly DNA.

Table 7.1   Major chemical carcinogens

Direct-acting carcinogens	Alkylating agents, acylating agents, anticancer drugs, etc.
Procarcinogens that require metabolic activation	Polycyclic and heterocyclic aromatic hydrocarbons, benz(α)anthracene, benzo(α)pyrene.
Aromatic amines, amides, azo dyes	2-naphthylamine, 2-acetylaminofluorene.
Natural plant and microbial products	Aflatoxin B, griseoflavin, betelnuts.
Others	Nitrosamine and amides, vinyl chloride, nickel, chromium, insecticides, fungicides, etc.

Radiations and heat act as physical carcinogens. Ionizing radiations and ultraviolet rays in direct sunlight are mutagenic and thereby carcinogenic. Radioactive material inhaled or injected accidentally may produce carcinoma. Workers engaged in preparing watch dials with luminous paint containing radioactive material developed osteogenic sarcoma of bones due to their practice of licking the paint brush with tongue and lips for making the tuft of bristles pointed, whereby, the materials entered the body and got selectively accumulated in bones. Exposure to atomic radiations increases the incidence of malignancy in the population affected. Heat as a physical carcinogen plays its role indirectly by inducing necrosis and regeneration.

Both RNA and DNA viruses can be transforming agents, such viruses are often called tumour viruses. Tumour viruses cause transformation as a consequence of their ability to integrate their genetic information into the host cell DNA; most often they cause the chronic production of one or more proteins called transforming proteins, which are responsible for maintaining the transformed state of the infected cells. These proteins are synthesized under the direction of transforming genes in an integrated viral genome.

Hormones are known to exert a trophic influence on their target organs by controlling their proliferation and secretory activities. Hormones particularly oestrogens, are important as tumour promoters in the development of some human cancers. The proliferation of cells of the uterine endometrium is stimulated by oestrogen, and there is a chance of spreading endometrial cancer. Tumours solely induced by excess hormonal stimulation are also called hormone-dependent tumours.

The role of increased hormonal levels in tumorigenesis has been illustrated by the technique of parabiosis by which two or more animals of the same genetic constitution are sutured side by side so that they possess a common blood supply. Thus, when a normal female is joined in parabiotic union with either a castrated male or a female, the gonadotrophic hormone excreted in excess by the castrate is transferred to the intact partner and stimulates its ovaries. Although more oestrogens are produced, they are rapidly broken down by the liver and consequently do not pass over to the castrate, whose pituitary gland remains uninhibited. Granulosa cell tumours have been reported to develop in this manner.

There are also presence of endogenous carcinogens and co-carcinogens. The structural carbon-ring nucleus of sex hormones, sterols, bile acids and cholesterol is similar to that of carcinogenic hydrocarbons. The endogenous carcinogens may originate through some error in synthesis or metabolism of sex hormones or bile acids; such errors may be genetically based appearing at various age periods or governed by mutations. Methyl cholanthrene has been synthesized from deoxycholic acid, a normal constituent of bile.

Diet, age, heredity, trauma and chronic irritation are considered as co-carcinogens. There are certain diseases or malformations, which are superimposed by malignant changes with great frequency so that detection of such diseases or malformations increases the chances of malignant tumours appearing in the patient. These include leukoplakia, syphilitic glossitis, xeroderma pigmentosum, etc.

DRUG DISCOVERY

A huge range of targets for drug discovery has been uncovered in cancer research. Among these the most successful are **tyrosine kinases**. Cancer drugs could be targeted to the molecular peculiarities of cancer cells. A major success for this approach is Herceptin, an antibody drug that targets a specific protein present on some breast cancer called *HER/neu*.

A related approach to the Herpatin is to treat cancer as an invasion by dangerous cells and to vaccinate the patient against them *HER/neu* and MUC-1, a protein that is common on pancreatic cancers as well as some others that act as suitable antigens. Cancer vaccines are most suited for melanomas. They can be single proteins, killed or engineered cancer cells or immune cells, such as dendritic cells which are involved in "presenting" the antigen to the immune system.

Immunotherapeutics is a related idea to boost the immune system to recognize cancer on its own. Drugs such as interleukin-2 and interferon have been tried with this, with indifferent success.

DIAGNOSTICS

Use of antibodies, DNA probes, and PCR have revolutionized cancer diagnosis to find out the molecular problems behind each disease. The problem is that, at the moment, the drugs are not here to do much about it. The exception is herceptin, where a specific diagnostic test can be used to tell, whether a breast cancer patient has *HER/neu* gene (or) not. If they do, then herceptin can be used.

Summary

- Neoplasm (tumour) is an abnormal mass of tissue, the growth of which exceeds and is uncoordinated with that of the normal tissues.

- The two types of tumours are benign and malignant tumours.

- Cancer cells fall into one of the three main groups namely carcinomas, sarcomas and leukemias.

- The development of cancer is a multi-step process involving mutation and selection for cells with progressively increasing capacity for proliferation, survival, invasion and metastasis.

- Basic fibroblast growth factor (BFGF), transferring growth factor-α (TGF-α) and vascular endothelial growth factor (VEGF) are the growth factors secreted by many tumours; all have angiogenic properties.

- The process by which a new clone of tumour cells has evolved on the basis of its increased growth rate or other properties that confer a selective advantage is called clonal selection.

- Study of the morphology and metabolism of cancer cells help in clear differentiation of cancer cells from normal cells.

- Invasion and metastasis are biological hallmarks of malignancy.

- Carcinogenesis is a process which induces cancer in normal tissues.

- Oncogenes are capable of inducing cell transformation.

- *ras* and *myc* oncogenes act in cooperation to induce cellular transformation.

- Chemical carcinogens are highly reactive electrophiles that react with the electron-rich atoms in RNA, cellular proteins and mainly DNA.

REVIEW QUESTIONS

1. Define:

    i. oncology

    ii. cancer

2. Explain the types of cancer.

3. Brief on initiation of tumour.

4. Explain clonal selection.

5. Discuss in detail, the morphology of the cancer cells.

6. Explain the process of transformation of normal cells to neoplastic cells.

7. Give a detailed account of tumour suppressor genes.

8. Give a brief account of exogenous carcinogens.

9. What are endogenous carcinogens?

10. Brief on drugs employed in cancer treatment.

8

GENE THERAPY

Retinoblastoma—The Two-hit Hypothesis

Our current understanding of tumour-suppressing genes began with observations by Alfred Knudson, who developed the two-mutation hypothesis of cancer causation.

Knudson was interested in retinoblastoma (RB), a rare childhood eye cancer. Distinct tumours, representing individual original cancerous cells, develop in the eye. Sometimes RB affects one eye, and sometimes both. Knudson examined the medical records of 48 children with RB between 1944 and 1969. He recorded the following information for each child:

1. Whether one or two eyes were affected
2. How old the child was at the time of diagnosis
3. Whether any other relatives had RB
4. Sex
5. Number of tumours per eye

The fact that RB occurred in boys and girls told Knudson that any genetic control was autosomal. Pooling data from families with more than one case of RB revealed that approximately 50 percent of the children of an affected parent were also affected, suggesting dominant inheritance. Knudson also noted that in some families, a child with two affected eyes would have an affected grandparent, but both parents had healthy eyes. A picture of autosomal dominant inheritance with incomplete penetrance began to emerge.

Knudson, however, proposed a different explanation. He hypothesized that an initial, inherited recessive mutation had to be followed by a second somatic mutation in the eye to trigger tumour formation. This idea became known as the "two-hit hypothesis" of cancer causation. Occasionally the second mutation would not occur, and this would explain the unaffected parents between two affected generations.

Knudson's two-mutation hypothesis explained another observation gleaned from his search of the medical records. Children with tumours in both eyes become affected much earlier than children

G enes are the blueprint of our bodies, governing factors such as growth, development and functioning. Excitement has come from the finding that genes that guide cell differentiation and pattern formation are linked to human disease. A genetic mutation means that a gene contains a fault or "spelling mistake" that disrupts the gene message. A fault in a gene can occur spontaneously or can be inherited. Genetic faults can cause a wide range of disorders and be involved in susceptibility to some cancers. For example, mutation in β-catenin contributes to human colon cancer and other cancers (Morin *et al.*, 1997).

The primary goal of human medical genetic research is to develop treatments for each of the many different genetic diseases. A few gene products, the characteristics of the diseases caused by mutations in the genes that encode these products, and the current treatments for these diseases are listed in Table 8.1.

with tumours in only one eye—generally before the age of five. This would make sense if a hereditary, bilateral (two-eye) form of RB requires germ-line mutation followed by a somatic mutation, but a nonhereditary, unilateral (one-eye) form results from two somatic mutations in the same gene in the same cell. In other words, in the inherited form of RB, a child is already born halfway on the road to tumour development—just one somatic mutation in the eye is needed. The unilateral, noninherited form appears later in childhood because it takes longer for the required two somatic mutations to occur in the same cell.

Next, Knudson used a mathematical expression called a Poisson distribution to estimate the number of events required to account for a certain pattern of observations. He found that the average number of tumours per eye—three—was consistent with a two-hit mechanism, according to the equation.

Although it would be another 15 years before researchers identified the RB gene on chromosome 13, and longer still before its role in controlling the cell cycle was identified, Knudson's insights paved the way for the particular discovery and for recognition of the widespread action of tumour suppressors in general.

Table 8.1 Current therapies for 10 single-gene human disorders

Gene product	Disease and symptoms	Frequency	Current therapy	Prognosis
Adenosine deaminase	Severe combined immunodeficiency Loss of T and B cells	1:1,000,000	Bone marrow transplant Adenosine deaminase replacement	Without therapy: fatal by 2 years of age With therapy: clinical improvement
LDL receptor	Familial hypercholesterolaemia Elevated blood serum cholesterol level Coronary artery disease	1:500 (heterozygotes)	Diet, drugs, liver transplant	Some clinical improvement
Glucocerebrosidase	Gaucher's disease Accumulation of glucocerebroside in macrophages causing liver, spleen, and bone damage	1:2,500 (Jewish population); rare in non-Jewish populations	Symptomatic treatment: removal of spleen, antibiotics, repairing bone damage, bone marrow ransplant, enzyme replacement	Some clinical improvement

Blood clotting factor VIII	Haemophilia A Altered plasma protein that causes defective blood clotting, chronic internal bleeding into joints, excessive bleeding after wounding	1:10,000 (males)	Concentrates of factor VIII by transfusion	Extended life expectancy, requires continual treatment, risk of viral infection from transfusions
Phenylalanine hydroxylase	Phenylketonuria Excess phenylalanine in the bloodstream of newborns causes mental retardation	1:10,000	Restricition of dietary phenylalanine	Good in many cases, if treatment is started early and maintained
α_1-Antitrypsin	Emphysema Deficiency of serum protein protease inhibitor, damage to the lungs, cirrhosis of the liver	1:3,500	Replacement therapy, lowering environmental risks	Progression of the disease is slowed but not stopped

(Contd.)

Table 8.1 (Continued)

Gene product	Disease and symptoms	Frequency	Current therapy	Prognosis
Cystic fibrosis transmembrane regulator	Cystic fibrosis Multisystem disease, pancreatic insufficiency in some cases, intestinal blockage, blocked airways of the lungs	1:2,500 (Caucasians)	Antibiotics, physical clearing of the lungs, upgrading diet	Fatal by late 20s
Ornithine transcarbamylase	Hyperammonaemia Urea cycle defect, ammonia accumulation arginine deficiency Early form: within 72 h of birth, lethargy, vomiting, coma, death, and, if survival, irreversible brain damage Late form: vomiting, lethargy, seizures	1:40,000	Restricted protein diet, arginine-supplemented diet, drugs, liver transplant	Late form: good Early (severe) form: symptoms diminished

Dystrophin	Duchenne muscular dystrophy Progressive muscle wasting	1 : 7,500 (males)	Only supportive treatments: good nutrition, aid in respiratory function, confinement to a wheelchair	Fatal by early 20s
β-globin	Sickle-cell disease Chronic anaemia, multisystem disease, damage to spleen, heart, kidneys, liver, and brain; heterozygotes have a mild form of the disease	1 : 500 (heterozygotes in populations of black African origin; less frequent in other populations)	Blood transfusions, drugs, analgesics, bone marrow transplant	Symptoms diminished; treatments suboptimal

Because of the physiological complexity of these genetic diseases, curative measures have consisted mainly of treating the symptoms by administering drugs or conducting surgery. For some of the genetic diseases that cause errors of metabolism, restrictive diets have been devised to control the high endogenous levels of the toxicant that are a consequence of the gene mutation. Unfortunately, the most effective treatment for many genetic diseases can be given only when the patient enters a crisis phase. Although these interventions are lifesaving, they are only temporary. Because genetic diseases often affect multiple systems and tend to be progressively debilitating, it has not been an easy task to develop effective therapies. Moreover, on the whole, the existing treatments tend to be inadequate, with very few of the patients having normal lifespans or offspring. Although patients benefit from treatment, it is in most cases suboptimal, recurring, costly, and time-consuming. Consequently, new forms of therapies are continually being sought.

With the discovery of the molecular basis of DNA transformation in bacteria, in which a gene can be transferred to another strain, researchers have speculated that human genetic diseases might be cured in an analogous way by introduction of the normal gene into the appropriate somatic cell type. Human gene therapy is defined as the deliberate alteration of the genetic material of living cells to prevent or treat disease. A distinction is made between somatic gene therapies, which alters the DNA of the body's differentiated cells, and germ-line gene intervention, which changes the DNA of reproductive cells. The term "human gene therapy" came into use because it sounded more benign than the term "human genetic engineering". The term "therapy" indicates that desirable ends are sought, but it has no necessary linkage to any universally agreed upon concept of disease.

Gene therapy is the treatment or prevention of disease, and involves replacement of a defective abnormal gene into the cells of the patient who is deficient in the normal gene product. The technical basis of gene therapy is gene delivery, i.e., introducing the desired gene into the appropriate cells of the patient. Gene therapy is regarded by many as a potential evolution in medicine. This is because gene therapies are aimed at treating or eliminating the cause of disease, whereas most current drugs treat the symptoms. This radical improvement is possible because the gene-based approach can provide superior targeting and prolonged duration of action. Hence, in comparison with other forms of therapy, it will permit biological effects that are more subtle and better localized to the most appropriate cells. Therapeutic gene therapy is still in its infancy, but has the potential to revolutionize treatment for all kinds of genetic diseases.

The first clinical studies involving gene transfer began in 1990 and since then gene therapy has become the focus of a whole new industry. All clinical studies involve gene addition rather than correction or replacement of defective genes, which is technically more difficult. All clinical protocols approved to date involve gene transfer only to somatic cells rather than germ-line cells (Table 8.2).

SOMATIC VERSUS GERM-LINE GENE THERAPY

Researchers distinguish two types of gene therapy, depending on whether it occurs in somatic tissue or in gametes or fertilized ova.

Correcting only the somatic cells that a genetic condition affects is called somatic gene therapy. This form of the technology is nonheritable, which means that a recipient does not pass the genetic correction to offspring. For example, a bone marrow transplant is somatic gene therapy because it replaces only bone marrow cells which are somatic. Clearing lungs congested from cystic fibrosis with a nasal spray containing functional CFTR genes is also somatic gene therapy; it does not alter the gametes.

Table 8.2 Gene therapy in phase II clinical studies

Disease	Vector	Gene transfer	Gene	Therapeutic strategy
Cancer	Cationic lipid	Tumour cells *in vivo*	HLAB7/IL-2/	Immunogenicity enhancement
	"	"	AdE1A	"
	Ad	"	p53	Apoptosis induction
	Ad	"	Ad death genes	Killing by lytic virus
	RV	"	*TK*	Killing by enzyme and/or prodrug
	RV	Fibroblasts *ex vivo*	IL-12	Immunogenicity enhancement
Limb ischaemia	Naked DNA	Muscle cells *in vivo*	VEGF	Angiogenesis stimulation
Cystic fibrosis	AAV	Airway cells *in vivo*	CFTR	Provision of functional protein
HIV	RV	T cells *ex vivo*	Chimeric TCR	Retargeting killer T cell

Somatic gene therapy helps an individual, but not, his or her descendants.

We must know a great deal about a disorder to develop a somatic gene therapy. Which cells, in which tissues and organs, function abnormally, and when in development do they do so? Which biochemical is abnormal or missing? What DNA sequence must we assess to correct the abnormality? How many cells must be sequence-altered to alleviate symptoms? Even when we have much of this information, designing a gene therapy is challenging. Some early gene therapy experiments have been disappointing because the correction was not sufficient to overcome symptoms or because the immune system attacks the altered cells.

FIRST APPROVED TRIAL

The first approved human experiments have begun in the USA using the technique of somatic cell gene insertion. It is still very much at the experimental stage but the scientists did have to meet very strict criteria in order to conduct their experiments (Culliton, 1989a), and further trials are under regulatory consideration (Gershon, 1990a). The first trial did not replace a defective gene, but inserted a marker gene into cells for tracking the cells involved in cancer therapy.

The therapy involves the use of cells which attack cancer, and called tumour-infiltrating lymphocytes (TILs). They are isolated from the patient's own tumour, then grown in large numbers *in vitro*. The cells are then given back to the patient, and stimulated by a naturally occurring hormone, interleukin-2. The procedure is known to help about half of the patients. In order to discover how this therapy works, the TILs were genetically marked to trace them in the patients. The initial trial involved ten patients, but this number has now been increased following the success of the preliminary group of patients.

The next trial will attempt to insert the gene to express the hormone, interleukin-2 themselves so it is self-sustaining and more targeted as the first trial has been successful (Culliton, 1989). A trial involving the insertion of the gene for tumour necrosis factor in TILs, which will be conducted in fifty patients with advanced melanoma, passed the final stages of approval in August 1990. The trial was led by the same researchers, Steven Rosenberg of the National Cancer Institute. It should have been easier to justify ethically because there is some hope of therapeutic value in this trial. Tumour necrosis factor has been shown to shrink tumours in mice, and it is hoped that the TILs will cluster around the tumours, releasing the factor which will kill the tumour, and then the TILs will die so that the production of the tumour necrosis factor is limited to such sites (a high blood level is toxic).

Foetal Gene Therapy

Another major purpose of somatic cell therapy will be during foetal life, as many diseases require treatment to begin during foetal life (Robertson, 1985). The alternative is selective abortion, at an early stage. The future situation may make early foetal screening a routine, by chorionic villi sampling at 6–8 weeks, allowing the screening of any genetic disease. If a disease is detected, then the decision will be whether to have selective abortion and have another foetus or to try to treat the disease. As the number of gene probes increases, it is possible that all embryos will be shown to carry the 5–10 harmful recessive alleles that they contain and one could imagine the situation where it is very hard to find the foetus free of a potentially harmful allele or that will not have a risky allele and the tendency for some disease, so therapy will eventually dominate.

Foetal surgery or therapy or dietary supplements have been used for many diseases already and the increasing use of conventional therapy will be supplemented by gene therapy. If there is a risk of harm to the foetus then it may be more ethical to use selective abortion at that early stage. It is seen by many that it is more ethical to end the life of a present human embryo than to let the embryo grow to become a person or to try experimental therapy on the embryo and risk the damage to the child. The question of selective abortion comes down to the status of the human embryo and the possibilities of early screening. The technology is here and already some parents have had to make these types of decisions. The choice will depend on the type of disease.

Germ-line gene therapy alters the DNA of gamete or fertilized ovum. As a result, all cells of the individual that develop harbour the change. Unlike somatic gene therapy, germ-line gene therapy is heritable and passes to offspring. The steps are the same but the goals are different. Germ-line gene therapy is technically more difficult, and raises more ethical challenges. The two main methods of performing germ-line gene therapy are:

- To treat pre-embryo that carries a serious genetic defect before implantation in the mother.

- To treat the germ cells (sperm or egg cells) of affected adults so that their genetic defects would not be passed on to their offspring.

Results of Animal Gene Therapy

There have been experiments on animal models as a prerequisite to experiments on humans. There have been several animal systems for testing human gene therapy made. Retroviruses have been successfully used to target genes into animal cells for several years (Thomas, 1986) and foreign genes have been expressed in mice using treated cells (Bernstein *et al.*, 1986; Keller and Wagner, 1986). Larger animal models are under study to determine the persistence of infected bone marrow cells in transplanted animals (Anderson *et al.*, 1986, Kwok *et al.*, 1986). Various types of retrovirus vectors have been developed. Human bone marrow cells have been infected and have expressed inserted drug-resistance genes (Miller *et al.,* 1986). There was expression of the human ADA gene in murine cells *in vivo* (Williams *et al.,* 1986). One group of scientists placed the human gene for the production of ADA into the bone marrow into the rhesus and macaque monkeys and foetal sheep in preliminary experiments (Anderson *et al.,* 1986). Several experiments using SCID-mice and human ADA-deficient lymphocytes expressing an inserted ADA gene have been successful, as described.

Other groups continued to work on procedures to infect haematopoietic stem cells, for diseases such as ADA deficiency (Cournoyer *et al.,* 1990). These are the reasons for optimism with the approved gene therapy trial; it is based on the years of preliminary work by many groups. The human globin gene has been expressed in mice that were treated with retrovirus-infected stem cells, so that the expression problems in stem cells may be soon overcome (Dzierzak *et al.,* 1988). Another experiment in mice was to test a model for treating retinoblastoma in tumour development. Both positive acting oncogenes needed to be switched on and negative acting tumour suppressor genes need to be mutated. Both alleles of a suppressor gene need to be turned off before cancer can proceed. In gene transfer studies the suppressor gene function for retinoblastoma could be reinserted and tumour cell formation repressed preventing the tumour formation (Freidmann, 1990b). This work may be extended to humans.

There is a contrast between diseases that affect the lymphoid cells and diseases that affect the haematopoietic system. The haematopoietic cells have a short life and so cannot be useful targets of gene therapy. One of these diseases is Gaucher's disease. Human cells have expressed the genes for glucocerebrosidases needed for the correction of Gaucher's disease (Choudary *et al.,* 1986; Sorge *et al.,* 1986). It is another disease with no known cure, which affects 0.2% of people born

(2% of Jewish population). The absence of glucocerebrosidase results in the accumulation of nondegraded membrane lipoprotein within the reticuloendothelial cells. An appropriate *in vivo* model for the detection and measurement of pluripotent haematopoietic stem cells does not yet exist as further experiments are needed. But there is some progress (Parkman and Kohn, 1990).

One of the positive outcomes of extensive animal trials has been the development of effective and stable retroviral vectors for gene transfer. There are alternatives to retroviral vectors being developed, such as methods of targeted modification of human genes, and these will soon be possible. Promising alternatives to retroviral vectors are being developed, such as the use of laser micropuncture of the cell membrane to facilitate direct gene transfer. This technique has been used on cultured human cells with an efficiency 100-fold more than the standard calcium phosphate method of DNA transfer in which cells are passively incubated in the DNA solution. The efficiency of the physical methods is about 1% of the target cells that incorporate the genes.

If there are no naturally occurring animal models for some human genetic diseases, targeted gene mutagenesis can be used to create diseased animals. Mice have been genetically engineered using embryonic stem cells to be HPRT-deficient mice. These were hoped to provide mouse equivalents of humans suffering from Lesch–Nyhan disease. Genetic therapy has been tried on them, as preliminary experiments for human gene therapy, correcting the gene deficiency.

There have been a variety of trials in animals. There is some progress on gene transfer methods. Lipoprotein vesicles using the substance lipofectin has been used to insert genes into the neuroepithelium of frog embryos, as well as into skeletal muscle cells of young mice. The biolistic approach used in plants for gene transfer is being applied to animals, and will also be used in the gene therapy trial on ADA gene insertions in humans, which has been approved.

After conception, the genotype may be normal, without a genetic disease, or abnormal, with a genetic disease. There are several stages at which therapy could occur, germ-line gene therapy must occur in the very early embryo. After this stage, somatic cell therapy can be performed, before or after birth. Symptomatic therapy usually is given after birth. But with modern medicine, symptomatic therapy may also be given before birth in some diseases.

The knowledge of the biochemical basis of genetic diseases is often essential in order to develop a reasonable therapy. In the majority of genetic diseases, there is some therapy available and in some cases the therapy can effectively restore normal health in spite of the continued presence of the abnormal genotype.

Protein replacement   Symptomatic therapy involves using the normal gene product as a substitute for the defective gene product. One of the first diseases to be treated this way was diabetes mellitus. Diabetes is due to inadequate production of the hormone insulin. The treatment varies from just a diet which is low in carbohydrate, to taking regular amounts of the hormone insulin to ensure a sufficient level in the body for normal function. Insulin is a relatively simple molecule so it has been produced in large amounts for some time. It can be obtained from animal tissue extraction, usually pigs, or the exact human protein can now be produced from genetically engineered bacteria.

Dietary treatment   Substitutional therapy is not successful in every case, as we may not be able to administer the defective gene product or enzyme from outside the body. In phenylketonuria (PKU), the defective enzyme is localized in the liver and a substitute can be inserted. However, the disease can be successfully treated by a dietary treatment involving a reduction in the intake of phenylalanine. This is possible because the disease only affects people due to the accumulation of an abnormal toxic product derived from a specific substance in the diet, so these people can live otherwise normal lives if they omit this substance from their diet.

GENE THERAPY PROCESS

The basic steps of gene therapy include:

- The faulty gene that causes a specific disease must be identified.
- The location of the affected cells must be pinpointed.
- A healthy version of the gene must be available.
- The healthy gene has to be delivered to the cell.

Gene Delivery

Gene therapy requires introduction of foreign DNA sequences with stable integration, gene expression and an appropriate regulation in the target tissue. There are two strategies used to deliver genes—*ex vivo* and *in vivo* transfer.

In *ex vivo* transfer, cells are removed from the patient, an appropriate gene is introduced in these cells and then these genetically engineered cells are transplanted back into the patient's body. This offers the advantage of more efficient gene transfer and the possibility of cell propagation to generate higher cell doses. It is patient-specific and more costly.

A promising technique is to put the healthy gene inside a deactivated virus which has limited genetic information of its own. A virus that causes diseases, such as the common cold, works by slipping into a cell, taking over its DNA and forcing it to produce more viruses. A deactivated virus can therefore enter the specific cell and deliver the healthy gene.

In *in vivo* approach, the desired gene is directly introduced into the target tissue.

Other techniques involve extracting stem cells from an individual's bone marrow, putting the healthy gene into them and re-inserting the stem cell into their bone marrow. These stem cells can develop into a range of cell types.

GENE TRANSFER TECHNIQUES

Transfer of the gene can be accomplished by the following methods:

1. Physical
2. Biological (viral vectors)

PHYSICAL TRANSFECTION METHODS

They include

- Liposome-mediated DNA transfer
- Receptor-mediated endocytosis

Liposome-mediated DNA Transfer

It involves complexing plasmid DNA (with foreign DNA) with liposomes and introducing it into the target cell (Figure 8.1).

Using this technique one can introduce larger amount of DNA into the target cells, than what is possible with viral vector systems. This can be as large as an artificially constructed minichromosome, which includes elements for regulation of gene expression apart from the particular structural gene. These elements regulate gene expression in a physiologically controlled manner. The disadvantage of this method is that the gene expression is transient; therefore the treatment has to be repeated.

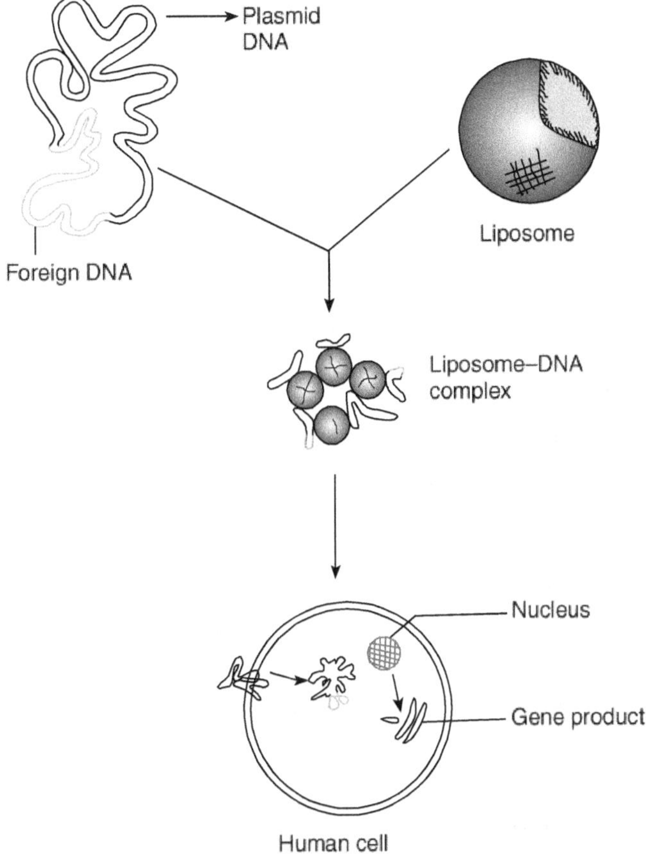

Figure 8.1   Liposome-mediated gene therapy

Receptor-mediated Endocytosis

In this method, a complex is made between plasmid DNA containing foreign DNA and specific polypeptide ligands for which the cell has receptors. The DNA is targeted to these receptors. For example, DNA is complexed to a glycoprotein containing galactose. This will be recognized by the receptors on liver cells, which are specific to glycoproteins with a terminal galactose.

This causes internalization of the complex by endocytosis. The endocytic vesicles fuse with lysosomes and here the complex is degraded and the foreign gene escapes from the lysosome, to be expressed. The rate at which it (foreign DNA) escapes from the lysosomes can be increased by inclusion of adenovirus or fusogenic influenza gene products.

The major advantage of the physical transfection method is that it is free from the risk of viruses. Liposome-mediated gene therapy is being actively tested on cystic fibrosis patients in U.K. and U.S.A.

BIOLOGICAL TRANSFECTION METHODS USING VIRAL VECTORS

In biological transfection methods, viral vectors form an efficient mode of gene delivery into the target cells. Various viruses are used namely retroviruses, adenoviruses, herpesvirus, adeno-associated virus, parvovirus, etc. The viruses are rendered replication-deficient by removing encapsidation (Ψ, psi) gene sequences.

Retroviruses

They are derived from the family of viruses that includes human immunodeficiency virus (HIV) and oncogenic viruses. The retroviruses integrate into the host DNA and make copies of their genome using reverse transcriptase enzyme. The provirus thus formed serves as a template for production of mRNA for various viral gene products as well as the new genomic RNA of the virus.

The virus must not only be efficient in transferring and expressing the transduced gene, but it must also be safe. It is possible that the retroviral vectors might activate endogenous retroviral genomes in target cells by recombining with them and expression of the genes inserted has proved difficult. The use of disabled viruses is probably essential to "ensure" safety. The study of gene expression is also essential, it has been found that the retroviral vector used to transfer genes into cells does alter the expression of the inserted gene. The chromosomal sites which retroviruses integrate may be fundamentally different from sites into which DNA segments integrate using DNA transfection without a viral vector.

The experience from a number of clinical trials with more than 200 patients has established that replication-defective retrovirus vectors are not responsible for any apparent adverse effects. Moreover, the safety features of retroviral vectors are continually being upgraded. For example, a plasmid that carries the retroviral *gag* and *pol* genes under the control of the 5′-LTR promoter, a therapeutic gene, and the *env* gene driven by the cytomegalovirus promoter has been constructed. After transfection, this plasmid directs the production of replication-defective virus particles. It is very unlikely that this construction, which is called a plasmovirus, will recombine to form replication-competent retroviruses. Although the DNA-carrying capacity of a plasmovirus is only 3.5 kb, most potential therapeutic cDNAs and anticancer genes range from 0.5 to 2 kb in length.

The retroviruses used in gene therapy need a couple of elements 1) packaging cell line and 2) helper virus.

Packaging cell line   It is the cell line that has been infected with the retrovirus, which is genetically engineered to lack region of proviral DNA called packaging sequence.

Helper virus/vector   It consists of a retroviral provirus with more than 90% of its viral genomic material removed, leaving only minimal sequences essential to produce copies of the viral RNA along with the sequences necessary for packaging of the viral genomic RNA. This is a vector backbone in which the foreign gene can be inserted. If this helper virus is introduced into the packaging cell line that contains provirus in which packaging sequences are missing, the RNA produced by the vector provirus can be packaged into viruses. These **virions** can be used to infect, or more precisely called transduce, the target cells.

The demerits of retroviruses as vectors in gene therapy are as follows:

1. Only smaller DNA sequences (less than 7 kb usually) can be introduced.

2. Retroviral vectors can transduce only dividing cells and hence CNS disorders are not emendable to it.

3. The retroviruses can only be used *in vitro.*

4. Another demerit of retroviruses is that they are unstable.

5. They cannot be purified for use in gene therapy without reducing their capability to transduce target cells. This means that contamination with the replication-competent retroviruses is inevitable. They could serve as oncogenes, causing malignant transformation.

6. Controlling the levels of expression of the introduced gene is yet another difficult task.

Retroviral gene transfer can be effectively used in the case of hepatocytes, haematopoietic stem cells, fibroblasts, myoblasts, endothelial cells, etc.

Adenoviruses

Adenoviruses infect a wide range of nondividing human cells and have been used extensively as live vaccines against respiratory infections and gastroenteritis without side effects. These features make adenovirus a likely prospect for delivering genes to target cells.

After infection of a target cell with a recombinant adenovirus, the DNA is passed into the cell nucleus, where the therapeutic gene is expressed. The recombinant DNA construct does not integrate into a chromosome, and consequently, it does not persist for long periods. Therefore, adenovirus-based gene therapy requires periodic administration with additional recombinant viruses.

Adenovirus gene delivery systems have been used in gene therapy trials for treating cystic fibrosis. The initial results were not encouraging. Very few of the patients' cells were transduced with the cystic fibrosis transmembrane regulating gene (CFTR), and multiple treatments with recombinant adenovirus triggered severe immunological responses in the patients. A low level of expression of some of the adenoviral genes after infection is responsible for the immunological response that destroys the transduced cells (Figure 8.2). A plasmid with a therapeutic gene (TG) that is inserted into a segment of the adenovirus genome

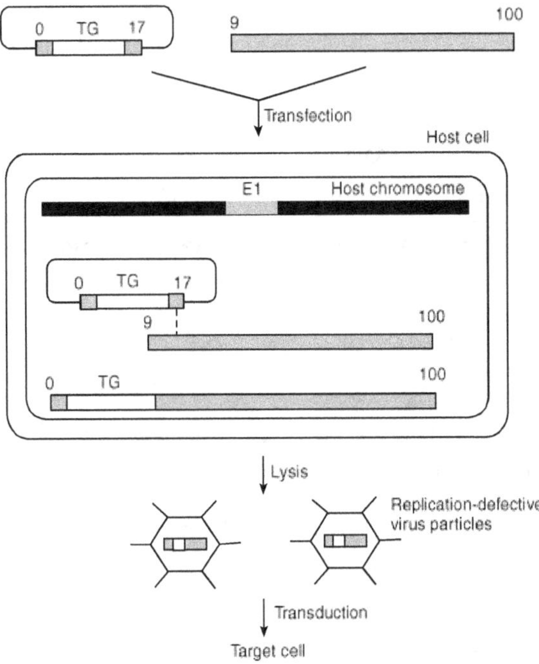

Figure 8.2   Adenovirus vector system

extending from map unit 0 to 17 and a linear adenovirus DNA molecule that consists of map units 9 to 100 are both introduced into a host cell that carries an integrated and functional adenovirus E1 gene (E1). The entire adenovirus genome is 100 map units. A recombination event (dotted line) between the shared DNA regions of the plasmid and linear adenovirus DNA molecule creates a DNA molecule that is equivalent to a full-length adenovirus genome. The recombinant DNA molecule with the therapeutic gene is packaged and released after cell lysis. The recombinant virus particles are replication-defective. Plasmid DNA which is part of the final construct does not impact on the packaging of the recombinant DNA molecule (not shown).

The merits of adenoviruses as vectors are as follows:

1. They are stable.
2. They can be easily purified.
3. They are suitable for targeted treatment of specific tissues, e.g. respiratory tract diseases.
4. They can infect/transduce non-dividing cells.
5. They can carry larger DNA segments as big as 36 kb long.

The demerits of adenoviruses as vectors in gene therapy are as follows:

1. They do not integrate into the host genome therefore expression of the introduced gene is unstable.
2. The gene expression is often transient.
3. By virtue of their infectivity, they can produce adverse effects secondary to infection.
4. Adenoviruses contain genes which can cause malignant transformation, hence their use as vectors carries the menace of inducing malignancy.

Herpesvirus

There are many different disorders that affect the central and peripheral nervous systems, including tumours, metabolic and immunological defects and neurodegenerative syndromes such as Alzheimer and Parkinson diseases. Neurological disorders are responsible for more hospitalizations and chronic care than almost all other diseases combined.

Herpesviruses are neurotropic viruses, which on suitable modification can be effectively used for gene therapy in central nervous system disorders.

Herpesvirus has a natural affinity for non-dividing cells and hence it is suitable for transfection of neurons. It can also be used in hepatocytes.

The genome of herpes simplex virus is a 152-kb long, double-stranded DNA molecule. The virus fuses to the membrane of a neuron and is transported to the nucleus. The reproductive cycle consists of a lytic phase and a latent phase. During the former phase, the virus replicates and virus particles are produced. During the latter phase, the viral genome condenses and at least two viral promoters called latency-associated promoters are active.

About 30 kb of the herpes simplex virus genome can be replaced by cloned DNA without significant effects on replication, packaging, or infectivity. However, the large size of the virus makes genetic manipulations difficult. To overcome this problem, a scaled-down version of the herpes simplex virus genome consisting of the herpes simplex virus origin of replication and packaging signal has been incorporated into an *E. coli* plasmid that can, in addition, carry up to 8 kb of cloned DNA. These herpes simplex virus-based constructs are called amplicons (amplicon plasmids). The herpes simplex virus amplicon plasmid vector system is shown in Figure 8.3. An HSV origin of replication (HSV *ori*), an HSV packaging signal, and a therapeutic gene (TG)

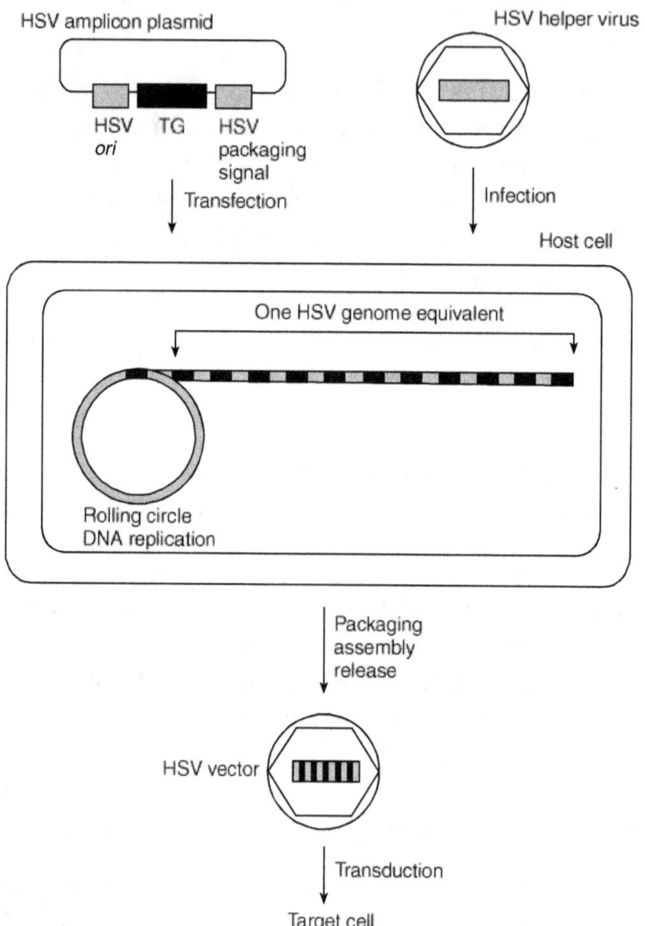

Figure 8.3 Herpes simplex virus (HSV) amplicon plasmid vector system

are inserted into an *Escherichia* coli plasmid (HSV amplicon plasmid). The HSV amplicon plasmid is transfected into a host cell that is infected with helper HSV. The HSV amplicon plasmid DNA undergoes rolling circle DNA replication. One HSV genome equivalent to about 10 amplicons is packaged as a unit into an HSV capsid that is produced by the HSV helper virus. The HSV helper genome is not packaged. HSV particles carrying multiple copies of the therapeutic gene are released after cell lysis and used to transduce neurons.

An immediate disadvantage of using herpesviruses as vectors is their toxic effects on the nerve cells and the following immune response. Since herpesviruses do not integrate into the host genome, the expression of the introduced gene may be unstable.

Parvovirus and simian virus 40 are being considered for gene therapy in CNS and smooth muscle cells respectively. Studies undertaken on adeno-associated virus (AAV) have revealed the following merits.

1. They form a stable preparation.
2. They lack pathogenicity.
3. They have high efficiency of integration.

The disadvantages of AAV as vectors include the possibility of an immune response.

Researchers are also experimenting with introducing a 47th (artificial human) chromosome into target cells. This chromosome could exist autonomously alongside the standard 46, not affecting their workings or causing any mutations. It would be a large vector capable of carrying substantial amounts of genetic code and scientists anticipate that, because of its construction and autonomy, the body's immune system would not attack it. A problem with this potential method is the difficulty in delivering such a large molecule into the nucleus of a target cell.

TARGET TISSUES

Insertion of a normal gene into the diseased tissue depends upon the proliferative state of the tissue, accessibility to gene manipulations and normal site of gene expression.

Liver   Hepatocytes are refractory to retrovirus *in vivo*, they are however more susceptible to transduction by retrovirus *in vitro*. Liver forms a suitable target organ owing to its rich vascularity. Hepatocytes are removed after partial hepatectomy. They are grown in culture, transduced with desired gene through retroviral vector and then returned via hepatic artery or portal nervous system. However there is a risk of portal venous thrombosis, which can lead to portal hypertension. This approach has been used for patients with familial hypercholesterolaemia caused by missense mutation of LDLR gene (low density lipoprotein receptor gene). This leads to reduction of LDL levels for a short term; the long-term benefit is still awaited. Other disorders involving liver in which similar approach can be considered are haemophilia A, α 1-antitrypsin deficiency and phenylketonuria.

Muscle   Direct injection of foreign DNA into the muscle has met with reasonable success in terms of retention and expression of the foreign gene. As an alternative, myoblasts can be injected into the muscle. This results in their incorporation into the recipient muscle fascicles. This approach can be used *in vitro* to insert genes into the myoblasts, which are totally unrelated to the muscle function, e.g. factor VIII and human growth hormone.

CNS   Vector systems are being developed for CNS disorders. They consist of replication-defective neurotropic adenoviruses lacking E1 region. They are then made infective by growing them in the cells engineered to express E1 genes.

Alternatively one can transplant cells which have been genetically modified *in vitro*, into specific regions of the brain like caudate nucleus, in Huntington's disease.

Bone marrow   Treatment of bone marrow disorders poses a problem because of the low frequency of stem cells. Haematopoietic stem cells (HSCs) form an ideal target for gene therapy. However currently many gene therapy efforts are aimed at more differentiated haematopoietic cells such as lymphocytes.

Harvesting and inserting gene directly into the stem cells is possible using monoclonal antibodies that recognize cell surface marker CD34. For ADA deficiency which leads to severe combined immunodeficiency (SCID), lymphocytes are isolated from the blood, grown under conditions promoting growth of T lymphocytes (culture medium containing antibody to T-cell receptor and T-cell growth factor IL-2). ADA gene is introduced in T cells using retroviral vector.

Other target tissues include fibroblasts, endothelial cells, airway epithelial cells, glial cells, etc.

DISEASES AMENABLE TO GENE THERAPY

Diseases in which gene therapy can be conceived include both genetic and nongenetic diseases. More than 2000 genetic diseases of man could be cured by gene therapy. However, it should meet the following prerequisites of candidate disease.

1. The gene that causes the genetic disorders should be responsible for a product that functions at the terminal stage of its formation.

2. The etiology of the disease should be restricted to a single tissue, so that the replacement is easy.

3. The clinical symptoms of the disease should result from the expression of the single pair of recessive genes (e.g. sickle cell anaemia). Under this condition, it would be necessary to replace in the target cell only one of alleles.

4. The genetic disorder should be such that by altering the DNA of a few target cells, amelioration of the disease could be achieved.

5. At the molecular level, the difference in both the normal and abnormal gene sequence should be as small as possible.

In other words, the abnormal base sequence that needs replacement by the normal sequence should be small. For instance, in sickle cell anaemia gene, only one base pair will have to be replaced in order to render it normal.

Some shortcomings in selecting candidate disease for gene therapy include the following difficulties.

- Disorders that involve complex gene regulation

- Diseases in relatively inaccessible tissues like CNS

Gene therapy in such situations (CNS) has to be instituted before irreversible damage occurs.

Although characterization of many hereditary disorders has been done at the gene or protein level, the genetics of most of the diseases are complex and still poorly understood. Hence genetic diseases like cystic fibrosis, sickle cell anaemia, thalassemia are not even tackled so far and target disease for gene therapy are only a few relatively (Table 8.3).

Table 8.3 Target diseases for gene therapy in human beings

Protein	Disease /state in absence of protein
Factor VIII	Haemophilia
Hypoxanthine-guanine phosphoribosyl transferase	Lesch–Nyhan disease
Adenosine deaminase	Combined immunodeficiency disease
Pyrimidine nucleoside phosphorylase	Immunodeficiency
–	Cystic fibrosis
–	Muscular dystrophy
–	Asthma
–	High blood pressure

Since 1984, it has become possible to target the appropriate cells for gene therapy. Calcium-phosphate-mediated uptake is ineffective here because a very small number of cells would be transfected. The use of retrovirus vectors for cloning of desired gene in the recipient's stem cells is being tried.

Medical researchers are painstakingly tracking down defective genes responsible for several common disorders such as asthma, high blood pressure, cystic fibrosis, haemophilia B, muscular dystrophy and cancer. Results are encouraging.

Scientists from the University of Michigan medical centre are trying to insert a gene in the liver cells of patients with high cholesterol to remove the excess of it.

British doctors have narrowed their hunt for culprit gene for asthma, the most common disorder, to just 100 of the many thousands of genes that are present in the human body. And identifying the exact gene may be just a couple of month's work. Such a gene if isolated, cloned and made to produce a protein, if any, then a drug to counter that can be made.

For hypertension also, scientists have identified two chromosome segments that contain genes regulating blood pressure in naturally hypertensive rats. Scientists have identified markers on chromosome 10 that are believed to be linked to the gene regulating blood pressure.

Another disorder, "muscular dystrophy", is the wasting disease of muscles, which kills 1 in 3000 boys early in life. The normal healthy gene makes dystrophin, a prothin that is found on the outer membrane of the muscle cells. This disorder is also on the list of research experiments.

Adenosine Deaminase (ADA) Deficiency

A person born with ADA deficiency lacks an important enzyme in his/her immune system. This means that infections are likely and can even be fatal. ADA deficiency was the first genetic disorder to undergo experimental gene therapy trials in 1990. It was chosen because a single, relatively uncomplicated gene causes it. The results were promising.

The evolution of a cure   From her birth in July 1982, Laura Cay fought infection after infection. Colds rapidly became pneumonia, landing her in hospital. Routine vaccines caused severe abscesses. In February 1983, doctors identified Laura's problem to severe combined immune deficiency (SCID) due to adenosine deaminase (ADA) deficiency. She has inherited the autosomal recessive disorder from two carrier parents.

Lack of ADA blocks a biochemical pathway that breaks down the metabolic toxin deoxyinosine into uric acid, which is excreted. Without ADA, the substance that ADA normally acts upon, builds up and destroys T cells. Without helper T cells to stimulate them, B cells cannot produce antibodies. Hence, both the branches of the immune system fail. The child becomes extremely prone to infections and cancer and despite medical treatment, usually does not live beyond a year in the outside environment.

In 1983 and again in 1984, she received bone marrow transplants from her father, which temporarily bolstered her immunity. Red blood cell transfusions also helped for a time. By the end of 1985, Laura was gravely ill. Then she was chosen to be the first recipient of a new treatment. She received her first injection of PEG-ADA. This is the missing enzyme, ADA, taken from a cow and modified by adding polyethylene glycol (PEG) chains to it. PEG is the major ingredient in antifreeze.

Laura began responding to PEG-ADA almost immediately. Within hours, her ADA level increased 20-fold. After 3 months, toxins were no longer in her blood, but her immunity was still suppressed. After 6 months, though, Laura's immune function neared normal for the first time ever and stayed that way, with weekly doses of PEG-ADA.

On September 14, 1990 researchers at the US National Institutes of Health, Maryland, performed the first gene therapy procedure on four-year-old Ashanthi De Silva. Born with the rare genetic disease, SCID, she was vulnerable to every passing germ. In her treatment, she began receiving her own white blood cells intravenously after modifications. That is, the doctors removed WBC from the child's body, let the cells grow in the laboratory, inserted the missing gene into the cells and then infused the genetically modified blood cells back into the patient's bloodstream. Laboratory tests showed that this gene therapy did not "heal" a sufficient percentage of the girl's cells but strengthened her immune system.

Cancer

Gene therapy for cancer involves the introduction of tumour suppressor gene or inactivating an oncogene or use of immune cells and so on. Potential strategies for gene therapy in the cancer treatment are as follows:

1. *Tumour suppressor gene*   Inserting a wild type tumour suppressor gene, for example, *p53* or the gene involved in Wilm's tumour.

2. *Blocking oncogenes*   Blocking expression of an oncogene, for example, by introducing the gene that encodes antisense K-*ras* message.

3. *Suicide gene*   Insertion of a sensitivity gene or suicide gene into the tumour, e.g. introducing a gene that encodes HSVT *tk* (thymidine kinase gene of herpes simplex virus).

4. *Promoting immunogenicity of tumour*   This is achieved by introducing genes that encode foreign antigens.

5. *Use of genes for cytokines*   Enhancing immune cells to increase anti-tumour activity by inducing genes that encode cytokines.

6. *Protecting stem cells from the toxic effects of chemotherapy*
    Introducing the gene that confers MDR-1 (multiple drug resistance-1).

About half of the current gene therapy trials target cancer. Molecular surgery and manipulation of the immune response are two promising strategies.

Molecular surgery targeting brain tumours   Viruses may treat a type of brain tumour called a glioma, which affects the glial cells that support and interact with neurons. Unlike neurons, glia can divide. Cancerous glias divide very fast, usually causing death within a year. Based on the difference between the dividing capacities of neurons and glia, researchers reasoned that an agent directed against only dividing cells might halt the cancer. One candidate was a "suicide" gene from the herpes simplex virus. In the presence of a certain drug, activation of the gene causes the cell containing it to die.

The herpes gene therapy system has several components. Mouse fibroblasts are infected with a retrovirus vector that contains the herpes gene encoding an enzyme, thymidine kinase. Any cell that produces thymidine kinase is susceptible to the anti-herpes drug ganciclovir. Because a retrovirus can only infect dividing cells, it should not harm nondividing, healthy brain neurons.

The booby-trapped mouse fibroblasts are injected into the brain tumour in very ill patients (a hole is drilled in the skull). There, the implanted cells produce the viruses, which infect neighbouring tumour cells, which then produce thymidine kinase. When the patient takes ganciclovir, the drug is changed into a toxin that kills the cell and nearby cells. A few people have greatly improved with this treatment.

Vaccines   Another genetic approach to battling cancer is to enable tumour cells to produce immune system biochemicals or to mark tumour cells so that the immune system recognizes them more easily. These approaches are called cancer vaccines.

In *ex vivo* trials, a patient's cancer cells are removed, altered in a way that attracts an immune response, then reimplanted into the patient. Specifically, the tumour cells are altered to overproduce cytokines (such as interleukins, interferons, or tumour neurosis factor) or HLA cell surface molecules.

Melanoma, an often fatal skin cancer, is amenable to both *ex vivo* and *in situ* treatment, because it is on the body's surface and therefore accessible. In one group of *ex vivo* experiments, melanoma cells are removed, given genes encoding interleukins, then reimplanted, where the genes are expressed and evoke an immune response. In such experiment, researchers inject liposomes bearing genes encoding HLA proteins directly into tumours. This protein, when displayed on tumour cell surfaces, stimulates the immune system to respond to the tumour as if it was foreign tissue.

Peripheral Vascular Disease

In persons with peripheral vascular disease the arterial segments or the vascular grafts can be resurfaced by endothelial cells or smooth muscle cells in which anticlotting agent genes have been incorporated.

Coronary Artery Disease

In persons having a family history of an early coronary artery disease, gene therapy could be used to introduce LDL receptor. A particularly important area under investigation is the prevention of reocclusion after angioplasty. One approach involves use of mutant forms of tissue type plasminogen activator (TPA) having thrombolytic effect. This can be delivered by adenovirus to the specific tissue and can quickly lyse a clot, the other alternative being modifying endothelial cells so that they can secrete TPA. These genetically engineered endothelial cells can be implanted in the graft to prevent clotting. Preventing smooth muscle proliferation through genes may also be rewarding because it is supposed to be the principal cause of reocclusion.

Haemophilia

Haemophilia is a genetic disorder caused by the deficiency of some blood vasculating factors—factor II, VIII, IX, etc. The deficiency of these factors can lead to profuse bleeding, both internally and externally, either spontaneously or with some injuries.

The current treatment is intravenous injections of factor VIII or IX prepared from human blood donations. Factor VIII is available in three forms: i) cryoprecipitate ii) freeze-dried concentrate, iii) genetically engineered recombinant DNA product. The last two can be stored by ordinary freezing. Complications of viral infections and inhibitors to the factor development are observed. Pig-originated factor VIII is also available but can create heterologous antibodies to pig proteins. Hence it is used only if other alternatives are not found useful. Genetically engineered factor VIII and IX are alternatives to human plasma-derived products. An example is the synthetic product developed by a company in San Francisco based on the work of a British research group. This product is purer, safer and has reached the licensing stage after successful clinical trials.

Hopes of incorporating healthy cloned genes into cells of people suffering from haemophilia defective gene have risen. Researchers at the Oxford University and the University of Washington, Seattle, have carried out the combined work, which creates this hope. With this, patients will make their own supplies of factor IX.

Cystic Fibrosis

In cystic fibrosis, gene therapy aims at delivering CFTR or α-1 antitrypsin gene directly into the epithelial cells lining airways. The α-1 antitrypsin gene can also be directed towards hepatocytes. As lung tissue proliferates very slowly, it is more suitable for adenoviral vectors. However this corrects only the pulmonary complications of cystic fibrosis. The other method used to deliver CFTR gene is liposome-mediated gene transfer. At NIH (National Institute of Health), the first human gene therapy trial on cystic fibrosis was initiated on April 17, 1993.

The Immune System

Current research is focusing on the immune system, which is a collection of special cells and chemicals that fight infection. If the immune system is not functioning in the right way, illness can result. One theory on cancer suggests that the immune system is failing to stop the overgrowth of cells that form a tumour. If the immune system could be "bolstered" with gene therapy, it may also be used as a form of immunization against particular infections, such as HIV/AIDS and malaria.

Another condition where gene therapy is promising is called X-linked severe combined immune deficiency (X-SCID). Children affected by X-SCID have a faulty gene that means they have no working immune system, so their bodies cannot fight infections. Only boys are affected due to the pattern of inheritance of the faulty gene. Until recently, unless they could be given a matched bone marrow transplant, boys with X-SCID faced a lifetime living in a sterile bubble. With gene therapy, bone marrow from the boy is first harvested. The white blood stem cells from the bone marrow are then infected with a virus carrying a working gene, before returning the cells to the boy's body. This treatment has worked to successfully restore the immune system in several children. However there have been a number of concerns related to the treatment's safety.

RISKS OF GENE THERAPY

Some of the risks of gene therapy include the following.

- The immune system may respond to the new gene and cause inflammation.

❋ The healthy gene might be slotted into the wrong spot.

❋ The healthy gene might produce too much missing enzyme or protein, causing other health problems.

❋ Other genes may be accidentally delivered to the cell.

❋ The deactivated virus might target other cells as well as the intended cells.

❋ The deactivated virus may be contagious.

While these risks have not been observed in gene therapy tested on animals, there are still concerns over the safety of the technique.

The factors that keep gene therapy from becoming an effective treatment for genetic disease are listed below:

1. *Short-lived nature of gene therapy*   Before gene therapy can become a permanent cure for any condition, the therapeutic DNA introduced into target cells must remain functional and the cells containing the therapeutic DNA must be long-lived and stable. Problems with integrating therapeutic DNA into the genome and the rapidly dividing nature of many cells prevent gene therapy from achieving any long-term benefits. Patients will have to undergo multiple rounds of gene therapy.

2. *Immune response*   Anytime a foreign object is introduced into human tissues, the immune system is designed to attack the invader. The risk of stimulating the immune system in a way that reduces gene therapy effectiveness is always a potential risk. Furthermore, the immune system's enhanced response to invaders makes it difficult for gene therapy to be repeated in patients.

3. *Problems with viral vectors*   Viruses present a variety of potential problems to the patient—toxicity, immune and inflammatory responses and gene control and targeting tissues. In addition, there is always the fear that

the viral vector, once inside the patient, may recover its ability to cause disease.

4. *Multigene disorders* Conditions or disorders that arise from mutations in a single gene are the best candidates for gene therapy. Unfortunately, the most commonly occurring disorders, such as heart disease, high blood pressure, Alzheimer's disease, arthritis and diabetes, are caused by the combined effects of variations in many genes. Multigene or multifactorial disorders such as these would be especially difficult to treat effectively using gene therapy.

RECENT DEVELOPMENTS IN GENE THERAPY

1. The research team at University of California, Los Angeles, gets genes into the brain using liposomes coated in polyethylene glycol (PEG), a polymer. This method has potential for treating Parkinson's disease.

2. RNA interference or gene silencing may be a new way to treat Huntington's short pieces of double-stranded RNA (short, interfering RNAs or siRNAs) which are used by cells to degrade RNA of a particular sequence. If an siRNA is designed to match the RNA copied from a faulty gene, then the abnormal protein product of that gene will not be produced.

3. New gene therapy approach repairs errors in mRNA derived from defective genes. The technique has potential to treat the blood disorder thalassaemia, cystic fibrosis and some cancers.

4. Sickle cell is successfully treated in mice.

ARGUMENTS IN FAVOUR OF GENE THERAPY

1. The general argument in favour of gene therapy is that it can be used to treat desperately ill patients or to prevent the

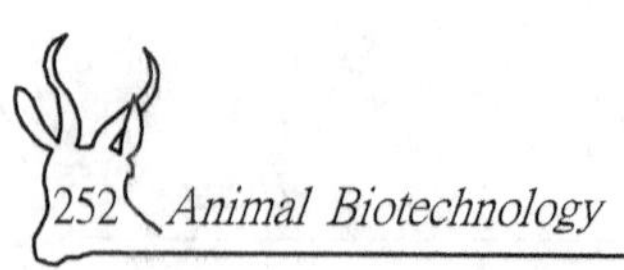

onset of horrible illnesses. Conventional treatment has failed for the candidate diseases for gene therapy and for these patients gene therapy is the only hope for a future.

2. Germ-line gene therapy offers a true cure, not simply palliative or symptomatic treatment. It may be the only effective way of addressing some genetic disease.

3. By preventing the transmission of disease genes, the expense and risk of somatic cell therapy for multiple generations is avoided.

4. Medicine should respond to the reproductive health needs of prospective parents at risk for transmitting serious genetic disease.

5. The scientific community has a right to free inquiry, within the bounds of acceptable human research.

ARGUMENTS AGAINST GENE THERAPY

Gene therapy patients would need to be under surveillance for decades to monitor long-term effects of the therapy on future generations. Arguments specifically against the development of germ-line therapy techniques include the following:

1. Germ-line therapy and experiments would involve too much scientific uncertainty and clinical risks and the long-term effects of such therapy are unknown.

2. Such gene therapy would open the door to attempts at altering human traits not associated with disease, which could exacerbate problems of social discrimination.

3. As germ-line gene therapy involves research on early embryos and affects their offspring, such research essentially creates generations of unconsenting research subjects.

4. Gene therapy is very expensive and will never be cost-effective enough to meet high social priority.

5. Germ-line gene therapy would violate the rights of subsequent generations to inherit a genetic endowment that has not been intentionally modified.

WHEN SHOULD WE USE GENE THERAPY?

The goal of biomedical research has always been to alleviate human suffering, and as we have seen, gene therapy is a proper part of that. The techniques are necessary and they provide new approaches. What remains undecided is whether they are ethically and socially acceptable. Just because a new technology becomes available it may not necessarily be the most rational one to be used. Gene therapy has been described as a preventive therapy, preventing diseases at the fundamental level.

Because of the doubts about success, the immediate prospect of gene therapy is limited to life-threatening diseases that do not have any other cure, and are due to a single gene defect whose effects can be corrected by the insertion of the normal gene into the bone marrow without the need for precise regulation of gene expression. The major disadvantage of gene therapy is the technical difficulty of expression and appropriate regulation of new genes in somatic cells. The main advantage is the compatibility of the reintroduced cells with the patient and the avoidance of immune rejection. Many patients with genetic disease do not have suitable donors for transplantation.

Gene therapy also offers the possibility of introducing novel genetic elements that, for example, may confer drug resistance to normal bone marrow cells to allow their survival during chemotherapeutic treatment for cancer. It is also another medical tool to help individuals overcome an illness, and somatic cell therapy has basically no new ethical problems for existing treatments. The point at which we stop using gene therapy is when it no longer is a treatment for a disease, but becomes enhancement.

FUTURE PROSPECTS

When the age of gene therapy dawned in the 1990s, expectations were high and for good reason. Work in the 1980s had clearly shown that abundant, pure, human biochemicals, useful as drugs, could come from recombinant single cells and from transgenic organisms. It was merely a matter of time, researchers speculated, until genetic altering of somatic tissue would routinely treat a variety of illnesses.

In reality, gene therapy progress has been slow. Boys with DMD who receive myoblasts with healthy dystrophin genes do not walk again, people with cystic fibrosis who sniff viruses bearing the CFTR gene do not enjoy vastly improved lung function. Genetic corrections for babies with SCID given normal ADA genes in umbilical cord stem cells take years, with only partial correction achieved.

There is no single vector, at present with genetic utility for all types of disease and it is unlikely that such a universal vector will emerge in the next few years. Over this period, adenovirus is likely to remain the most suitable vector for *in vivo* therapeutic concepts and requires only short-term expression, especially where a single dose will be sufficient for clinical benefit. Most of gene therapy that has progressed into phase II clinical studies involved *in vivo* gene transfer. It is likely that there will be renewal of interest in *ex vivo* therapy with stem cells.

Preparation of haematopoietic progenitor cells is becoming increasingly routine in medical centres. Considerable advance is being made in identifying and propagating new types of human stem cells, including totipotent stem cells derived from embryonic or foetal tissue.

The approach of developing a non-viral vector with large insert capacity, which exploits the site-specific integration machinery of AAV for safe and sustained transgene expression is very attractive, but is at an early stage of development.

For gene therapy to become a widely used medical therapy, few hurdles need to be overcome. Firstly, we need gene transfer vectors that can be injected directly into the patient. Secondly, the vector should integrate safely into a noncritical site on chromosomes or homologous recombination replacing defective gene.

Finally the introduced gene should have the ability to respond to physiological changes in blood or cellular metabolites. For example, in gene therapy for diabetes, rise or fall in blood glucose levels should be sensed and responded to by an appropriately engineered insulin gene.

To conclude, the most important recent development in gene therapy is the clear demonstration of efficacy in clinical studies, which might lead to a restoration of public and investor confidence in gene therapy.

Summary

- Gene therapy involves the treatment or prevention of disease replacement of a defective abnormal gene into the cells of the patient who is deficient in the normal gene product.

- Researchers distinguish two types of gene therapy namely somatic and germ-line gene therapy.

- *Ex vivo* and *in vivo* are the two strategies to deliver genes.

- Insertion of a normal gene into the diseased tissue depends upon proliferative state of the tissue, accessibility to gene manipulations and normal site of gene expression.

- More than 2000 genetic diseases of man could be cured by gene therapy.

- Molecular surgery and manipulation of the immune response are the two promising strategies in gene therapy against cancer.

REVIEW QUESTIONS

1. Define gene therapy.
2. State the clinical significance of gene therapy.
3. Explain the steps involved in gene therapy.
4. Discuss in detail the gene delivery methods.
5. How do you perform gene therapy in bone marrow?
6. Discuss briefly the diseases amenable to gene therapy.
7. What are the risks of gene therapy?
8. Explain the differences between *ex vivo* and *in vivo* gene therapy with an example each.
9. Would somatic gene therapy or germ-line gene therapy have the potential to affect evolution? Cite a reason.
10. Discuss the factors to be considered in selecting a viral vector for gene therapy.
11. What are the two challenges in providing gene therapy for Duchenne muscular dystrophy?
12. Discuss the recent developments in gene therapy.

9

BIOTECHNOLOGY OF AQUACULTURE

INTRODUCTION

Nearly seventy percent of the earth's cover is made up of oceans. These oceans house millions of marine organisms, which have rich sources of various products and processes, and they hold a valuable bank of genetic information and gene pools. Reports suggest that marine biotechnology can help utilize more than thirty thousand known species of marine organisms to develop new products. This technology can also boost the development of vaccines, diagnostic and analytical agents and genetically modified organisms for aquaculture and the seafood industry. Biotechnology is also providing new and innovative approaches for improving animal stock like fish and prawn. It has even paved the way for developing ecological relationships of marine organisms and improving these resources.

The technique of cultivation of commercially important aquatic animals and plants is called as "Aquaculture". Otherwise, it is the manipulation and improvement of the production of aquatic beings. This practice has a significant bearing on the seafood industry.

The world's seafood demand is all set to shoot up by seventy per cent in the next thirty-five years. And with the seafood harvest from fisheries being on a gradual decline, the industry is being threatened by a major shortage in the coming years. The use of modern biotechnological tools for rearing and enhancing the production of aquatic species cannot only help meet the global demands of seafood, but also enhance aquaculture farming per se. These techniques also improve the health, reproduction, development of environmentally sensitive and sustainable systems. This in turn will lead to substantial commercialization of aquaculture.

GOOD PRACTICES IN AQUACULTURE

In order to have a successful aquaculture we should practise the following:

- The reproductive cycle of the fish species in culture has to be totally controlled.

- Genetic background of the brood stock should be excellent.

- The optimal physiological, environmental, and nutritional conditions for growth and development should be thoroughly understood.

- Supply of quality water should be sufficient and excellent.

- Innovative management techniques should be applied.

- Following this practice, novel or medicinal substances could be produced with genetic engineering of aquatic organisms.

GENETIC ENGINEERING OF AQUACULTURE

The DNA Construct

The process of creation of genetically engineered organisms is called as transgenesis. The initial step in transgenesis is to design and build a DNA construct. The aim is to insert DNA, which enables the GMO to produce a protein in a specific tissue at appropriate times and at a sufficient concentration. Therefore the prototype of a transgene is usually constructed in a plasmid to contain an appropriate promoter–enhancer element and a structural gene sequence.

The Transgene

The most commonly employed genes so far in fish transgenesis are growth hormone (GH) genes. The specific GH genes of humans, rat, cattle, salmon, rainbow trout and tilapia have all been used in attempts to increase growth rates and thus reduce time to market. Another gene of special relevance to aquaculture in polar regions is the antifreeze protein gene, which is found in fish such as ocean, pond or winter flound that live in sub-zero temperatures. If other species could be genetically modified to produce an antifreeze protein, then this could extend the area available for aquaculture.

A critical requirement of transgenesis is that the gene inserted into the GMO be expressed. Hence, any gene sequence in a

Table 9.1 Major aquaculture species which have been genetically engineered, with examples of the construct used

Species	Construct	
	Gene	**Promoter**
Salmon	hGH, bGH, rGH	mMT-1
	csGH	opAFP
	sbGH	wfAFP
	coIGHF	csMT-1
Trout	hGH	SV40
	rGH	mMT-1
	sbGH	Cβ-actin
	INHV-G protein	CMV-tk
	cα-globin	cα-globin
Tilapia	hGH, rGH	mMT-1
	rGH	cβ-actin
	tiGH	RSV
	tiGH	CMV

construct is accompanied by a promoter, the DNA sequence that directs transcription and translation of the gene into protein. The promoter sequence regulates when, where and how much of expression of the gene will occur. Early experimental work on transgenesis used promoters from avian (Rous sarcoma virus, RSV) or primate viruses (simian virus, SV40; cytomegalovirus, CMV) but recently, efforts have been made to design an "all fish" promoter from fish genomes rather than the genomes of other groups. Examples of the genes and promoters used for transgenesis of aquatic organisms are given in Table 9.1.

Once the DNA construct has been formed by splicing together the gene and the promoter, it must be amplified to produce the billions or trillions of copies needed for each transgenesis attempt. The method used is to insert the construct into a plasmid, which is replicated in *E. coli* by standard cloning methodology. Normally, the plasmid DNA is removed before transgenesis because this prevents the insertion of prokaryotic plasmid DNA into the target organism. Once the DNA construct has been created, amplified and purified out of the plasmid, it can be delivered into the eggs of the target organism.

Reporter Function

Special reporter genes have been used to identify and measure the strength of a promoter–enhancer element. Generally, the structural gene of the CAT, β-galactosidase, or luciferase gene is fused to a promoter–enhancer element. These are simply genes that are very obvious if they have successfully been incorporated into the host genome. They code for proteins which cause the embryo to turn blue or fluoresce green.

Loss of Function

The loss-of-function transgene is constructed for interfering with the expression of host genes. These genes might encode an antisense RNA to interfere with the post-transcriptional process or translation of endogenous mRNAs. Alternatively, these genes might encode a catalytic RNA (a ribozyme). It can cleave specific mRNAs and thereby cancel the production of the normal gene product. These genes could be potentially employed to produce disease-resistant transgenic brood stock for aquaculture or transgenic model fish defective in a particular gene product for basic research.

TRANSGENE DELIVERY

Microinjection

It is the most common transgene delivery method used for fish. Microinjection in fish is accomplished by one of the following two procedures, depending upon the ease of removal or softening of the egg chorion.

Intranuclear microinjection is a direct physical approach that overcomes many of the biological and other obstacles. It simply involves using remote-control levers to operate a micropipette tipped with a very fine glass needle. Individual fertilized eggs are held steady by suction against a blunt-ended tube and a mircropipette needle is inserted into the egg.

A small volume of buffer (usually 1–2 nl) containing a high copy number (10^6–10^7 copies) of the transgene construct is then injected into the egg, the needle is removed and the egg incubated in the normal way. Many fish eggs have an opaque chorion, which limits the ability to visualize the target for the injection of the DNA. Moreover, most fish eggs have a chorion that hardens upon fertilization and exposure to water. Ability to locate the injected DNA as close as possible to the region where the chromosomes are located is an important criterion, because otherwise the integration of the novel DNA into the chromosome of the GMO is far less likely.

The time for spermatozoa activation to first cleavage in fish is quite short and early embryonic cell division is rapid, therefore, there is only a short window of opportunity for microinjection into the fertilized egg or early embryo. Ideally the transgene should be injected at the single cell stage to ensure integration in all the cells of the GMO. However, even when this is done, integration often takes place after some cleavage divisions have occurred. This produced mosaic embryos where the transgene is present in some cells but not in others. Mosaicism is a very common outcome of transgene delivery by the microinjection method.

Advantages

- The amount of DNA delivered per cell is not limited by the technique and can be optimized. This improves the chance for integrative transformation.

- The delivery is precise, again increasing the chance of integrative transformation.

- The small structures can be injected containing only a few cells and with high regeneration potential.

- Since it is a direct physical approach, it is host-range-independent.

Applications of microinjection technique Different applications of microinjection technique in plant science have been reported. Some experiments deal with the introduction of fluorescent dyes into differentiated cells to study intracellular transport. Efforts have been made to microinject microorganisms into suspension cells. Attempts to transfer chromosomes or even cell organelles have been made. A common feature of these experiments is low numbers of manipulated cells.

DNA is microinjected in fish. The injected zygotes are cultured *in vitro* to blastocysts and then placed inside a foster mother. This technique has been used to introduce rat growth hormone gene. Microinjection of foreign DNA into newly fertilized eggs was first developed for the production of transgenic mice in the early 1980s. Since 1985, this technique has been adopted for introducing transgenes into Atlantic salmon, common carp, catfish, goldfish, loach, and zebra fish. As the male pronucleus is difficult to locate in fish zygote, the DNA is microinjected into cytoplasm of fertilized ovule within first few hours after fertilization. As the fishes undergo external fertilization, transfer of embryos into foster mothers is not needed. Transformed embryos are instead hatched in trays. Various gene transferred into fish are human or rat growth hormone, chicken delta-crystalline protein, *E. coli* β-galactosidase gene, *E. coli* neomycin-resistance gene and *E. coli* hygromycin-resistance gene. Cytoplasmic microinjection is a tedious and time-consuming procedure and is unsuitable for mass transfer.

Steps in gene transfer by microinjection   The transfer of foreign DNA into fish by direct microinjection is conducted as follows.

1. Eggs and sperms are collected in separate, dry containers.
2. Fertilization is initiated by adding water and sperms to the eggs, with stirring to enhance fertilization.
3. Fertilized eggs are microinjected within the first few hours after sterilization.
4. The injection apparatus consists of a dissecting stereomicroscope and two micromanipulators, one with a glass microneedle for delivering transgenes and the other with a micropipette for holding fish embryos in place.
5. In general, about 10^6–10^8 molecules of a linearized transgene (with or without plasmid DNA) in about 20 nL is injected into egg cytoplasm.
6. Following injection, the embryos are incubated in water until hatching.
7. Natural spawning in zebra fish or medaka can be induced by adjusting the photoperiod and water temperature. The

micropyle on the fertilized eggs will remain visible for at least 2 hours if precisely staged at 4°C immediately after fertilization. The DNA solution can be easily delivered into embryos by injection through this opening.

8. Depending on the fish species, the survival rate of injected fish embryos ranges from 35 to 80% while the rate of DNA integration ranges from 10 to 70% in the survivors.

9. The tough chorions of the fertilized eggs in some fish species (e.g. rainbow trout and Atlantic salmon) can frequently make insertion of glass needles difficult. This difficulty can be overcome by any one of the following methods.

   i. By inserting the injection needles through the micropyle.

   ii. By making an opening on the egg chorions by microsurgery.

   iii. By removing the chorion by mechanical or enzymatic means.

   iv. By reducing chorion hardening by initiating fertilization in a solution containing 1mM glutathione.

   v. By injecting the unfertilized eggs directly.

Disadvantages

- Injection can cause damage that affects embryonic survival and can result in quite high mortalities.

- Only one cell is targeted per injection.

- The handling requires specialized skill and instrumentation.

- Has low transformation rate.

Electroporation

Electroporation is a method which was first developed for work with cells in tissue culture and it involves subjecting the cells to a short burst of electrical impulse. When the cells are treated like this, their cell walls become temporarily much more porous and larger molecules can pass through. Cells to be treated are suspended at high concentration in a solution of high copy number DNA construct and held in a 1–2-ml container with flat electrodes on each side.

The transient current is passed and DNA construct molecules pass through the cell membrane. The electrical impulse and rates of decay are varied in order to produce optimum electroporation results. A major advantage of electroporation over microinjection is that there is no need to handle and manipulate eggs individually.

Electroporation has been tried on fish eggs, but the difficulty is that the eggs are quite large and have a chorion. The stresses on embryos seem greater with this electrical method than the cell puncture used in the injection method, since embryo survival following electroporation is poor. While there are problems with fish eggs, the method has great potential for transgenesis in molluscs and successful experimental trials have been conducted with abalones and oysters.

Sperm-mediated Transfer

The DNA binds readily to the outer coat of spermatozoa; hence this looks like a perfect method of getting novel DNA into the egg. But there are a number of pitfalls with this method.

1. In most of the cases it is only a single spermatozoan that enters the egg and reaches the pronucleus. Therefore, the number of copies of the DNA construct entering the egg or reaching the target is quite small when compared with the number of copies employed in electroporation/ microinjection.

2. There are some mechanisms in place in the host to prevent the ingress of foreign DNA by this process. Spermatozoa will often come into contact with extraneous DNA (from

cellular debris or bacteria) before fertilization so it is not surprising that mechanisms have evolved to prevent such DNA from integrating into host genome.

Attempts have been made to electroporate spermatozoa with DNA construct before fertilization; although this approach has not been extensively explored, early work suggests that the difficulties of the methodology outweigh any benefits.

Biolistics

In plants where there is a tough cell wall, experimenters have resorted to brute force to get foreign DNA into the cells. The method is called biological ballistics or biolistics, and involves coating microscopic particles, usually of gold, with DNA construct and explosively firing these particles into the cells through the cell membrane. This method has been tried with fish, sea urchins and oysters and results suggest that viable transgenic embryos can be produced. However, there is currently no evidence that biolistics offers a more effective or more efficient method of producing transgenic fish than microinjection.

Viral Vectors

Burns *et al.,* (1993) developed an alternative gene transfer method using a viral vector that has been rendered defective. The transgene is spliced into the defective virus that is nevertheless able to infect host virus and induce the replication of the transgene within those cells. A defective pantropic retroviral vector was used and this vector contains long terminal repeat (LTR) sequence of Moloney murine leukemia virus (Mo MLV) and transgenes packaged in a viral envelope with the G protein of vesicular stomatitis virus (VSV). Because the entry of VSV into cells is mediated by interaction of VSV G protein with a phospholipid component of the cell, this pseudotyped retroviral vector has a very broad host range and is able to transfer transgenes into many different cell types.

Using pantropic pseudotyped defective retrovirus as a gene transfer vector, transgene containing neomycin-resistant gene and β-galactosidase has been introduced into zebra fish and Medaka. However, there are obvious dangers involved in engineering any virus to be too good at getting into a wide range of species.

Lipofection

Synthetic lipid vesicles containing encapsulated DNA can be taken up directly by animal cells and this lipofection method has been tried with fish eggs. As with sperm-mediated transfer, initial expression of foreign DNA was followed by a rapid loss of expression suggesting that the novel DNA may have been destroyed by the host embryos.

TRANSGENE INTEGRATION

All that the delivery methods do is to get the construct into the egg. It must then become incorporated into one or more of the chromosomes of he host. If it just remains floating around in the cytoplasm of cells, or as a fragment of DNA in the nucleus not attached to a chromosome, it will probably not be replicated and will be lost during cell division. A very small proportion of inserted construct becomes integrated spontaneously—which is what is currently relied upon in commercial transgenesis—but methods are being developed to increase transgene integration success.

A major difficulty with the integration of the transgene is targeting its position on a chromosome. Chromosomes are not homogeneous throughout their length but consists of heterochromatic and euchromatic regions and genes in heterochromatic regions are seldom switched on. If the transgene gets incorporated in such regions then it is unlikely to be expressed due to this position effect.

Three main experimental approaches are being used to try to get transgene into the nucleus and to locate them more precisely on the chromosomes. They are as follows:

1. Many viruses have proteins containing special amino acid sequences that assist the viruses to get into the nucleus where they can replicate. These are called nuclear-localizing sequences (NLS) and such sequences can be added to the medium containing the transgene construct to assist with entry into the nucleus.

2. A pseudotyped retrovirus (murine leukemia retrovirus with one of its genes replaced by one from the vesicular stomatitis virus) that encodes a special integrase protein can be used. Once inside the cell the integrase protein is produced and this can assist with integration of the transgene DNA into the chromosome. Since these viruses can infect any cell, the risk working with it is high.

3. The enzymes responsible for the insertion of the highly mobile DNA elements known as jumping genes or transposons may be harnessed. These transposes act in a fairly straightforward manner, cutting a DNA segment from one place in a chromosome and pasting it in at another on the same or a different chromosome. Natural transposes isolated from a number of organisms failed to work well, so a synthetic transpose has been manufactured and the result—the Sleeping Beauty (SB) transposon—effectively enhances integration (up to about 20-fold in zebra fish embryo).

None of these approaches is yet used commercially, but work in this area is increasing knowledge of the effects of positioning on gene function.

DETECTING INTEGRATION AND EXPRESSION OF TRANSGENE

The production of fish or shellfish GMOs by any method requires confirmation. We need to establish that the DNA construct is present in the GMO, that it has become integrated into the host genome and that the expected gene product is being expressed at the right level and in the right tissues for the particular commercial production for which the GMO was designed.

Dot blots and PCR help us to identify those individuals in which the transgene is present. Restriction digestion and Southern blotting provide evidence of transgene incorporation.

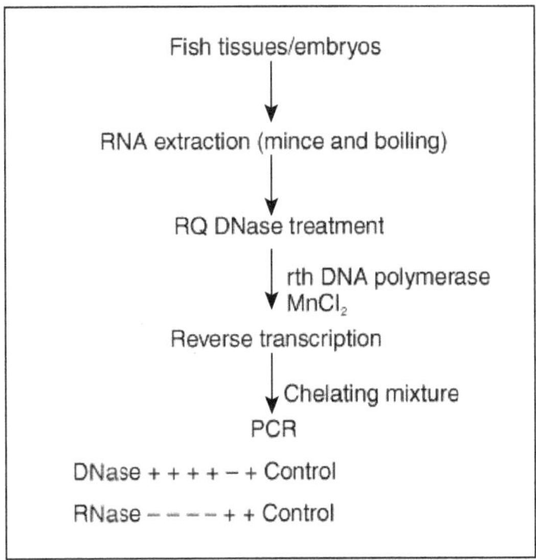

Figure 9.1 Strategy of detecting rtGH transgene expression by reverse transcription (RT)–PCR Assay

Expression of the transgene can be identified by evaluating the level of production of the protein itself. The following methods are used to know about expression of transgene:

1. RNA northern/dot-blot hybridization
2. RNase protection assay
3. Reverse transcription-PCR (Figure 9.1)
4. Immunoblotting assay
5. Other biochemical assays

FISH ANTIFREEZE PROTEIN GENE PROMOTER

The studies on biochemistry and molecular biology of fish antifreeze proteins are important aspects of fish biotechnology. It includes studies of their structure and function, protein engineering, gene structure and expression, and gene transfer for the production of freeze-resistant transgenic fish.

The sea water temperature in the polar regions during winter can be as low as $-1.9°C$. To avoid freezing, several marine teleosts inhabiting these regions produce antifreeze proteins (AFPs) in their blood, with a concentration of 10–25 mg/ml. These proteins inhibit ice crystal formation and effectively lower their freezing temperature. The trout and salmon metallothionein, carp β-actin, salmon histone, protomine and AFP genes from several fish species are well-characterized promoters from fish genes.

FUTURE PROSPECTS

Superior transgenic fish strains can be produced for aquaculture. These traits may include elevated growth enhancement, improved food conversion efficiency, resistance to some known diseases, tolerance to low oxygen concentrations, and tolerance to subzero temperature.

Summary

☑ The technique of cultivation of commercially important aquatic animals and plants is called as "aquaculture".

☑ The prototype of a transgene is usually constructed in a plasmid to contain an appropriate promoter–enhancer element and structural gene sequence.

☑ A critical requirement of transgenesis is that the gene inserted into the GMO be expressed.

☑ Reporter genes used to identify and measure the strength of a promoter–enhancer element.

☑ Expression of the transgene can be identified by evaluating the level of production of the protein itself.

REVIEW QUESTIONS

1. Define aquaculture.

2. Define transgene.

3. What is transgenesis?

4. Explain the function of reporter genes.

5. State the gene delivery methods.

6. Explain the applications of microinjection technique.

7. Brief on electroporation.

8. How is the transgene integrated?

9. Explain the role of fish antifreeze gene promoter in the production of growth hormone.

10. Explain the screening techniques used to detect transgene.

GENETIC ENGINEERING AS APPLIED TO SPECIFIC AREAS

Following the discovery of cloning strategies and gene transfer techniques (discussed in chapters 3 and 4), thousands of DNA fragments of prokaryotes and eukaryotes have been cloned during the last two decades. The cloned genes are utilized commercially in medicine, agriculture and industry, and the production of valuable products. In this connection, some of the cloned genes and the application of genetic engineering in the production of regulatory proteins and blood products are described here.

Genetic recombination techniques are in successful use, because of their wide applications to human life. Genetic engineering is also employed for the production of some valuable regulatory proteins/factors such as thyrotropin hormone releasing factor, somatostatin, etc.

REGULATORY PROTEINS

Thyrotropin Hormone Releasing Factor (THRF)

Guillemin *et al.* (1982) stated that the transfer of genes from the pancreas of human beings to microbes cause the production of THRF from the genetically modified microbes. THRF is used to regulate the thyroid gland of human beings.

Growth Hormone Releasing Factor (GRF)

Genetically engineered microorganisms also have the capacity to produce GRF in culture. The GRF thus produced is used in the treatment of pituitary dwarfness.

Somatostatin

Scientists have stated that the chemical synthesis of somatostatin is somewhat difficult and lengthier than the biosynthesis of somatostatin in bacteria. It is a 14-residue polypeptide hormone synthesized in the hypothalamus. It is the first polypeptide, which was expressed in *Escherichia coli* cells as a part of the fusion peptide, which inhibited the secretion of growth hormone, glucagon and insulin (Itakura *et al.*, 1977). It does not contain any internal methionine. Eight single-stranded DNA segments were synthesized chemically which were annealed in an overlapping manner to form a double-stranded DNA (synthetic gene). It has single-stranded projections at each end as the same are formed by *Eco* RI. The synthesized gene contained 51 base pairs, which were terminated by two nonsense codons and preceded by a methionine codon as below:

ATG — (42 base pairs encoding somatostatin) — TGATAG

Two plasmids pSOM I and pSOM II-3, were constructed. The synthetic gene was introduced into *E. coli* β-galactosidase gene at different sites. The chemically synthesized gene was inserted downstream from the *lac* promoter in such a way that the gene fusion should have specified a polypeptide in which the first 7 amino acids of β-galactosidase were fused to somatostatin. But the somatostatin was not detected in transformed bacteria, possibly because it was degraded in *E. coli* (Glover, 1994). An alternative plasmid, pSOM II-3, was constructed, that had both promoter/operator and *lacZ* regions. The *lacZ* gene encodes peptide of β-galactosidase. If the resulting frame of *lacZ* is maintained after inserting a DNA, a fusion peptide is produced. The plasmid was cleaved in the *lacZ* region by restriction enzyme (Figure 10.1). The synthesized gene was inserted into the plasmid at *Eco* RI site near C-terminus of β-galactosidase. The plasmid in the transformed bacterium directs the synthesis of a fused protein consisting of NH_2-termined segment of β-galactosidase fragment coupled by methionine to somatostatin. It was stabilized from proteolytic degradation of β-galactosidase moiety. This fused protein was purified and treated with cyanogen bromide (CnBr), which cleaves only protein at the carboxyl terminal of the methionine. Thus the methionine linker remains attached to β-galactosidase fragment, and somatostatin was released. Now it has become possible to inhibit the degradation of foreign protein in *E. coli* by introduction of protein inhibition (*PIN*) gene of T4 phage.

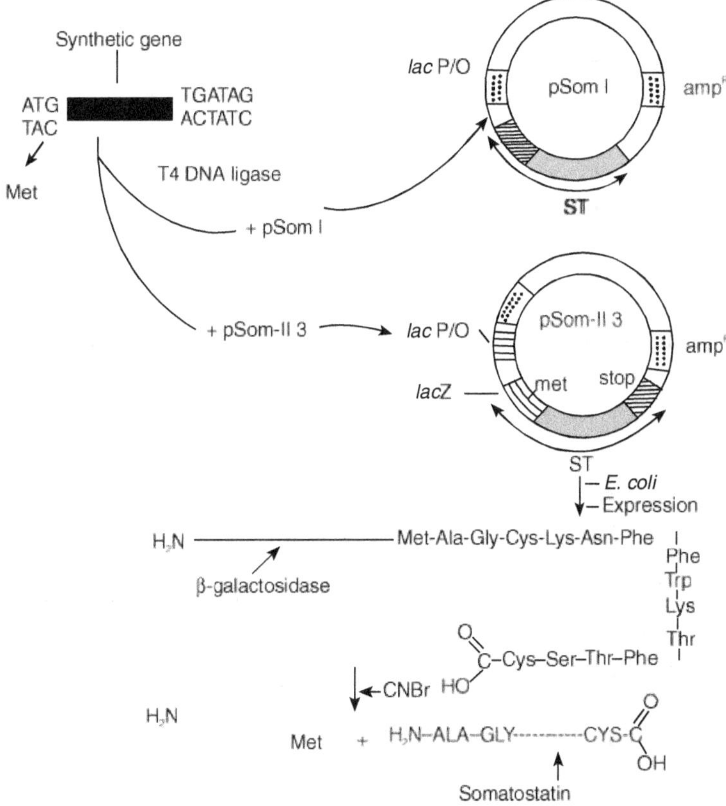

Figure 10.1 Synthesis of somatostatin, Lac P/O—lactose promoter/operator, ST—somatostatin; Amp—ampicillin resistance

Somatostatin finds application in regulating the release of several hormones from the pituitary gland.

Somatotropin

It is a human growth hormone secreted by the anterior lobe of pituitary glands which consists of 191 amino acid units. Its secretion is regulated by two other hormones (somatostatin and growth hormone releasing hormone) produced by hypothalamus. Deficiency of somatotropin in about 3% cases is due to heredity. The extraction of somatotropin pharmaceutically from the pituitary glands could not meet the annual demand of this hormone. Biosynthesis of somatotropin was achieved through gene cloning procedures.

Somatomedin

Hall (1972) explained the genetic engineering of bacterial cells for producing somatomedin A, which stimulates the fixation of sulphur in cartilages. Somatomedin B, a variant, can be produced by modifying bacterial cells using recombinant DNA technology. It regulates the incorporation of protein in collagen.

Human Granulocyte Macrophage Colony Stimulating Factor (GM-CSF)

GM-CSF is a protein that stimulates the growth of cells important for the body's response to infection and cancer. It promotes the growth of granulocytes and macrophages.

Beta Endorphin

Beta endorphin is a 30-amino acid long neuropeptide with opiate activity. It is another growth hormone which was expressed in genetically engineered *E. coli* cells. Shine *et al.* (1980) isolated the genes encoding endorphin from mRNA and inserted them adjacent to β-galactosidase gene on a plasmid. The mRNA contained large precursor of protein that consisted of, besides β-endorphin, the hormones α-melanotropin, corticotropin, β-lipoprotein and β-melanotropin. The β-endorphin is cleaved from the C-terminus of the precursor peptide. In this way, the transformed bacteria produced an insoluble fusion protein between β-galactosidase and β-endorphin. β-endorphin can be removed from the hybrid protein by trypsin which cleaves only at arginine residue. Before doing so, internal lysines are protected from trypsinization by citraconylation as below:

$$NH_3\text{–}\beta\text{-gal–}\beta\text{-melanotropin–}\beta\text{-endorphin–COOH}$$

$$\downarrow$$

Citraconylation

$$\downarrow$$

Trypsin

$$\downarrow$$

$$\beta\text{-gal} + \beta\text{-endorphin}$$

The β-endorphin produced by the transformed *E. coli* cells are used in the treatment of severe pains and a few diseases.

Epidermal Growth Factor (EGF)

Stanley Cohen discovered this growth factor during animal cell culture experiments. This consists of 53-amino acid polypeptide chains, and the submaxillary gland produces this growth factor. The EGF is known to stimulate the epidermal cell proliferation and nerve cell regeneration. EGF has major action on stem cells, leucocytes, macrophages and tissue cells. More number of animal cells show response to the EGF in the medium. This is also required for embryo development.

Lipocortin

Lipocortin is a protein produced by our body cells to fight inflammation. Biogen Company has been successful in making

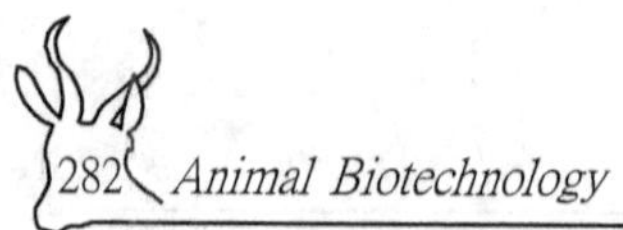

lipocortin by rDNA technology but it is on clinical trials. It can be used against arthritis and asthma without any undesirable side effects. Corticosteroids work by stimulating the body to produce lipocortin, which is less toxic and more specific.

Thymic Hormones

These are also produced in large amounts from genetically engineered microbes.

Erythropoietin

It is a glycoprotein which stimulates the division and differentiation of myeloid stem cells to red blood cells. This was produced by kidney and foetal liver cells. The molecular weight is approximately 30,400 daltons and it is a 166-amino acid polypeptide chain. The protein contains two disulphide bonds and four sites of glycosylation. The disulphide bonds and glycosylation are very much essential for the biological activity. It acts as a mitogenic substance and enhances survival of cells in culture condition and prevents programmed cell death.

Tropomyosin

Bailey discovered tropomyosin (TM) in 1946. Its solubility and structural properties show a similarity to those of myosin, hence its name. Tropomyosin is a rod-shaped molecule (about 400Å long and 20Å wide) with a molecular weight of 65,000–70,000. It consists of two α-helical chains arranged in a parallel coiled-coil configuration. Tropomyosin molecules are bonded head to tail (Figure 10.2). In skeletal muscle, TM accounts for about 3% of the total muscle protein.

On SDS-PAGE, TM shows two bands, corresponding to the α (fast) and β (slow) chains. The molecular weight of the chains does not differ significantly; it is in the range of 33,000–35,000. Both chains have 284 amino acid residues, but they differ in

39 residues. The amino acid sequence shows a repeating pattern of non-polar and polar residues, totalling 7 residues. These are called tropomyosin isoforms.

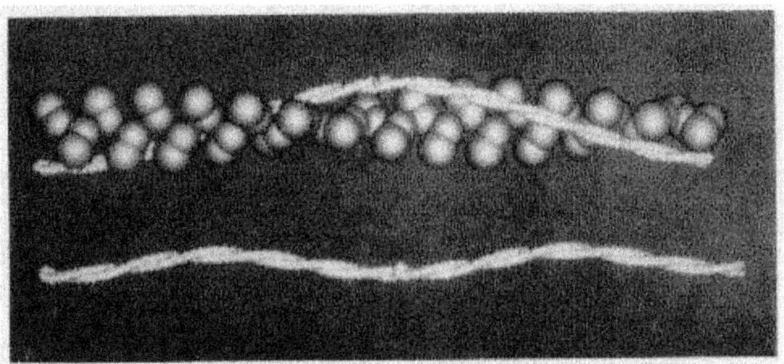

Figure 10.2 Head to tail joint in TM (lower part) and the interaction of TM with each of seven actin monomers (upper part).

The molecular length of a chain calculated from sequence, $284 \times 1.49\text{Å} = 423\text{Å}$ (Number of residues multiplied by the effective residue translation in a coiled coil), is larger than the length of 400Å calculated from X-ray studies. This difference can be accounted for by assuming an overlap of 8–9 residues between the ends of TM molecules.

The ratio of the concentration of α and β subunit varies with the muscle type. Slow skeletal muscle and foetal muscle contain a larger portion of the β subunit than fast skeletal muscle, whereas rabbit and avian cardiac muscle contains only the α subunit.

The amino acid sequences of the α and β chains of rabbit skeletal muscle TM showed a repeating pattern of non-polar and polar amino acids. This is characteristic for a coiled-coil structure in which two α-helices interact along their length by the "knobs into holes" packing of non-polar residues to form a hydrophobic core (Figure 10.3). Polar and ionic side chains are directed towards the exterior of the 2-stranded rope-like arrangement, where they can interact with solvent and/or other protein molecules.

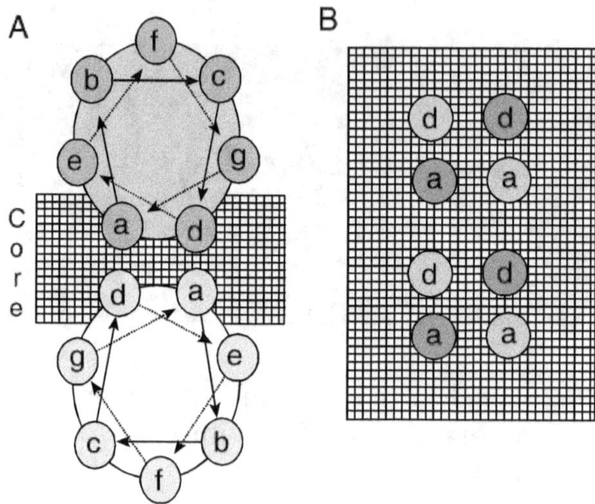

Figure 10.3 Interaction between the two α-helices of TM coiled-coil. Each α-helix is shown with seven residues (a–g) in two turns. (A) End of view looking from N-terminus. The interface between the α-helices derives primarily from hydrophobic residues in core positions a and d. (B) The core interface viewed parallel to the coiled-coil axis shows how residues from one chain occupy the spaces between the corresponding residues from the second chain to give "knobs in holes" packing.

Tropomyosin has high affinity to actin, as evidenced by the difficulty in removing TM in the course of actin purification. Each TM molecule binds 7 actin monomers in F-actin. When bound to actin, each TM is believed to be supercoiled with a radius of about 40Å and the molecular length is reduced to about 385Å. Since both actin and TM are abundant in charged residues, it is reasonable to assume that electrostatic interaction plays a major role in binding of TM to actin.

Electron microscopic image reconstruction and X-ray diffraction studies suggest that during muscle activation TM moves from its lateral position on the actin filament by a distance of 10–15Å towards the centre of the groove in the actin double helix. It is postulated that in the resting muscle, TM occupies the site of actin necessary for combination with the myosin head. The movement of TM, at the beginning of contraction, liberates the myosin-binding site, thus actomyosin can be formed and the muscle can contract. The tropomyosin-binding component of troponin (TN-T) binds to TM (Figure 10.4).

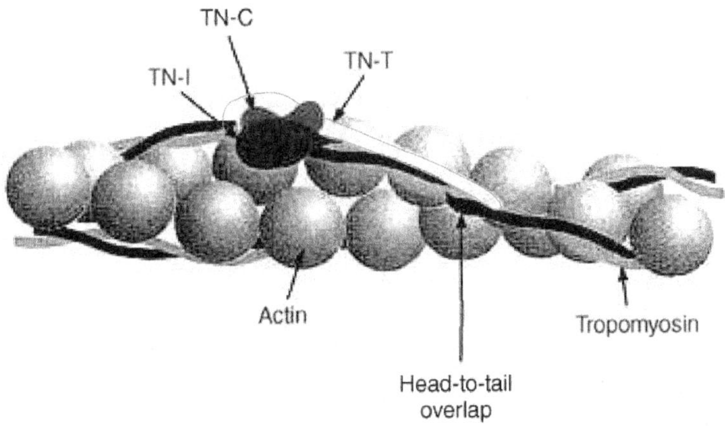

Figure 10.4 A model of the molecular arrangement of TN subunits, TM, and actin in skeletal muscle thin filament (From Gordon *et al.*, 2000). Note the adjacent TM molecules overlap head to tail with the N-terminus of TN-T lying along the overlap region. The C-terminus of TN-T interacts with TN-C and TN-I, and TN-I also interacts with actin.

Troponin C is the Ca^{2+} receptor in the thin filament. It has been crystallized and its three-dimensional structure determined.

Ebashi discovered troponin (TN) in 1963. Subsequently, Greaser and Gergely (1971) showed that troponin consists of 3 components. These protein subunits differ in their molecular weights (based on early electrophoretic studies), and each possesses a specific function (Figure 10.5).

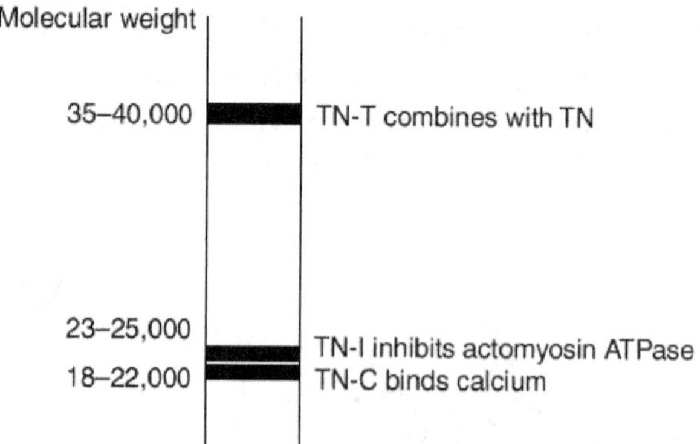

Figure 10.5　SDS-PAGE of troponin

Troponin T is an asymmetric molecule that interacts with TM. Evidence obtained from co-crystals of TN-T with TM suggests that TN-T is an asymmetric molecule that contacts the C-terminal region of TM along much of its length. Skeletal TN-T extends 160Å along the TM molecules whereas the cardiac isoform with a slightly longer polypeptide chain occupies 180Å.

The TM–TN-T interaction serves to fix the position of the entire TN complex within the thin filament, so that subtle changes in the conformation of these proteins may regulate contraction. In this respect, it is important that TN-T also binds TN-C, thus the Ca^{2+}-induced conformational changes in TN-C are transmitted through TN-T to TM. A highly conserved protein domain was described in the amino acid sequence of TN-T, which is characterized by a heptad repeat motif with a potential for α-helical coiled coil formation. A similar, potentially coiled-coil-forming domain is also conserved in all known TN-I sequences, suggesting that these protein domains play a role in TN-T–TN-I interaction. This is another example of how specific interactions of the TN subunits build the entire TN molecule.

Troponin T, similar to TN-C and TN-I, has several isoforms in the molecular mass range of 30–35 kDa. Variations in TN isoforms may modulate muscle performance. TN-C belongs to the Ca^{2+}-binding protein family. Its Ca^{2+}-binding domains comprise

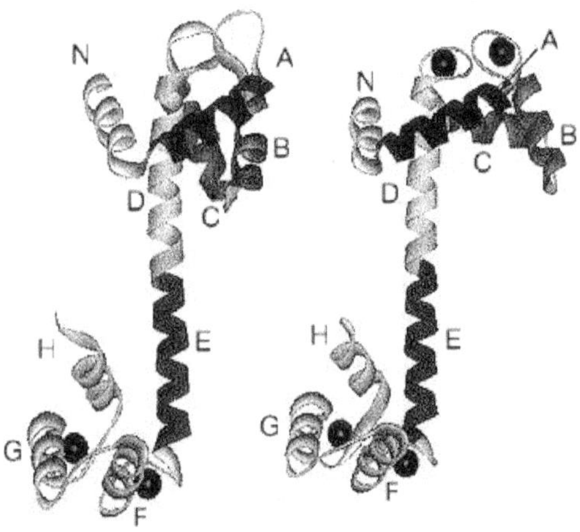

Figure 10.6 Ribbon diagram of skeletal muscle TN-C with one molecule of Ca^{2+} bound (left) and two molecules of Ca^{2+} bound (right). Solid circles represent Ca^{2+}, the letters A–H correspond to the helices, N is the amino terminal end and C is the carboxyl terminal end.

a short α-helix, a Ca^{2+}-binding loop and a second short α-helix. The domain folds around the Ca^{2+} ion in a manner of a clenched right hand, with the extended forefinger and thumb representing the helices and the middle finger forming the loop structure. This domain is termed an E-F hand. Two E-F hand domains form a stable pair in many Ca^{2+}-binding proteins. In TN-C, the two E-F hand domains are connected by a long α-helix resulting in four

Ca^{2+}-binding sites per molecule. Two of the four sites have low affinity to Ca^{2+} and are located in the amino-terminal domain. Only the low affinity Ca^{2+}-binding sites of TN-C are involved in the regulation of muscle contraction (Figure 10.6).

TN-C forms specific complexes with both TN-I and TN-T. The crystal structure of TN-C complexed with the N-terminal fragment of TN-I has been determined. In the complex, TN-C has a compact globular shape in contrast to the elongated dumbbell shaped molecules of uncomplexed TN-C. The TN-I fragment binds to the N-terminal domain of TN-C in its Ca^{2+}-bound conformation. Apparently, the complex formation between the two proteins requires a specific sterical configuration, and, therefore, the complex has functional significance.

Troponin-I inhibits the Mg^{2+}-activated ATPase of actomyosin (identical with the actin and Mg^{2+}-activated myosin ATPase). TN-I is a basic protein that readily complexes with the acidic TN-C; importantly Ca^{2+} strengthens the complex formation. Thus, upon muscle stimulation, TN-C binds Ca^{2+} and then complexes with TN-I. This provides a simple mechanism to relieve the inhibition of the actomyosin Mg^{2+}-ATPase by TN-I.

BLOOD PRODUCTS

Blood and blood products constitute a major group of traditional biologics. The main components of blood are the red and white blood cells, along with platelets and the plasma in which these cellular elements are suspended. Whole blood remains in routine therapeutic use, as do red blood cells and platelet concentrates. A variety of therapeutically important blood proteins also continue to be purified from plasma. These include various clotting factors and immunoglobulins. Such blood products include the following

- Whole blood
- Red blood cells

- ❋ Platelet concentrate
- ❋ Plasma and plasma protein fraction
- ❋ Albumin
- ❋ Clotting factors (particularly factors VII, VIII, IX and XIII)
- ❋ Haemoglobin

Whole Blood

Whole blood is blood which has been aseptically withdrawn from humans. A suitable anticoagulant is added (often heparin or a citrate dextrose-based substance), although no preservative is present. The blood is usually stored at temperatures ranging from 1 to 8°C, and has a short shelf-life (48 hours after collection if heparin is used as the anticoagulant; or up to 35 days if citrate phosphate dextrose with adenine is employed). In addition to being screened for likely pathogens, the ABO blood group and the Rh group are also determined.

The blood is generally warmed to 37°C just prior to transfusion. Whole blood is often used to replace blood lost due to injury or surgery. The number of units (1 unit equals approximately 510 ml) administered depends upon the health and age of the recipient, along with the therapeutic indication. Administration of whole blood may also be undertaken to supply a recipient with a particular blood constituent (e.g. a clotting factor, immunoglobulin, platelets or red blood cells). However, this practice is minimized in favour of direct administration of the specific blood constituent needed. Whole blood is occasionally used for treating acute loss or for transfusion.

Red Blood Cells

Concentrated red blood cells are prepared from the whole blood of a single donor, from which the anticoagulant and some of the plasma has been removed.

Packed (plasma-depleted) red cells having packed cell volume greater than 70% is the treatment of choice for most transfusions. A diuretic is often given simultaneously and the infusion should be sufficiently slow to avoid circulation overload. Iron chelation therapy should be considered with the patients on a regular transfusion programme to avoid iron overload.

Red blood cell preparations are usually stored at 2–8°C. If stored unfrozen, its useful shelf life must not exceed the shelf life of the whole blood from which it was derived. If stored frozen, the shelf life is extended to three years (although the product must subsequently be used within 24 hours after thawing).

Platelet Concentrates

Platelets play a central role in the blood clotting process. They are often administered prophylactically or therapeutically in order to prevent/minimize blood loss due to haemorrhage in persons suffering from thrombocytopenia (low blood platelet levels).

Platelet concentrate consists of platelets obtained from whole blood. The platelets obtained from a unit of blood are usually resuspended in 20–50 ml of the original plasma and contain not less than 5.5×10^{10} platelets per unit.

For prophylaxis, the platelet count should be kept above $5–10 \times 10^9/l$ unless, there are additional risk factors such as sepsis, drug use or coagulation disorders for which the threshold should be higher. For minor invasive procedures, for example, liver biopsy (or) lumbar puncture, the platelet count should be raised to above $50 \times 10^9/l$. Therapeutic use is indicated in bleeding associated with platelet.

Plasma and Plasma Protein Fraction

The plasma protein contains a euglobulin called plasminogen or profibrinolysin which, when activated, becomes a substance called plasmin or fibrinolysin. **Plasmin** is a proteolytic enzyme that resembles trypsin, the most important proteolytic digestive enzyme of pancreatic secretion. It digests the fibrin fibres as well as other substances in the surrounding blood such as fibrinogen, factor VI, factor VIII, prothrombin and factor XII. Therefore, whenever plasmin is formed in a blood clot, it can cause lysis of the clot and destruction of many of the clotting factors, thereby sometimes even causing hypocoagulability of the blood.

Prothrombin   It is a plasma protein, an α_2-globulin having a molecular weight of 68,700. It is present in normal plasma in a concentration of about 15 ng/dl. It is an unstable protein that can split easily into smaller compounds, one of which is thrombin, which has a molecular weight of 33,700, almost exactly one half of that of prothrombin. The liver forms prothrombin continually and it is continually being used throughout the body for blood clotting. If the liver fails to produce prothrombin, its concentration in the plasma falls too low to provide normal blood coagulation within one to several days.

Fibrinogen   It is a high molecular weight protein (3,40,000) that occurs in the plasma in quantities of 100–700 mg/dl. Fibrinogen is formed in the liver, and liver diseases occasionally decrease the concentration of circulating fibrinogen. Because of its large molecular size, little fibrinogen normally leaks into the interstitial fluids and because it is one of the essential factors in the coagulation process, interstitial fluids ordinarily coagulate poorly, if at all. Yet, when the permeability of the capillaries becomes pathologically increased, fibrinogen does then leak into the tissue fluids in sufficient quantities to allow clotting of these fluids in much the same way as plasma and whole blood.

Albumin

Human serum albumin (HSA) is the single most abundant protein in blood (Table 10.1). Its normal concentration is approximately

42 g per litre, representing 60% of total plasma protein. The vascular system of an average adult thus contains in the region of 150 g of albumin. HSA is responsible for over 80% of the colloidal osmotic pressure of human blood, more than any other plasma constituent. HSA is thus responsible for retaining sufficient fluid within blood vessels. It has been aptly described as the protein that makes blood thicker than water.

Albumin molecules also temporarily leave the circulation and enter the lymphatic system which harbours a large pool of this protein (up to 230 g in an adult). Lower quantities of albumin are also present in the skin.

In addition to its osmoregulatory function, HSA serves a transport function. Various metabolites, predominantly bound to HSA travel throughout the vascular system. These include fatty acids, amino acids, steroid hormones and heavy metals (e.g. copper and zinc), as well as many drugs.

HSA is a 585-amino acid, 65.5-kDa polypeptide. It is one of the few plasma proteins, which is unglycosylated. A prominent feature is the presence of 17 disulphide bonds, which help stabilize the molecule's three-dimensional structure. HSA is synthesized and secreted from the liver, and its gene is present on human chromosome 4.

HSA is used therapeutically as an aqueous solution, and is available in concentrated form (15–25% protein) or as an isotonic solution (4–5% protein). In both cases, more than 95% of the protein present is albumin. It can be prepared by fractionation from normal plasma or serum, or purified from placentae. The source material must first be screened for the presence of indicator pathogens. After purification, a suitable stabilizer (often sodium caprylate) is added, but no preservative. The solution is then sterilized by filtration and aseptically filled into final sterile containers. The relative heat stability of HAS allows a measure of subsequent heat treatment, which further reduces the risk of accidental transmission of viable pathogens (particularly viruses). This treatment normally entails heating the product to 60°C for 10 hours. It is then normally incubated at 30–32°C for a further 14 days and subsequently examined for any signs of microbial growth.

Table 10.1 The major plasma proteins of known function found in human blood

Protein	Normal plasma concentration (g/litre)	Molecular mass (kDa)	Function
Albumin	35–45	66.5	Osmoregulation, transport
Retinol-binding protein	0.03–0.06	21	Retinol transport
Thyroxine-binding globulin	0.01–0.02	58	Binds/transports thyroxine
Transcortin	0.03–0.04	52	Cortisol and corticosterone transport
Ceruloplasmin	0.1–0.6	151	Copper transport
Haptoglobin			
Type 1-1	1.0–2.2	100	Binds and helps conserve haemoglobin
Type 2-1	1.6–3.0	200	
Type 2-2	1.2–2.6	400	
Transferrin	2.0–3.2	76.5	Iron transport
Haemopexin	0.5–1.0	57	Binds haem destined for disposal
β_2-microglobulin	0.002	11.8	Associated with HLA histocompatibility antigen
γ-globulins	7.0–15.0	150	Antibodies
Tranthyretin	0.1–0.4	55	Binds thyroxine

HSA is used as a plasma expander in the treatment of haemorrhage, shock, burns and oedema, as well as administered to some patients after surgery. For adults, an initial infusion containing at least 25 g of albumin is used. The annual world demand for HSA exceeds 300 tonnes, representing a market value of the order of $1 billion.

Despite screening of raw material and heat treatment of final product, HSA derived from naive blood sometimes (though rarely) will harbour pathogens. rDNA technology provides a way of overcoming such concerns and the HSA gene and cDNA have been expressed in a wide variety of microbial systems, including *E. coli, Bacillus subtilis, Saccharomyces cerevisiae, Pichia pastoris* and *Aspergillus niger.* (Its lack of glycosylation renders possible production of native HSA in prokaryotic as well as eukaryotic systems.) However, HSA's relatively large size, as well as the presence of so many disulphide bonds, can complicate recombinant production of high levels of correctly folded products in some production systems. The main stumbling block in replacing native HSA with a recombinant version, however, is an economic one. Unlike most biopharmaceuticals, HSA can be produced in large quantities and inexpensively by direct extraction from its native source. Native HSA currently sells at Rs. 90– Rs. 140 per gram. Although it can be guaranteed blood-pathogen-free, no recombinant HSA products will likely be able to compete with this price.

Clotting Factors

The process of blood coagulation is dependent upon a large number of blood clotting factors which act in a sequential manner. At least 12 distinct factors participate in the coagulation cascade, along with several macromolecular cofactors. The clotting factors are all designated by roman numerals (Table 10.2) and with the exception of factor IV, all are proteins. Most factors are proteolytic zymogens, which become sequentially activated. An activated factor is indicated by inclusion of a subscript "a" (e.g. factor XII_a = activated factor XII).

Although the final steps of the blood-clotting cascade are identical, the initial steps can occur via two distinct pathways: the extrinsic and intrinsic pathways. Both pathways are initiated when specific clotting proteins make contact with specific surface molecules exposed only upon damage to a blood vessel. Clotting occurs much more rapidly when initiated via the extrinsic pathway.

Two coagulation factors function uniquely in the extrinsic pathway: factor III (tissue factor) and factor VII. Tissue factor is an integral membrane protein present in a wide variety of tissue types (particularly lung and brain). This protein is exposed to blood constituents only upon rupture of a blood vessel, and it initiates the extrinsic coagulation cascade at the site of damage as described below.

Factor VII contains a number of γ-carboxyglutamate residues (as do factors II, IX and X), which play an essential role in facilitating their binding of Ca^{2+} ions. The events initiating the extrinsic pathway entail the interaction of factor VII with Ca^{2+} and tissue factor. In this associated form, factor VII becomes proteolytically active. It displays both binding affinity form and catalytic activity against factor X by proteolytic processing, and factor X_2, which initiates the terminal stages of clot formation, remains attached to the tissue-factor–Ca^{2+} complex at the site of damage. This ensures that clot formation only occurs at the point where it is needed.

The initial steps of the intrinsic pathway are somewhat more complicated. This system requires the presence of clotting factors VIII, IX, and XII, all of which, except for factor VIII, are endo-acting proteases. As in the case of the extrinsic pathway, the intrinsic pathway is triggered upon exposure of the clotting factors to proteins present on the surface of body tissue exposed by

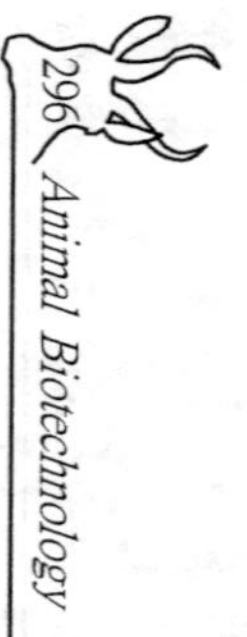

Table 10.2 Coagulation factors that promote blood clotting process (Note that the factor originally designated as VI was later shown to be factor V)

Factor number	Common name	Pathway in which it functions	Function
I	Fibrinogen	Both	Forms structural basis of clot after its conversion to fibrin
II	Prothrombin	Both	Precursor of thrombin, which activates factors I, V, VII, VIII and XIII
III	Tissue factor (thromboplastin)	Extrinsic	Accessory tissue protein which initiates extrinsic pathway
IV	Calcium ions	Both	Required for activation of factor XIII and stabilizes some factors
V	Proaccelerin	Both	Accessory protein, enhances rate of activation of X
VII	Proconvertin	Extrinsic	Precursor of convertin (VII$_a$) which activates X (extrinsic system)
VIII	Antihaemophilic factor	Intrinsic	Accessory protein, enhances activation of X (intrinsic system)

IX	Christmas factor	Intrinsic	Activated IX directly activates X (intrinsic system)
X	Stuart factor	Both	Activated form (X) converts prothrombin to thrombin
XI	Plasma thromboplastin antecedent	Intrinsic	Activated form (XIa) serves to activate IX
XII	Hageman factor	Intrinsic	Activated by surface contact or the kallikrein system. XII helps initiate intrinsic system
XII	Fibrin-stabilizing factor	Both	Activated form cross-links fibrin, forming a hard clot

vascular injury. These protein-binding/activation sites probably include collagen.

Additional protein constituents of the intrinsic cascade include prekallikrein, an 88-kDa protein zymogen of the protease kallikrein, and high-molecular-mass kininogen (HMK), a 150-kDa plasma glycoprotein which serves as an accessory factor.

The intrinsic pathway appears to be initiated when factor XII is activated on contact with surface proteins exposed at the site of damage. HMK also appears to form part of this initial activating complex (Figure 10.7).

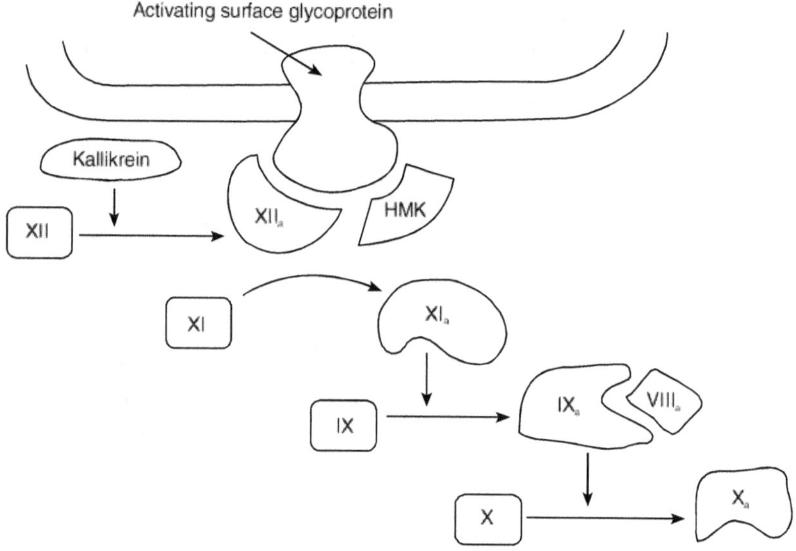

Figure 10.7 The steps unique to the intrinsic coagulation pathway. Factor XII$_a$ can also convert prekallikrein to kallikrein by proteolysis, but this is omitted from the figure for the sake of clarity.

Factor XII$_a$ can proteolytically cleave and hence activate two substrates:

- Prekallikrein, yielding kallikrein (which, in turn, can directly activate more factor XII_a)

- Factor XI, forming factor XI_a

Factor XIa, in turn, activates factor IX. Factor IX_a then promotes the activation of factor X, but only when it (i.e., factor IX_a) is associated with factor VIII_a. Factor VIII_a is formed by the direct action of thrombin on factor VIII. The thrombin will be present at this stage because of prior activation of the intrinsic pathway.

Both intrinsic and extrinsic pathways generate activated factor X. This protease, in turn, catalyses the proteolytic conversion of prothrombin (factor II) into thrombin (factor II_a). Thrombin, in turn, catalyses the proteolytic conversion of fibrinogen (I) into fibrin (I_a). Individual fibrin molecules aggregate forming a soft clot. Factor XIII_a catalyses the formation of covalent cross-links between individual fibrin molecules, forming a hard clot.

Prothrombin (factor II) is a 582-amino acid, 72.5-kDa glycoprotein, which represents the circulating zymogen of thrombin (II_a). It contains up to six γ-carboxyglutamate residues towards its N-terminal end, via which it binds several Ca^{2+} ions. Binding of Ca^{2+} facilitates prothrombin binding to factor X_a at the site of vascular injury. The factor X_a complex the proteolytically cleaves prothrombin at two sites ($Arg^{274} - Thr^{275}$ and $Arg^{323} - Ile^{324}$), yielding active thrombin and an inactive polypeptide fragment.

Fibrinogen (factor I) is a large (340 kDa) glycoprotein consisting of two identical tripeptide units, α, β and γ. Its overall structural composition may thus be represented as $(\alpha\beta\gamma)_2$. Interchain disulphide bonds hold together not only the $\alpha - \beta$ and $\beta - \gamma$ chains, but also the two tripeptide subunits. Slight heterogeneity of plasma fibrinogen molecules has been observed. This is apparently not only caused by variations in its carbohydrate content, but also slight proteolytic degradation at its C- and N-termini.

The overall molecule which exhibits a pI in the region of 5.5, displays an α-helical content of approximately 33%. Its quaternary structure is characterized by three compact domain-like regions (one each on either ends and one in the middle), separated by two extended stretches.

The N-terminal regions of the α and β fibrinogen chains are rich in charged amino acids which, via charge repulsion, play an important role in preventing aggregation of individual fibrinogen molecules. Thrombin, which catalyses the proteolytic activation of fibrinogen, hydrolyses these N-terminal peptides. This renders individual fibrin molecules more conducive to aggregation, therefore promoting soft clot formation (Figure 10.8).

The soft clot is stabilized by the introduction of covalent cross-linkages between individual participating fibrin molecules. This reaction is catalysed by factor XIII_a, as shown in Figure 10.8. Factor XIII is present in both plasma and platelets. Plasma factor XIII is a 320-kDa tetramer, composed of two (70 kDa) α chains, and two (90 kDa) β chains. The platelet form of factor XIII is composed solely of the two α chains (i.e., a 140-kDa α dimer). Both forms of factor XIII are activated upon proteolytic cleavage by thrombin (factor IIa), which hydrolyses a single glycine bond in the N-terminal region of α chains. This generates a 73-amino acid peptide and a modified α chain, α'.

In the case of platelet-derived factor XIII, the resultant product $(\alpha')_2$ is the activated form. Thrombin action on plasma-derived factor XIII generates an $\alpha'_2\beta_2$ dimer, which is devoid of trans glutaminase activity. However in the presence of Ca^{2+}, the $\alpha'\beta$ chains dissociate, yielding the biologically active α'_2.

Haemoglobin

The most important component of red blood corpuscles is haemoglobin. It is a complex of protein and iron. Protein contains the globulin, which constitutes 94% and iron contains haem, which constitutes 6% of haemoglobin.

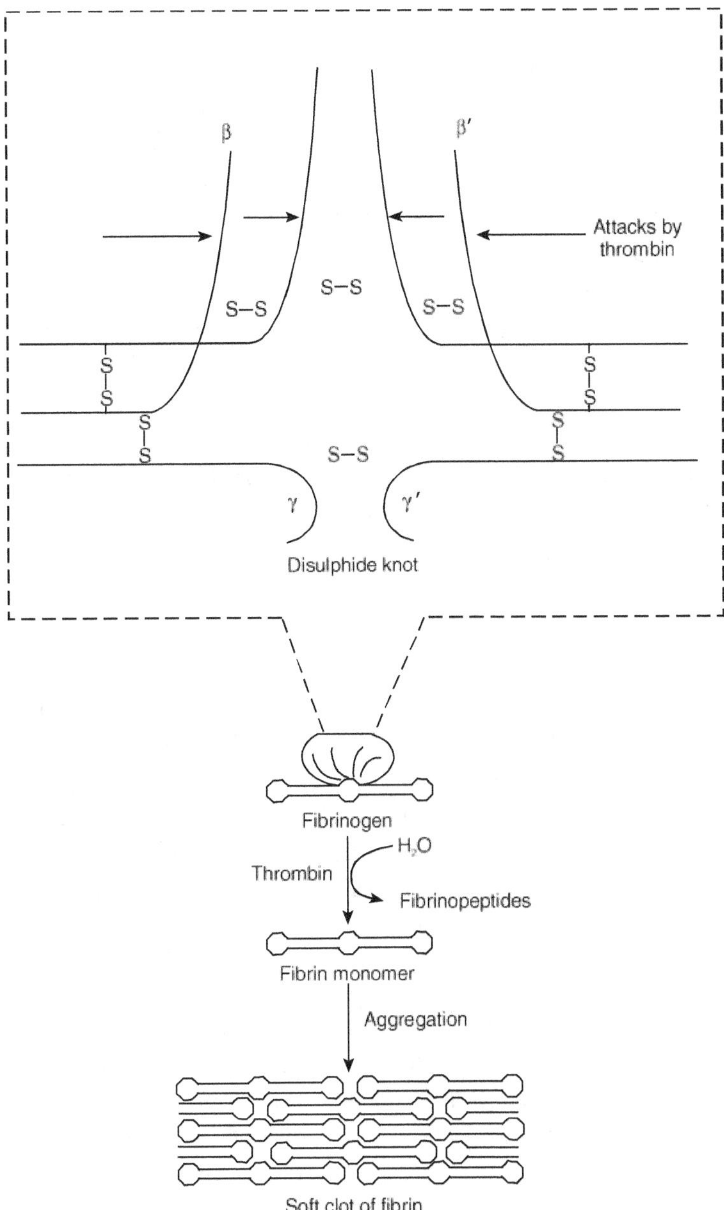

Figure 10.8 Diagrammatic representation of the fibrinogen molecule and its conversion to the soft clot of fibrin

Haemoglobin is a tetramer with two alpha and two beta polypeptide chains. Each of the four polypeptide chains has haem as prosthetic group. This is responsible for the oxygen-binding properties of haemoglobin. Modified haemoglobin molecules called as HBOC are prepared from donated blood. Bovine blood is poor in oxygen transport. Haemoglobin has been successfully produced by recombinant DNA technology in *E. coli* and yeast systems. It is also produced in transgenic animals. Although recombinant haemoglobin is a technical feasibility, outdated stocks of human blood still remain an expensive source of haemoglobin. Contaminating proteins of nonhuman origin in recombinant haemoglobin are immunogenic. Recombinant DNA technology may be helpful in research to develop human haemoglobin mutants with desirable characteristics such as improved stability or decreased oxygen affinity.

ANTICOAGULANTS

Haemostasis is the process of rapid arrest of blood loss upon vascular damage, in order to maintain a relatively constant blood volume. Blood clot formation is essential to maintain haemostasis, but inappropriate clotting can give rise to serious, if not fatal, medical conditions. The formation of a blood clot (a thrombus) occurs inappropriately within diseased blood vessels. This partially or completely obstructs the flow of blood (and hence oxygen) to the tissues normally served by that blood vessel.

Thrombus formation in a coronary artery (the arteries that supply the heart muscle itself with oxygen and nutrients) is termed coronary thrombosis. This results in a heart attack, characterized by the death (infarction) of oxygen-deprived heart muscle—hence the term myocardial infarction. The development of a thrombus in a vessel supplying blood to the brain can result in development of a stroke. In addition, a thrombus (or part thereof) which has formed at a particular site in the vascular system may become detached. After travelling through the blood, this may lodge in another blood vessel, obstructing blood flow at that point. This

Anticoagulants are substances, which can prevent blood from clotting and hence, are of therapeutic use in cases where a high risk of coagulation is diagnosed. They are often administered to patients with coronary heart disease and to patients who have experienced a heart attack or stroke (in an effort to prevent recurrent episodes). The major anticoagulants used for therapeutic purposes are listed in Table 10.3.

Table 10.3 Anticoagulants used therapeutically or that display therapeutic potential

Anticoagulant	Structure	Source	Molecular mass (Da)
Heparin	Glycosaminoglycan	Beef lung, pig gastric mucosa	3000–40000
Dicoumarol	Coumarin-based	Chemical manufacture	336.3
Warfarin	Coumarin-based	Chemical manufacture	308.4
Hirudin	Polypeptide	Leech saliva, genetic engineering	7000
Ancrod	Polypeptide	Snake venom, genetic engineering	35000

Dicoumarol and related molecules are generally used over prolonged periods, while heparin is used over shorter periods. Hirudin has recently been approved for general medical use, and ancrod remains under clinical investigation.

Heparin

Proteoglycans are widely distributed throughout the body, being most abundant in connective tissue, where they may contribute up to 30% of that tissue's dry weight. They consist of a polypeptide backbone to which heteropolysaccharide chains are attached. However, unlike glycoproteins, proteoglycans consist of up to, or in excess of, 95% carbohydrate and their properties resemble those of polysaccharides more than those of proteins.

Proteolytic digestion of proteoglycans liberates the carbohydrate side chains, which are known as glycosaminoglycans (also known as mucopolysaccharides). All the glycosaminoglycans contain derivatives of glucosamine or galactosamine. Six major groups are known, one of which is heparin.

Heparin is associated with many tissues but is found mainly stored intracellularly as granules in mast cells, which line the endothelium of blood vessels. Upon release into the bloodstream, heparin binds to, and activates, an additional plasma protein: antithrombin III.

Antithrombin III displays a heparin-binding site, as well as a binding site for a number of activated proteolytic clotting factors, including II_a, IX_a, X_a, XI_a and XII_a. Binding of heparin to antithrombin III activates the latter, facilitating complex formation between the heparin, anti-thrombin III_a, and the activated clotting factor. This binding, which is of very high affinity, inactivates the clotting factor. Moreover, the heparin now dissociates from the complex and combines with another antithrombin III molecule, thereby initiating another turn of this cycle.

Heparin represents one of the few carbohydrate-based biomolecules to find therapeutic application. It was originally extracted from liver (hence its name), but commercial preparations are now obtained by extraction from beef lung or porcine gastric mucosa.

While the product has proven to be an effective (and relatively inexpensive) anticoagulant, it does suffer from a number of clinical disadvantages. Its requirement for a cofactor (i.e., antithrombin III) can lead to unpredictable clinical responses. Activated platelets also secrete at least two factors (platelet factor 4 and histidine-rich glycoprotein), which can bind and neutralize heparin. However, perhaps this molecule's most serious clinical limitation is its inability to inhibit thrombin bound to fibrin (i.e., thrombin functioning on the surface of a newly developing clot). Despite such disadvantages, heparin, however, still enjoys widespread clinical use.

Vitamin K Antimetabolites

Dicoumarol and warfarin are related, coumarin–based anticoagulants which, unlike heparin, may be administered orally. These compounds induce their anticoagulant effect by preventing the vitamin K-dependent γ carboxylation of several of their glutamate residues; for example, 10 of the first 33 residues present in prothrombin are γ carboxyglutamate. This post-translational modification is required in order to allow these factors to bind Ca^{2+} ions, a prerequisite to their effective functioning. Vitamin K is an essential cofactor for the carboxylase enzyme, and its replacement with the antimetabolite dicoumarol renders the enzyme inactive. As a consequence, defective blood factors are produced which hinder effective functioning of the coagulation cascade.

The only major side effect of these oral anticoagulants is prolonged bleeding, thus the dosage levels are chosen with care. Dicoumarol was first isolated from spoiled sweet clover hay that was the agent, which promoted haemorrhage disease in cattle. Both dicoumarol and warfarin have also been utilized (at high doses) as rat poisons.

Hirudin

Hirudin is a leech–derived anticoagulant, which functions by directly inhibiting thrombin. A range of bloodsucking animals contain substances in their saliva, which specifically inhibit some element of the blood coagulation system.

A bite from any such parasite is characterized by prolonged host bleeding. This property led to the documented use of leeches an aid to bloodletting as far back as several hundred years BC. The method was particularly fashionable in Europe at the beginning of the 19th century. Many doctors at that time believed that most illnesses were related in some way to blood composition and bloodletting was a common, if ineffective, therapy. The Napoleonic army surgeons, for example, used leeches to withdraw blood from soldiers suffering from conditions as diverse as infections and mental disease.

With the advent of modern medical principles, the medical usage of leeches waned somewhat. In more recent years, however, they have staged a limited comeback. They are occasionally used to drain blood from inflamed tissue, and in procedures associated with plastic surgery.

The presence of an anticoagulant in the saliva of the leech, *Hirudo medicinalis,* was first described in 1884. However, it was not until 1957 that the major anticoagulant activity present was purified and named hirudin. Hirudin is a short (65-amino acid) polypeptide, of molecular mass 7000 Da. The tyrosine residue at position 63 is unusual in that it contains a sulphate group. The molecule appears to have two domains. The globular N-terminal domain is stabilized by three disulphide linkages, while the C-terminal domain is more elongated and exhibits a high content of acidic amino acids.

Hirudin exhibits its anticoagulant effect by tightly binding thrombin, thus inactivating it. In addition to its critical role in the production of a fibrin clot, thrombin displays several other (non-enzymatic) biological activities important in sustaining haemostasis. These include the following:

- It is a potent inducer of platelet activation and aggregation.
- It functions as a chemoattractant for monocytes and neutrophils.
- It stimulates endothelial transport.

One molecule of hirudin binds a single molecule of thrombin with great affinity ($K_d \sim 10^{-12}$ M). Binding and inactivation occur as a two-step mechanism. The C-terminal region of hirudin first binds along a groove on the surface of thrombin, resulting in a small conformational change of the enzyme. This then facilitates binding of the N-terminal region to the active site area. Binding of hirudin inhibits all the major functions of thrombin. Fragments of hirudin can also bind thrombin, but will generally only inhibit some of thrombin's range of activities. For example, binding of an N-terminal hirudin fragment to thrombin inhibits only the thrombin catalytic activity.

Hirudin displays several potential therapeutic advantages as an anticoagulant. These include the following:

- It acts directly upon thrombin.
- It does not require a cofactor to exist in its inhibitory effect.
- It is less likely than many other anticoagulants to induce unintentional haemorrhage.
- It is a weak immunogen.

Although the therapeutic potential of hirudin was appreciated for many years, insufficient material could be purified from the native source to support clinical trials, never mind its widespread medical application. The hirudin gene was cloned in the 1980s, and it has subsequently been expressed in a number of recombinant systems including *E. coli*, *Bacillus subtilis* and *Saccharomyces cerevisiae.* A recombinant form of hirudin (given the trade name Refludan) has recently gained approval for general medical use. The recombinant production system was constructed by insertion (via a plasmid) of a synthetic hirudin gene into a strain of *Saccharomyces cerevisiae.* The yeast cells secrete the product, which is then purified by various fractionation techniques. The recombinant molecule displays a slightly altered amino acid sequence when compared to the native product. Its first two amino acids, leucine and threonine, replace two valines of native hirudin. It is also devoid of the sulphate group normally present on tyrosine 63. Clinical trials, however, have proven this slightly altered product to be both safe and effective. The final product is presented in freeze-dried form with the sugar, mannitol, representing the major added excipient. The product, which displays a useful shelf life of two years when stored at room temperature, is reconstituted with saline or water for injections immediately prior to its i.v. administration.

Ancrod serine protease is purified from the venom of the Malaysian pit viper. This enzyme cleaves fibrin molecules prior to clot formation. However, it has no effect once clots are formed.

THROMBOLYTIC AGENTS

The natural process of thrombosis functions to plug a damaged blood vessel, thus maintaining haemostasis until the damaged vessel can be repaired. Subsequent to this repair, the clot is removed, viz. an enzymatic degradative process known as fibrinolysis. Fibrinolysis normally depends upon the serine protease plasmin, which is capable of degrading the fibrin strands present in the clot.

In situations where inappropriate clot formation results in the blockage of a blood vessel, the tissue damage that ensues depends, to a point, upon how long the clot blocks blood flow.

Rapid removal of the clot can often minimize the severity of tissue damage. Thus, several thrombolytic (clot-degrading) agents have found medical application. The market for an effective thrombolytic agent is substantial. In the USA alone, it is estimated that 1.5 million people suffer acute myocardial infarction each year, while another 0.5 million suffer strokes.

Table 10.4 Thrombolytic agents approved for general medical use, or which are currently undergoing clinical evaluation

Agent	Source
Tissue plasminogen activator (tPA)	Recombinant: CHO cell line (Alteplase)
Streptokinase	*Streptococcus haemolyticus*
Urokinase	Human urine Tissue culture using human kidney cells
Staphylokinase	*Streptococcus aureus* Various recombinant sources including *E. coli*

Tissue Plasminogen Activator

Plasmin is a protease which catalyses the proteolytic degradation of fibrin present in clots, thus effectively dissolving the clot. Plasmin is derived from plasminogen, its circulating zymogen. Plasminogen is synthesized in and released from the kidneys. It is a single-chain 90-kDa glycoprotein, which is stabilized by several disulphide linkages.

Tissue plasminogen activator (tPA, also known as fibrinokinase) represents the most important physiological activator of plasminogen. tPA is a 527-amino acid serine protease. It is synthesized predominantly in vascular endothelial cells (cells lining the inside of blood vessels). tPA displays four potential glycosylation sites, three of which are normally glycosylated (residues 117, 184 and 448). It is normally found in the blood in two forms: a single-chain polypeptide (type I tPA) and a two-chain structure (type II) proteolytically derived from the single-chain structure. The two-chain form is the one predominantly associated with clots undergoing lysis, but both forms display fibrinolytic activity.

Fibrin contains binding sites for both plasminogen and tPA, thus bringing these into close proximity. This facilitates direct activation of the plasminogen at the clot surface. This activation process is potentiated by the fact that binding of tPA to fibrin (a) enhances the subsequent binding of plasminogen and (b) increases tPA's activity towards plasminogen by up to 600-fold.

Overall therefore, activation of the thrombolytic cascade occurs exactly where it is needed on the surface of the clot. This is important as the substrate specificity of plasmin is poor, and circulating plasmin displays the catalytic potential to proteolyse fibrinogen, factor V and factor VIII. Although soluble serum tPA displays a much reduced activity towards plasminogen, some freely circulating plasmin is produced by this reaction. If uncontrolled, this could increase the risk of subsequent haemorrhage. This scenario is usually averted as circulating plasmin is rapidly neutralized by another plasma protein, α_2-antiplasmin. (α_2-antiplasmin, a 70-kDa, single-chain glycoprotein, binds plasmin very tightly in a 1:1 complex). In contrast to free plasmin, plasmin present on a clot surface is very slowly inactivated by α_1-antiplasmin. The thrombolytic system has thus evolved in a self-regulating fashion, which facilitates efficient clot degradation with minimal potential disruption to other elements of the haemostatic mechanism.

Production of tPA   Although tPA was first studied in the late

1940s, its extensive characterization was hampered by the low levels at which it is normally synthesized. Detailed studies were facilitated in the 1980s after the discovery that the Bowes melanoma cell line produces and secretes large quantities of this protein. This also facilitated its initial clinical appraisal. The tPA gene was cloned from the melanoma cell line in 1983, and this facilitated subsequent large-scale production in CHO cell lines by recombinant DNA technology. The tPA cDNA contains 2530 nucleotides and encodes a mature protein of 527 amino acids. The glycosylation pattern was similar, though not identical, to the native human molecule. A marketing license for the product was first issued in the USA to Genentech in 1987 (under the trade name Alteplase). The therapeutic indication was for the treatment of acute myocardial infarction. The production process entails an initial (10,000-litre) formation step, during which the cultured CHO cells produce and secrete tPA into the fermentation medium. After removal of the cells by submicron filtration and initial concentration, the product is purified by a combination of several chromatographic steps. The final product has been shown to be greater than 99% pure by several analytical techniques, including HPLC, SDS-PAGE, tryptic mapping and N-terminal sequencing.

Alteplase has proved to be effective in the early treatment of patients with acute myocardial infarction (i.e., those treated within 12 hours after the first symptoms occur), and has significantly increased rates of patient survival (as measured one day and 30 days after the initial event). tPA has thus established itself as a first line option in the management of acute myocardial infarction. A therapeutic dose of 90–100 mg (often administered by infusion over 90 minutes) results in a steady-state Alteplase concentration of 3–4 mg/litre during that period. The product is, however, cleared rapidly by the liver, displaying a serum half-life of minutes. As is the case for most thrombolytic agents, the most significant risk associated with tPA administration is the possible induction of severe haemorrhage.

Modified forms of tPA have also been generated in an effort

to develop a product with an improved therapeutic profile (e.g. faster acting or exhibiting a prolonged plasma half-life). Ecokinase is the trade name given to one such modified human tPA produced in recombinant *E. coli* cells. It gained marketing approval in Europe in 1996. Its development was based upon the generation of a synthetic nucleotide sequence encoding a shortened (355-amino acid) tPA molecule. This analogue contained only the tPA domains responsible for fibrin selectivity and catalytic activity. The nucleotide sequence was integrated into an expression vector subsequently introduced into *E. coli* (strain K12), by treatment with calcium chloride. The protein is expressed intracellularly, where it accumulates in the form of an inclusion body. Due to the prokaryotic production system, the product is non-glycosylated. The final sterile freeze-dried product is equally as effective as Alteplase and exhibits a two-year shelf life when stored at temperatures below 25°C. An overview of the production process is presented here.

Streptokinase

Streptokinase is an extracellular bacterial protein produced by several strains of *Streptococcus haemolyticus* group C. It displays a molecular mass in the region of 48 kDa and an isoelectric point of 4.7. Its ability to induce lysis of blood clots was first demonstrated in 1933. Early therapeutic preparations administered to patients often caused immunological and other complications, usually prompted by impurities present in these products. Chromatographic purification (particularly using gel-filtration and ion-exchange columns) overcame many of these initial difficulties. Modern chromatographically pure streptokinase preparations are usually supplied in freeze-dried form. These preparations often contain albumin as an excipient. The albumin prevents flocculation of the streptokinase upon its reconstitution.

Streptokinase is a widely employed thrombolytic agent. It is administered to treat a variety of thromboembolic disorders including

- pulmonary embolism (blockage of the pulmonary artery by an embolism), which can cause acute heart failure and sudden death. The pulmonary artery carries blood from the heart to the lungs for oxygenation,

- deep-vein thrombosis (thrombus formation in deep veins, usually in the legs),

- arterial occlusions (obstruction of an artery) and

- acute myocardial infarction.

Streptokinase induces its thrombolytic effect by binding specifically and tightly to plasminogen. This induces a conformational change in the plasminogen molecule, which renders it proteolytically active. In this way, the streptokinase–plasminogen complex catalyses the proteolytic conversion of plasminogen to active plasmin.

As a bacterial protein, streptokinase is viewed by the human immune system as an antigenic substance. In some cases, its administration has elicited allergic responses, which ranges from mild rashes to more serious anaphylactic shock. (Anaphylactic shock represents an extreme and generalized allergic response characterized by swelling, constriction of the bronchioles, circulatory collapse and heart failure).

Another disadvantage of streptokinase administration is the associated increased risk of haemorrhage. Streptokinase-activated plasminogen is capable of lysing not only clot-associated fibrin, but also free plasma fibrinogen. This can result in low serum fibrinogen levels and hence compromise haemostatic ability. It should not be administered to, for example, patients suffering from coagulation disorders or bleeding conditions such as ulcers. Despite such potential clinical complications, careful administration of streptokinase has saved countless thousands of lives.

Urokinase

The ability of some component of human urine to dissolve fibrin clots was first noted in 1885. It was not, however, until the 1950s that the active substance was isolated and named urokinase.

Urokinase is a serine protease produced by the kidney and is found in both the plasma and urine. It is capable of proteolytically converting plasminogen into plasmin. Two variants of the enzyme have been isolated—a 54-kDa species and a lower-molecular-mass moiety by proteolytic processing. Both forms exhibit enzymatic activity against plasminogen.

Urokinase is used clinically under the same circumstances as streptokinase and, because of its human origin, adverse immunological responses are less likely. Following acute medical events such as pulmonary embolism, the product is normally administered to the patient at initial high doses (by infusion) for several minutes. This is followed by hourly i.v. injections for up to 12 hours.

Urokinase utilized medically is generally purified directly from human urine. It binds to a range of adsorbents such as silica gel and, especially, kaolin (hydrated aluminium silicate), which can be used to initially concentrate and partially purify the product. It may also be concentrated and partially purified by precipitation using sodium chloride, ammonium sulphate or ethanol as precipitants.

Various chromatographic techniques may be utilized to further purify urokinase. Commonly employed methods include anion (DEAE-based) exchange chromatography, gel filtration on Sephadex G-100 and chromatography on hydroxyapatite columns. Urokinase is a relatively stable molecule. It remains active subsequent to incubation at 60°C for several hours, or brief incubation at a pH as low as 1.0 or as high as 10.0.

After its purification, sterile filtration and aseptic filling, human urokinase is normally freeze-dried. Because of its heat stability, the final product may also be heated to 60°C for up to 10 hours in an effort to inactivate any undeleted viral particles present. The product utilized clinically contains both molecular mass forms, with the higher-molecular-mass moiety predominating. Urokinase can also be produced by techniques of animal cell culture utilizing human kidney cells or by recombinant DNA technology. However, the product isolated from urine is the one used clinically thus far.

Staphylokinase

Staphylokinase is a protein produced by a number of strains of *Staphylococcus aureus*, which also displays therapeutic potential as a thrombolytic agent. The protein has been purified from its natural source by a combination of ammonium sulphate precipitation and cation exchange chromatography on CM cellulose. Affinity chromatography using plasmin or plasminogen immobilized to sepharose beads has also been used. The pure product is a 136-amino acid polypeptide displaying a molecular mass in the region of 16.5 kDa. Lower-molecular-mass derivatives lacking the first 6 or 10 N-terminal amino acids have also been characterized. All three appear to display similar thrombolytic activity at least *in vitro*.

The staphylokinase gene has been cloned in *E. coli*, as well as various other recombinant systems. The protein is expressed intracellularly in *E. coli* at high levels, representing 10–15% of total cellular protein. It can be purified directly from the clarified cellular homogenate by a combination of ion-exchange and hydrophobic interaction chromatography.

Although staphylokinase shows no significant homology with streptokinase, it induces a thrombolytic effect by a somewhat similar mechanism—it also forms a 1 : 1 stoichiometric complex with plasminogen. The proposed mechanism by which staphylokinase induces plasminogen activation is outlined in Figure 10.9. Binding of the staphylokinase to plasminogen appears to initially yield an inactive staphylokinase–plasminogen complex. However, complex formation somehow induces subsequent proteolytic cleavage of the bound plasminogen, forming plasmin

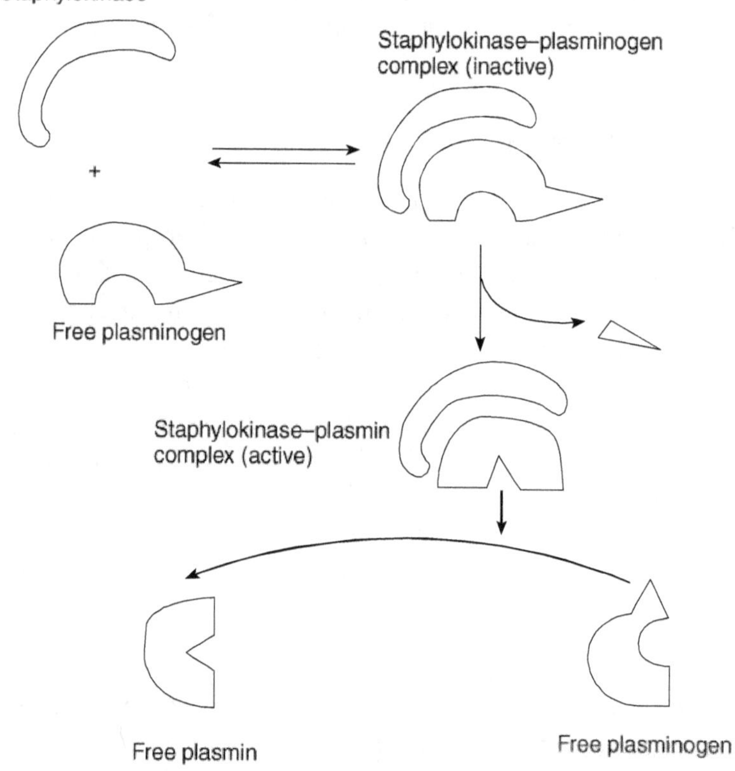

Figure 10.9 Schematic representation of the mechanism by which staphylokinase appears to activate the thrombolytic process via the generation of plasmin.

which remains complexed to the staphylokinase. This complex, via the plasmin, then appears to catalyse the conversion of free plasminogen to plasmin, and may even accelerate the process of conversion of ther staphylokinase–plasminogen complexes into staphylokinase–plasmin complexes. The net effect is the generation of active plasmin, which displays a direct thrombolytic effect by degrading clot-based fibrin.

The serum protein α_2-antiplasmin can inhibit the activated plasmin–staphylokinase complex. It appears that the α_2-antiplasmin can interact with the active plasmin moiety of the complex, resulting in dissociation of staphylokinase, and consequent formation of an inactive plasmin–α_2-antiplasmin complex.

The thrombolytic ability of (recombinant) staphylokinase has been evaluated in initial clinical trials with encouraging results. Eighty per cent of patients suffering from acute myocardial infarction who received staphylokinase responded positively (10 mg staphylokinase was administered by infusion over 30 minutes). Coupled to this is the finding that, in animal systems at least, staphylokinase appeared to elicit a less aggressive immunological response than streptokinase. Initial studies therefore suggest that staphylokinase may yet find therapeutic favour as a thrombolytic agent.

α_1-Antitrypsin

The respiratory tract is protected by a number of defence mechanisms, which include

- particle removal in the nostril/nasopharynx,

- particle expulsion (e.g. by coughing),

- upward removal of substances via mucociliary transport,

- ✺ presence in the lungs of immune cells, such as alveolar macrophages and

- ✺ production/presence of soluble protective factors, including α_2-antitrypsin, lysozyme, lactoferrin and interferon.

Failure/ineffective functioning of one or more of these mechanisms can impair normal respiratory function. Emphysema, for example, is a condition in which the alveoli of the lungs are damaged. This compromises the lung's capacity to exchange gases, and breathlessness often results. This condition is often promoted by smoking, respiratory infections or a deficiency in the production of serum α_1-antitrypsin.

α_1-antitrypsin is a 394-amino acid, 52-kDa serum glycoprotein. It is synthesized in the liver and secreted into the blood, where it is normally present at concentrations of 2–4 g/litre. It constitutes in excess of 90% of the α_1-globulin fraction of blood.

The α_1-antitrypsin gene is located on chromosome 14. A number of α_1-antitrypsin gene variants have been described. Their gene products can be distinguished by their differential mobility upon gel electrophoresis. The normal form is termed M, while point mutations in the gene have generated two major additional forms, S and Z. These mutations result in a greatly reduced level of synthesis and secretion into the blood of the mature α_1-antitrypsin. Persons inheriting two copies of the z gene, in particular, display greatly reduced levels of serum α_1-antitrypsin activity. This is often associated with the development of emphysema (particularly in smokers). The condition may be treated by the administration of purified α_1-antitrypsin. This protein constitutes the major serine protease inhibitor present in blood. It is a potent inhibitor of the protease, elastase, and serves to protect the lung from proteolytic damage by inhibiting neutrophil elastase. The product is administered on an ongoing basis to sufferers, who receive up to 200 g of the inhibitor each year. It is normally prepared by limited fractionation of whole human blood, although the large quantities required by patients heighten the risk of accidental transmission of blood-borne pathogens. The α_1-antitrypsin gene has been expressed in a number of recombinant systems, including milk of transgenic sheep. While use of the recombinant product would all but preclude blood pathogen transmission, it appears significantly more costly to produce.

Summary

- Genetic engineering is also employed for the production of some valuable regulatory proteins.

- Somatostatin is the first polypeptide expressed in *Escherichia coli* as a part of fusion peptide.

- β-endorphin produced by the transformed *E. coli* cells is used in the treatment of severe pains and a few diseases.

- Blood and blood products constitute a major group of traditional biologics.

- Some of the therapeutically important blood proteins are human serum albumin.

- Haemostasis is the process of rapid arrest of blood loss upon vascular damage.

- Heparin represents one of the few carbohydrate-based biomolecules to find therapeutic application.

- Hirudin is a leech-derived anticoagulant, which functions by directly inhibiting thrombin.

- Tissue plasminogen activator, also known as fibrinokinase, represents the most important physiological activator of plasminogen.

REVIEW QUESTIONS

1. Define regulatory proteins.
2. Give a few examples of regulatory proteins.
3. Write a short note on somatostatin.
4. Describe the process of β-endorphin expression in *E. coli*.
5. Give a detailed account of tropomyosin.
6. Write short notes on plasma and plasma protein fraction.
7. Give a brief account of human serum albumin.
8. Discuss in detail clotting factors.
9. Write in detail about anticoagulants.
10. Discuss the role of staphylokinase as a thrombolytic agent.

11

BIOMEDICAL APPLICATIONS

The Ethics of using a Recombinant Drug

Erythropoietin (EPO) is a hormone produced in the kidneys that consists of a 165-amino-acid protein portion plus four carbohydrate chains. When the oxygen level in the blood dips too low, cells in the kidneys produce EPO which travels to the bone marrow and binds to receptors on cells that give rise to red blood cell precursors. Soon, more red blood cells enter the circulation, and they carry more oxygen to the tissues.

This treatment also causes severe anaemia, because dialysis removes EPO from the blood. To counteract dialysis-induced anaemia, it was necessary to boost the patients' EPO levels.

Levels of EPO in human plasma are too low to make pooling from donors feasible. A more likely potential source was people suffering from disorders such as aplastic anemia and hookworm infection, which cause them to secrete large amounts of EPO.

Recombinant DNA technology solved the EPO problem. The hormone is produced in Hamster kidney cells, which can attach EPO's four-carbohydrate groups. Today EPO is sold under various names, including Epogen, Repotin, and Procrit.

EPO's ability to increase the oxygen-carrying capacity of the blood, and thereby increase physical endurance, has attracted the attention of competitive athletes. Training at high altitudes increases endurance, and the reason is EPO. The scarcer oxygen in the air at high altitudes stimulates the kidneys to produce EPO, which stimulates the production of more red blood cells. Athletes have attempted to reproduce this effect by abusing EPO.

The development of treatments, preventive procedures, and cures for human diseases was the outstanding contribution of medicine and science to human well-being in the 20th century. It is difficult to summarize the whole gamut of contributions in a text of limited space; these could be grouped under the following broad heads:

1. Disease prevention (vaccines)
2. Disease diagnosis (monoclonal antibodies and probes)
3. Forensic medicine (DNA fingerprinting)

Vaccine protects a recipient from pathogenic agents by establishing and immunological resistance to infection. The effectiveness of vaccines may be appreciated from the fact that smallpox, once a dreaded disease the world over, has been completely eradicated from the world; the last case of smallpox was reported in 1977.

An accurate diagnosis of the disease and its causal organism is critical to its effective management and cure. Conventionally, disease diagnosis is based on the following:

i. Microscopic examination
ii. Culture of specimen
iii. Immunological assays
iv. Detection and measurement of the pathogen-specific antibodies

These tests are often tedious and time-consuming, and may yield ambiguous results and some of them cannot be applied in certain cases. Hence novel diagnostic approaches have been developed by biotechnology, which are precise and very rapid, viz. monoclonal antibodies and probes.

The knowledge and techniques of medical science applied to assist in the resolution of crimes, legal disputes, etc. constitute forensic medicine. DNA fingerprinting or DNA profiling is a highly sensitive and extremely versatile approach to solve these problems.

VACCINES

Prevention of diseases is the most desirable, most convenient and highly effective approach to health; this is achieved by vaccination or immunization using biological preparations called **vaccines**. A vaccine is a preparation containing a pathogen (disease-producing organism) either in attenuated or inactivated state. This preparation is introduced into an individual to induce adequate antibody production against the pathogen, in particular, so that the individual becomes protected against infection, at a later date, by that pathogen. The introduction of a vaccine in an individual is called vaccination/immunization as it leads to the development of immunity in the vaccinated individuals to the concerned pathogen. Any molecule that induces production of antibodies specific to itself when introduced in the body of an animal is called **antigen**. Usually antigenic function is confined to a rather small portion of the antigen molecules known as **antigenic determinant** or **epitope**.

When an individual is vaccinated, the antigens of pathogen origin stimulate antibody production against themselves. This increase in antibody production takes some time, but since the vaccine does not have virulent pathogens there is no danger of disease development. When live virulent form of the same pathogen later enters into the system of an immunized animal, the high level of antibodies specific to the pathogen inactivate the pathogen, and thereby protect the animal against the disease.

The various vaccines can be grouped into 2 categories:

1. Vaccines containing killed or inactivated pathogens
2. Those containing live but attenuated pathogens

Attenuation means a drastic reduction in the virulence of a pathogen; this is achieved in one of the following ways.

- Several consecutive passages through an animal which is not the usual host of the pathogen, e.g. smallpox virus in calf.

- Several passages through cultured cells of the host, e.g. rabies virus in human diploid cell culture, or of a different species; yellow fever viruses in chick embryo cell culture.

- Selection of less virulent strains of pathogens, e.g. a mutant strain of polio virus.

- Treatment of the pathogen with some chemicals, e.g. BCG vaccines produced by culturing the bacteria on a medium containing bile.

- Culturing pathogens under unfavourable conditions like high temperature, e.g. anthrax vaccine obtained by cultivation of the bacterium, *Bacillus anthracis,* at 40–50°C.

In general, inactivation of viruses is always coupled with attenuation to minimize the accidental presence in vaccines of active virulent particles, which could cause disease in the vaccinated individuals.

The different vaccines differ in their composition, efficacy and duration of effective protection to the vaccinated individuals. Vaccines are one of the earliest examples of biotechnological intervention in human and animal health care. Vaccines offer the cheapest and most effective protection against diseases, and for some diseases, e.g. hepatitis B, AIDS, etc., they are the only means of protection. The effectiveness of vaccines may be highlighted by the success of WHO-sponsored mass vaccination against smallpox in completely wiping out this once ravaging disease from the face of earth.

HISTORY AND PROGRESS OF VACCINE DEVELOPMENT

Vaccine production and immunization or vaccination is the most important application of immunotechnology, as far as human welfare is concerned. The first experiment in immunization was performed by Louis Pasteur, a French chemist-turned-biologist, on July 6, 1885, when he treated a young boy against rabies. He extracted fluid from the spinal cord of a rabid dog and injected it in small amounts to the boy, who was bitten several times by a rabid dog. Apparently the extract from the spinal cord had stimulated the production of antibodies against the rabies virus. The science and application of the vaccine technology, however, has made significant progress in recent decades, so that the period 1950–1970 is considered to be golden age for vaccine development. During this period, vaccines for poliomyelitis, measles, mumps, and rubella were developed. In a majority of the vaccine recipients there are no significant ill effects and mortality rate among vaccines varies from about one per 2,00,000 to one in 3,000,000. Thus vaccines are quite safe and cost-effective.

Since the dramatic global eradication of smallpox, vaccination has also been remarkably successful in reducing morbidity and mortality due to yellow fever virus infections in Africa and Central America. In the United States, vaccination programs have nearly eliminated deaths due to diphtheria, measles, mumps, pertussis, paralytic poliomyelitis and rubella; the incidence of tetanus, moreover, has declined by about 90 percent.

But vaccination programs are still needed in certain regions of the world and the need for new and improved vaccines remains as great as ever, especially for AIDS, malaria (with drug-resistant strains of parasites spreading) and malignant diseases, such as virus-induced and certain other tumours.

Though a number of vaccines are produced in cultured animal cells, the procedure has several inherent problems. The problems involved in the production of vaccine in such cases include the following:

- Antigenic variation or presence of many serotypes as in influenza, rhinovirus and in HIV.
- Integration of viral DNA/cDNA in the host cell genome as in hepatitis B and retroviral infections.
- Large animal reservoirs as in influenza and human HIV.
- Infection transmitted by cells which may or may not express viral antigen as in HIV.
- Crucial cells of the immune system are infected as in HIV.
- Rigorous quality control is essential particularly to exclude the possibility of presence of infectious agents in the materials and cells used.

Therefore, alternative approaches of vaccine production and delivery have been/are being developed. They include

1. Production of antigenic proteins or only their antigenic determinants in GEMS (genetically engineered microbes).
2. Use of DNA as vaccines.
3. Production of antigenic proteins in vegetables/fruits to minimize storage costs and problems.
4. Minicells as vaccines.

Whatever the method of biotechnology used, vaccines produced by these new techniques have definite advantages over the current methods of vaccine production. These are:

- More effective vaccines with increased response and high specificity.
- Long-lasting immunity.

- No side effects, less toxicity.
- Production methods easier and cheap.
- Novel vaccines.

AN IDEAL VACCINE

An ideal vaccine or vaccination protocol should have the following features:

- It should not be tumorigenic, toxic or pathogenic.
- It should have very low levels of side effects in normal individuals.
- It should not spread either within the vaccinated individual or to other individuals (live vaccines).
- It should not contaminate the environment.
- It should be effective in producing long-lasting humoral and cellular immunities.
- It should not cause problems in individuals with impaired immune system.
- The technique of vaccination should be simple.
- The vaccine should be cheap so that it is generally affordable. So far, such an ideal vaccine has not been developed.

PRODUCTION OF VACCINES

For the production of vaccines against viral diseases, strains of the virus are often grown by using embryonated eggs. Individuals who are allergic to eggs cannot be given such vaccine preparations. Viral vaccines are produced by tissue culture. For example, the older rabies vaccine which was produced in embryonated duck eggs and had painful side effects, has been replaced with a vaccine produced in human fibroblast tissue cultures and has fewer side effects. The production of vaccines by bacteria, fungi and protozoa generally involves growing the microbial strain on an artificial medium, which minimizes problems with allergic response. Vaccines should be tested and standardized before use (Figure 11.1).

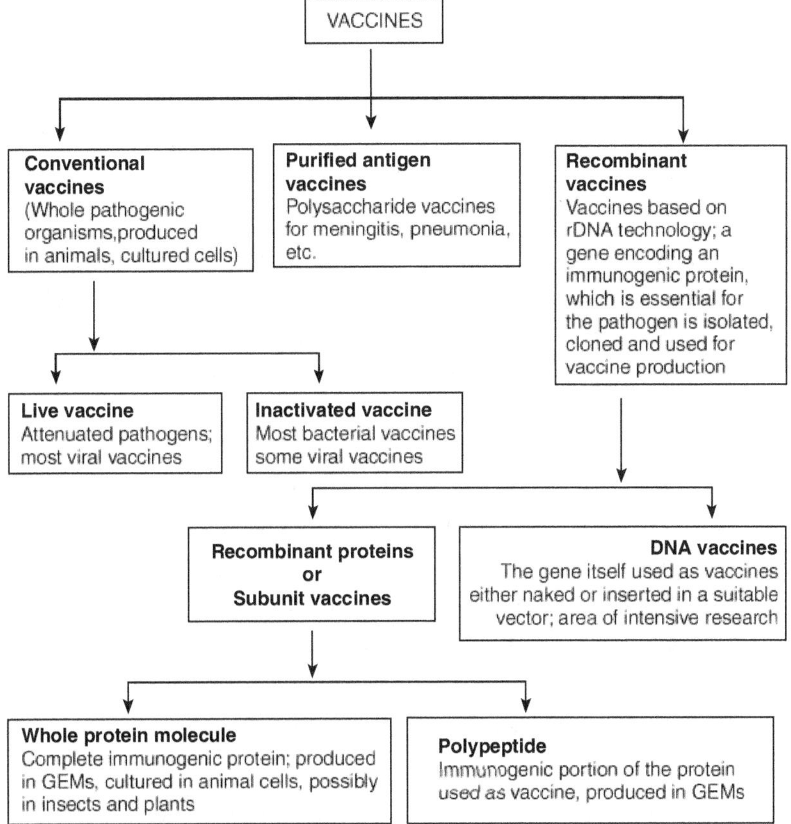

Figure 11.1 Summary of different types of vaccines either in commercial use or in various stages of development (DNA vaccines)

The majority of licensed vaccines for humans and animals presently in use are produced by conventional methods by manipulating the genes of bacteria causing diseases, and using the restructured genes for making antibodies. Conventional vaccines consist of whole pathogenic organisms which may either be killed, or live (attenuated). These vaccines, although highly effective and relatively easy to produce at low cost, suffer from many limitations.

Advantages of killed vaccines

* One of the major advantages of killed vaccines is that they are relatively stable under environmental conditions; therefore, it is not so crucial to maintain a cold chain to ensure efficacy of the vaccines.

* In specific disease situations such as rabies virus, the personnel involved are reluctant to use live viral vaccines, because of the fear that they may inject themselves with the vaccine and there may be some adverse side effects.

Disadvantages of killed vaccines

* The disadvantages of killed vaccines are that they do not replicate within the host and, therefore, large amounts of antigen are required for injection before induction of immunity. Since these vaccines are often produced in foreign tissue, there is also the possibility of reactions developing against foreign proteins.

* Since the vaccines are killed, they are generally injected intramuscularly.

* A killed vaccine is more effective against systemic viruses, which replicate in local mucosal sites. This latter disadvantage has lead to the development of a large number of attenuated vaccines.

Advantages of attenuated vaccines

* The main advantage of attenuated vaccines results primarily from their ability to replicate in the host.

- Since their mode of action is similar to natural infections, immunity is generally of a broader spectrum than it is with killed virus vaccines. Thus they can induce a range of immune response both locally as well as systemically.

- Attenuated vaccines develop immunity for longer duration than killed virus vaccines.

- Since the virus replicates in the host and produces large quantities of proteins to which the host responds, the possibility of injecting foreign proteins is dramatically reduced with attenuated virus vaccines.

Disadvantages of attenuated vaccines

- Since the vaccines are produced by passage in culture, to induce random mutation(s) or mutated with a specific agent and thereby reduce virulence, it is possible that passage in the natural host may result in reversion back to virulence, e.g. attenuated polio virus. In the case of polio, reversion can occur within a few days of oral immunization.

- Interference is also a potential problem. When viruses are grown in culture, there is a possibility to have other contaminating viruses present in it. For example, the presence of BVD (a ubiquitous virus) in viral vaccines grown for immunizing cattle is very common. This virus is present in many of the cell lines and foetal bovine sera than are used for growing bovine viruses. Such interference may result in reduced replication of the attenuated virus vaccine and thus, reduced immunity.

- Live attenuated virus vaccine induces latent infections and abortions if not administered properly or if administered at the wrong time in the animal's life.

- These limitations have prompted the successful development of vaccines, which are rather costly, at least for the present.

PURIFIED ANTIGEN VACCINES

These vaccines are based on purified antigens isolated from the concerned pathogens, i.e., these are non-recombinant. Since they do not contain the organism, the risk of pathogenicity is avoided. However, their cost is higher due to the steps involved in purification and vaccine preparation, and many of the isolated antigens are poorly immunogenic, e.g. polysaccharide antigens from the bacterial cell wall capsules of *Neisseria meningitis and Streptococcus pneumoniae.*

Many bacteria produce exotoxins which are highly immunogenic. But these toxins produce toxic effects, the intensity of which decreases with storage and heat, and formaldehyde and other chemicals accelerate this decline. However, most exotoxins that have lost their toxicity retain their immunogenicity; they are called **toxoids** and are used as effective vaccines, e.g. toxoids of the pathogens causing tetanus, diphtheria, gangrene, etc. Alum-precipitated toxoids are highly antigenic. The toxoid vaccines are quite effective and cheap.

Some toxoids are very good adjuvants, i.e., they increase the immunogenicity of other antigens, e.g. diphtheria toxoid. However, the B polysaccharide of *Haemophilus influenzae* is poorly immunogenic. But when the B polysaccharide is combined with diphtheria toxoid, its immunogenicity is greatly increased. In many cases, such adjuvant activities can be used to great advantage since most of the isolated antigens from pathogens are poorly immunogenic.

An immunizing serum (plural "sera") is the actual blood serum of a person or animal and contains an immunizing material or antibody. It is generally obtained by injecting a toxoid into an animal, permitting establishment of active immunity in the animal, and then withdrawing some of its blood. The corpuscles are removed from serum, which is then standardized as to the strength of antibodies and stored in sterile tubes or bottles.

Immunizing sera are used mostly for cure rather than prevention. They may be given for prevention after exposure has taken place, but their immunizing effects are too short-lived to give benefit if administered much in advance of the disease. As a curative agent, the action of immunizing serum is very rapid. Immunizing serum has been used in the cure of diphtheria and tetanus.

RECOMBINANT VACCINES

A recombinant vaccine contains either a protein or a gene encoding protein of pathogen origin that is immunogenic and critical to the pathogen function; the vaccine is produced using recombinant DNA technology. The vaccines based on recombinant proteins (= proteins produced by recombinant DNA technology) are also called subunit vaccines. Genes that encode desired antigen in the immunization process are introduced into the non-pathogenic organism for a mass production of the proteins. The concerned proteins are then purified and mixed with suitable stabilizers and adjuvants, if required, and used for immunization. A subunit vaccine has no potential for reverting to a pathogenic organism because only limited number of components from the original pathogen is present in the vaccine. A subunit vaccine should also have fewer side effects, as it does not have any infectious agents.

Steps in subunit vaccine production

- Identify protective proteins or epitopes on the proteins. Once this is done, an individual can either produce a subunit vaccine by rDNA technology or by synthetic peptide technology.

- Identify gene coding for the protein.

- Clone the gene coding for the specific protein and express it in a suitable expression system.

* Purify the protective protein to homogeneity using bovine herpes virus-1. BHV-1 has four glycoproteins—GVPI, GVPII, GVPIII and GVPIV.

* Immunosorbent columns with monoclonal antibodies are prepared and used for purification of large quantities of the BHV-I glycoproteins. These glycoproteins are then mixed with the adjuvant avidine and used to immunize animals against BHV-I virus.

The host organism used for expression of immunogenic proteins to be used as vaccines may be a genetically engineered microorganism.

Hepatitis B virus-core protein is planted in *E. coli*, yeast, *Bacillus*, etc. AIDS viral coat protein is engineered in vaccinia virus; foot-and-mouth disease (FMD) core protein is planted in *E. coli*. Malarial parasites and syphilis organism which are difficult to be grown has been planted in *E. coli*.

Myrin *et al.* have developed a more effective and less toxic cholera vaccine. It gives immunity equivalent to infection with natural bacterium. It was genetically altered *Vibrio cholerae* cells of E1T/Incaba strain. Side effects like diarrhoea can be eliminated by scaling the dose.

Malarial vaccine from Hoffmann La Roche of Switzerland is on the horizon. Surface antigen in sporozoite stage has been identified and isolated to give protection. Antigens and concerned genes from organisms in other parts of parasitic life cycle are under search, for the preparation of a multiple vaccine, which may give a broader spectrum protection against malaria.

DNA of spirochaete (*Treponema pallidum*, the causative agent of syphilis) has been isolated from infected rabbit testicles and is cloned in *E. coli* using vector coliphage. This has enabled development of a more specific diagnostic screening test for syphilis and manufacturing of an effective vaccine. Need for such vaccine is due to the difficulty of extraction and purification of antigen by conventional techniques, on the one hand, and *Treponema pallidum* becoming resistant to antibiotics in current use and, therefore, difficulty faced in the treatment of syphilis, on the other.

Foot-and-mouth disease (FMD), a highly contagious disease in cattle, sheep and swine worldwide, is caused by the virus with a single-stranded RNA. German scientists have successfully cloned double-stranded DNA copies of viral RNA into plasmid pBR 322 and grown in *E. coli*. It was found that more than 100 molecules of viral proteins were synthesized per bacterial cell with a positive serological test.

FMD virus is having a capsid with 60 copies of 4 polypeptides GVP I, GVP II, GVP III and GVP IV. Treatment with trypsin results in the cleavage of VP I and considerable decrease in infectivity and loss of immunizing activity. Synthesis of VP I protein precursor is being attempted to give long-lasting immunity.

Rabies is a zoonosis endemic in animals and humans in several countries of Africa, South America and Asia and is increasing in Europe. Virus propagation is difficult, so the cost of vaccine production is high. The virus has to be propagated in human cells. Rabies virus-protein-encoding gene is now cloned in *E. coli*, thus producing cheaper vaccine will be possible.

Hepatitis B virus-antigen-producing gene can be cloned in *E. coli* to produce antigen and the vaccine. But with *E. coli*, complications can be there due to the risk of contamination by endotoxins. Hardy *et al.* (1981) have described the production of *Bacillus subtilis* with both core antigen of hepatitis B virus and major antigen (VP I) of FMD virus.

Gene coding for a hepatitis B virus surface antigen is introduced into baker's yeast also. This antigen can provide a strong immune response. Yeast cells are then cultured in large fermentors for several days during which time the cells rupture. A vital part of the process is proprietary chemical process, which assembles the antigenic protein into a circular particle and resembles the natural molecule closely.

About 12 million people suffer from leprosy. In India, 300,000 new cases are added each year. To eliminate the disease, an effective vaccine in large quantity will be required. *Mycobacterium leprae*, the causative agent of leprosy is difficult for cultivation in laboratory and so difficult is vaccine production. Scientists are on the way to identify important antigens of *M. leprae* and introduce them in *E. coli* by genetic engineering. Three leprosy vaccines are already on trial in India.

There has been significant progress on four of the five vaccines against agents that cause diarrhoea—rotavirus, *Salmonella* sp., cholera organisms and *Shigella dysentriae*. Other diseases for which vaccines based on antigens produced through genetic engineering are being attempted and are in the various stages of development and/or trial, are AIDS, influenza, meningitis, herpes, rabies, pertussis, measles and poliomyelitis, tuberculosis, rheumatic fever, pneumococcal infections.

Vaccines designed to control fertility, based on genetically engineered products such as beta-chain of the hormone human chorionic gonadotrophin (HCG), are also under investigation. The antibodies produced inactivate the hormone and the uterus does not accept the fertilized egg. Birth control vaccine developed in India is proved to be safe and devoid of any side effects. HCG currently used for vaccine production is from the trimester urine of a pregnant woman through a procedure that is difficult and expensive.

Cheap vaccine for schistosomiasis, which affects about 250 million people in the tropical regions of the world every year, is on the horizon. There is also progress in making a genetically engineered vaccine for dengue fever. Dengue fever is a mosquito-borne viral disease endemic in Asia, South Africa and South and Central America.

The discovery of a unique HB antigen called Australia antigen (HBsAg) in 1965 in the blood of infected persons has lead to the development of two effective vaccines against this dreaded disease: a plasma-derived vaccine in 1981, and a more acceptable recombinant DNA technology-based yeast-derived vaccine in 1986. Dr. Baruch Blumberg received Nobel Prize in 1976 for his pioneering contributions to HB. The HBsAg is present in the serum of infected people as small particles of 22 nm size. These nucleic acid-free particles have no infectious properties. The blood serum taken from an acute case of HB contained enough HBsAg to accord at least this partial protection. In 1981, through this concept, HB vaccine was made available by Merck and Co., USA and Institute Pasture Production, Paris as Hepatavax-B.

The second generation of HB vaccine has been developed by employing the techniques of genetic engineering. The fragments of DNA are joined to the DNA of a suitable vector. The vector may be a plasmid, phage or a cosmid. This is done by the use of restriction endonucleases. Artificial linkers containing restriction sites can be attached to the foreign DNA fragments to allow insertion into the vector. Once joined, the composite DNA is introduced into the bacteria or eukaryotic cell system either by transformation or transfection. After successful integration of the composite genome into that of the host cell, it affects the metabolism of the host cell. This further directs the translation of additional genetic information packed in vector DNA, and consequently leads to the synthesis of foreign proteins by the host cells. In this way, the host cell acts as a mini-factory for the production of specific protein which is foreign to it.

Control of Hepatitis-B Virus through Vaccines

Hepatitis-B is a major public health problem. An estimated 210 million hepatitis-B virus carriers are in the world. India is known to be the second largest reservoir of this virus. The disease caused by this virus is characterized by higher morbidity, mortality and an amazing speed of spread of infection. About 80% liver cancer is attributed to hepatitis-B infection. Till date no chemotherapeutic agents are available for its treatment. Prophylaxis against this has come in practice in 1981 with the introduction of a plasma-derived vaccine. The scientists feared to introduce this vaccine because of the possible danger of spread of other infections including AIDS. In 1986, a genetically engineered yeast-derived vaccine has been introduced that is of non-plasma origin and has been widely accepted by the public.

After infection, HBV fails to grow and even in cultured cells it does not grow. This property has been explained to be due to inhibition of its molecular expression. A recombinant vaccine for HBV was produced by cloning HBsAg gene of the virus in yeast cells. The yeast system has its complex membranes and ability of secreting glycosylate protein. This has made it possible to build an autonomously replicating plasmid containing HBsAg gene near the yeast alcohol dehydrogenase (ADH) I promoter (Figure 11.2). The HBsAg gene contains 6-bp long sequence preceding the AUG that synthesizes N-terminal methionine. This is joined to ADH promoter cloned in the yeast vector PMA-56. The recombinant plasmid is inserted into yeast cells. The transformed yeast cells are multiplied in tryptophan-free medium. The transformed cells are selected. The cloned yeasts are cultured for expression of HBsAg gene. This inserted gene sequence expresses and produces particles similar to the 22 μm particle of HBV as these particles are produced in the serum of HBV patients. The expressed HBsAg particles have similarity in structure and immunogenicity with those isolated from HBV-infected cells of patients. Its high immunogenicity has made it possible to market the recombinant product as vaccine against HBV infection.

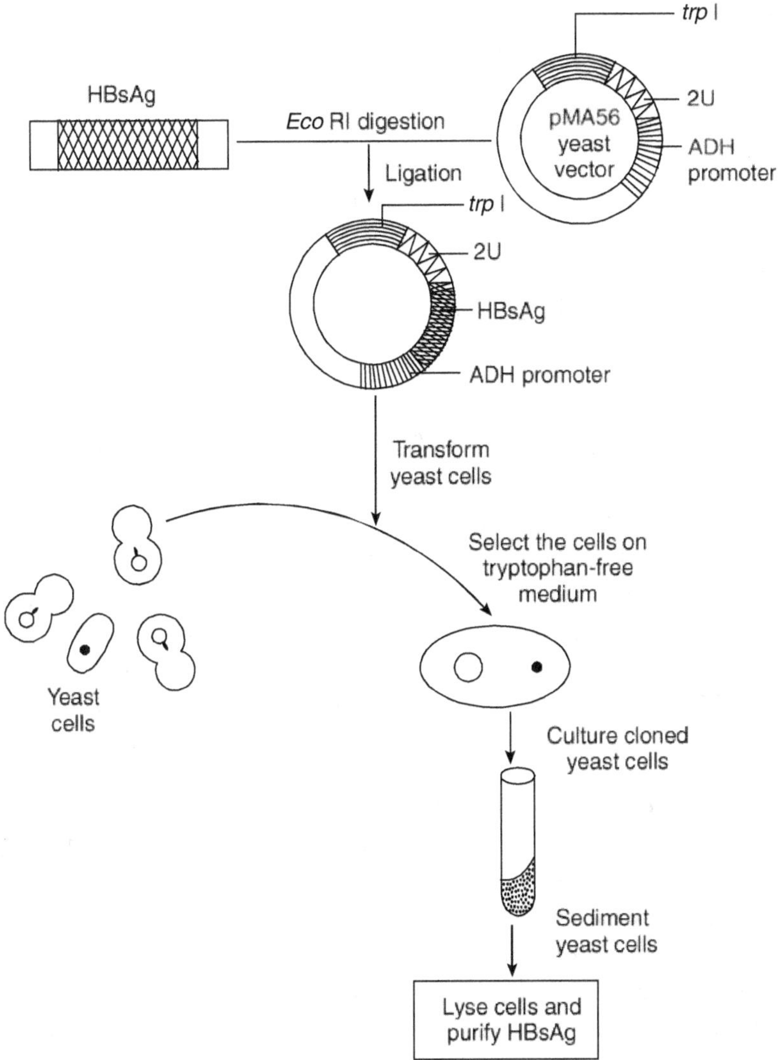

Figure 11.2   Expression of HBsAg gene in yeast

RECOMBINANT POLYPEPTIDE VACCINES

Generally, the whole protein molecule is not necessary for immunogenicity; the immunogenic property is usually confined to a small portion of the protein molecule. For example, the immunogenicity of foot-and-mouth disease virus coat protein is due to its amino acids 114–160, and also 210–213. Segments of proteins containing either of these two amino acid sequences are effective in immunization; they induce antibodies which neutralize the virus and thereby provide protection against the FMD. Similarly, the immunogenicity of coat protein of feline leukemia virus (FLV) is due to a 14-amino acid long segment; this segment produces a partial immunogenic response in guinea pigs.

In some cases, the immunogenic protein may be composed of 2 or more distinct polypeptides. In such cases, it may be desirable to use only one of the polypeptides as a vaccine for various reasons. For example, the cholera enterotoxin consists of 3 polypeptides, viz. A_1, A_2 and B polypeptides. The A polypeptides are toxic, while the B polypeptide is non-toxic but immunogenic. The gene encoding B polypeptide has been cloned, and the recombinant B polypeptide thus produced is being used as a vaccine against cholera; the recombinant B polypeptide is used in combination with inactivated cholera cells, as an oral vaccine in place of the conventional injectable vaccine.

Recombinant protein or polypeptide vaccines are safe since whole organisms are not involved. They are also of high efficacy. But (i) their cost is very high and often prohibitive, since they are produced by either bacterial fermentation or in animal cell cultures, (ii) they have to be stored in low temperatures since heat destabilizes the proteins, (iii) this makes their storage and transportation, especially in developing countries, problematic and often limiting.

Recombinant peptides corresponding to species-specific **circumsporozoite protein** (CSP) of *Plasmodium* have been obtained and show promising results as antimalarial vaccines. The introduction of the specific protein into various expression systems help us to identify the specific epitopes involved in inducing protective immunity and synthesis of the peptide. Using monoclonal antibodies against different proteins of bovine rotavirus, one can identify a immunodominant neutralizing epitope on the outer coat glycoprotein of bovine rotavirus. This epitope is identified by the ability of monoclonal antibody directed against it to neutralize virus *in vitro* as well as to prevent diarrhoea in animal *in vivo*. The advantage of peptide vaccines is that it is not possible to identify crucial epitopes on all viruses. Synthetic polypeptides generally are not very immunogenic unless they are linked to specific carriers and incorporated in strong adjuvants. In addition to incorporation of adjuvants into peptide vaccines, the peptides can be engineered in such a way that they are linked to specific carriers, which can act as ideal delivery systems for presenting the important epitopes on the peptide.

MINICELLS AS VACCINES

Minicells derived from bacteria (*E. coli* or *B. subtilis*) carrying cloned genes encoding immunogenic antigen or from minicell producing (**min**) mutants of pathogens might be used as vaccines. Minicells do not replicate and will not survive long in the body. Thus they may provide antigenic stimulus to the host by inoculation via natural routes of infection (most probably the gut) without the problem of administering infective live organisms. Since production of minicells is yet not problem-free, contaminating nucleated cells of pathogens may be present. This limits the use of minicells as vaccine.

RECOMBINANT-VECTOR VACCINES

It is possible to introduce genes that encode major antigens of especially virulent pathogens into attenuated viruses or bacteria. The attenuated organism serves as a vector, replicating within the host and expressing the gene product of the pathogen. A number of organisms have been used for vector vaccines, including vaccinia virus, the canary poxvirus, attenuated poliovirus,

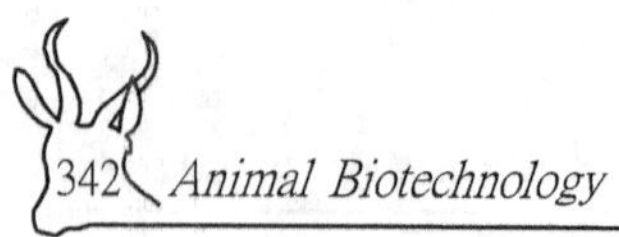

adenoviruses, attenuated strains of *Salmonella,* and the BCG strain of *Mycobacterium bovis.*

Vaccinia vector, the attenuated vaccine used to eradicate smallpox, has widely been employed as a vector vaccine. This large, complex virus, with a genome of 200 genes, can be engineered to carry several dozen foreign genes without impairing its capacity to infect host cell and replicate. The procedure for producing a vaccinia vector that carries a foreign gene from a pathogen is outlined in Figure 11.3. The genetically engineered vaccinia expresses high levels of the inserted gene product, which can then serve as a potent immunogen in an inoculated host. Like the smallpox vaccine, genetically engineered vaccinia vector vaccines can be administered simply by scratching the skin, causing a localized infection in host cells. If the foreign gene product expressed by the vaccinia is a viral envelope protein, it is inserted into the membrane of the infected host cell, inducing development of cell-mediated immunity as well as antibody-mediated immunity.

Other attenuated vector vaccines may prove to be safer than the vaccinia vaccine. The canary poxvirus has recently been tried as a vector vaccine. Like its relative vaccinia, the canary poxvirus is a large virus that can be easily engineered to carry multiple genes. Unlike vaccinia, the canary poxvirus does not appear to be virulent even in individuals with severe immune suppression. Another possible vector is an attenuated strain of *Salmonella typhimurium,* which has been engineered with genes from the bacterium that causes cholera. The advantage of this vector vaccine is that *Salmonella* infects cells of the mucosal lining of the gut and therefore will induce secretory IgA production. Effective immunity against a number of diseases, including cholera and gonorrhoea, depends on increased production of secretory IgA at mucous membrane surfaces. One of the poliovirus strain used in the Sabin vaccine is another candidate for a safe and effective vector vaccine. In this case, the poliovirus vector is engineered so that a portion of the gene that encodes the outer capsid protein

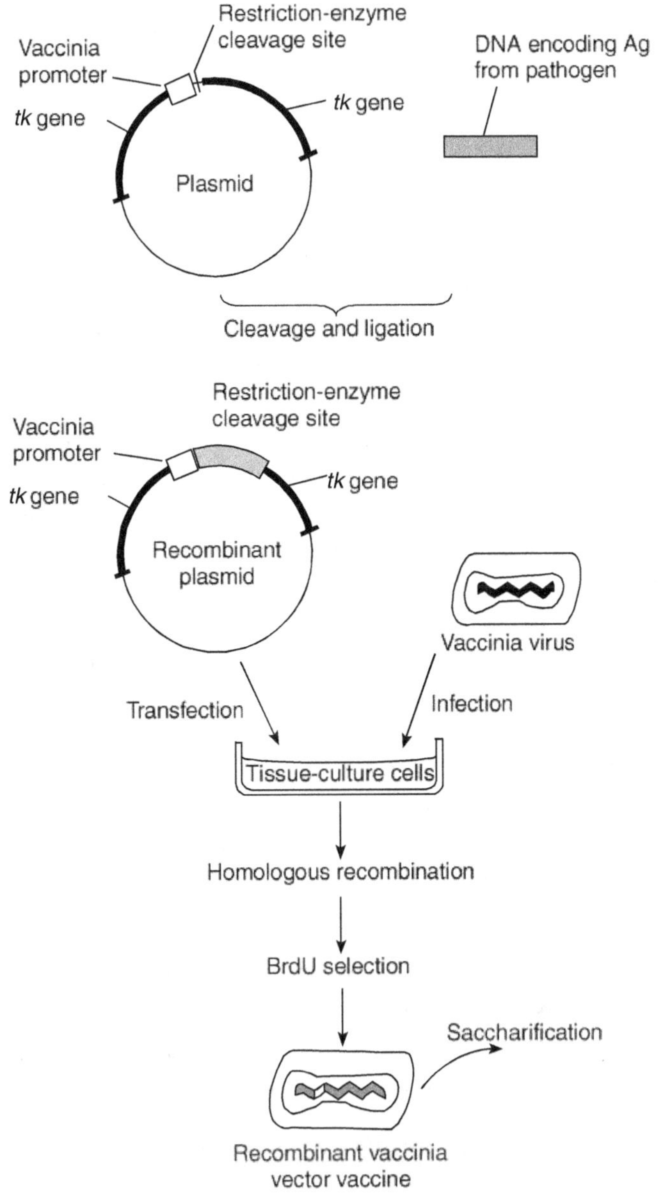

Figure 11.3   Production of vaccinia vector vaccine

of poliovirus is replaced by the DNA that encodes the epitope of choice. The resulting poliovirus chimera will express the desired epitope in a highly accessible presentation that protrudes from the nucleocapsid.

DNA VACCINES

Recently, vaccines based on DNA are being developed, and the results obtained with the influenza virus are quite exciting; these are regarded as the third revolution in vaccines. The strategy of DNA vaccines is as follows: the gene encoding the relevant immunogenic protein is isolated, cloned and then integrated into a suitable expression vector. This preparation is introduced into the individual to be immunized. The gene is ultimately expressed in the vaccinated individual and the immunogenic protein is expressed in sufficient quantities to invoke both humoral and cell-mediated immunities. It may be pointed out cell-mediated immune response is essential for recovery from infectious diseases. The various approaches for DNA vaccines are as follows:

1. Injection of pure DNA (or RNA) preparation into muscle.

2. Use of vectors (e.g. vaccinia viruses, adenoviruses, retroviruses, *E. coli*, *Salmonella typhimurium*, herpesviruses, etc.)

3. Reimplantation of autologous cells (cells of the individual to be vaccinated) into which the gene has been transferred, and

4. Particle gun delivery of plasmid DNA that contains the gene in an expression cassette.

Injection of pure DNA/RNA into the skeletal muscle leads to the uptake and expression of the DNA in the muscle cells, leading to both a humoral antibody response and a cell-mediated response. It is surprising to note that the injected DNA is taken up and expressed by the muscle cells with much greater efficiency than in tissue culture. The DNA appears either to integrate into the chromosomal DNA or to be maintained for long periods in an episomal form. The viral antigen is expressed not only by the muscle cells but also by dendritic cells in the area that take up the plasmid DNA and express the viral antigen. The fact that muscles express rather low-level of class I MHC molecules and do not express co-stimulatory molecules suggests that dendritic cells in the area may be crucial to the development of antigenic responses to DNA vaccines.

Another approach is to remove cells from the body of an individual into which the concerned immunogen-encoding gene is introduced and expressed. These cells are then reintroduced into the body of the individual in a variety of ways, e.g. simple infusion, implantation, encapsulation, etc. This approach although more cumbersome, has the advantage of enabling control of the modified cells within containment.

The immunogen-encoding gene may be integrated into an expression plasmid, which is purified, coated on gold or tungsten particles and introduced into skin cells by a particle gun. Antigen genes introduced into the skin of mice and guinea pigs elicited humoral immune response; it is not known if cellular immunity is also induced. Plasmid DNA is non-infectious, heat-stable and offers other advantages over viral/bacterial vectors. The skin cells are usually shed off in a few days after the inoculation, so that there is no long-term persistence of the modified cells.

DNA vaccine offers advantages over many of the existing vaccines. The encoded protein is expressed in the host in its natural form—there is no denaturation or modification. The immune response will be directed to the antigen exactly as it is expressed by the pathogen. DNA vaccines induce both humoral and cell-mediated immunity. This stimulation of both arms of the immune response normally requires immunization with a live attenuated vaccine. DNA vaccines cause prolonged expression of the antigen, which generates significant immunological memory.

The practical aspects of DNA vaccines are also very promising. Refrigeration is not required for the handling and storage of the plasmid DNA, a feature that greatly reduces the cost complexity of delivery. The same plasmid can be custom-tailored to make a variety of proteins, so that the same manufacturing techniques can be used for different DNA vaccines, each encoding an antigen from a different pathogen.

An improved method for administering these vaccines is to coat microscopic gold beads with the plasmid DNA and then deliver the coated particles through the skin into the underlying muscle with an air gun (called a **gene gun**). This will allow rapid delivery of a vaccine to large populations without the requirement for massive quantities of needles and syringes.

Tests of DNA vaccines in animal models have shown that the vaccines are able to induce protective immunity against a number of pathogens, including the influenza virus. At present, there are human trials underway with several different DNA vaccines, including those for malaria, AIDS, influenza, and herpes virus. Future experimental trials of DNA vaccines will mix genes for antigenic proteins with those for cytokines or chemokines that direct the immune response to the optimum pathway. For example, the IL-12 gene may be included in a DNA vaccine. The expression of IL-12 at the site of immunization will stimulate TH_1 type immunity induced by the vaccine.

Results to date with DNA vaccines are very promising, and they are likely to be used for human immunization within the next few years. Some drawbacks may prelude their universal application; for example, only protein antigens can be encoded—certain vaccines, such as those for pneumococcal and meningococcal infections, use protective polysaccharide antigens. Another shortcoming is the present inability to use DNA immunization with vaccines, such as the oral vaccines or those given as a nasal spray, that are applied to mucosal surfaces.

The first published report from India indicates modest success in the development of DNA vaccines against rabies and Japanese encephalitis virus (JEV) in experimental animals. The efficacy of DNA vaccine (G protein) against rabies is correlated to levels of neutralizing antibodies, whereas in the case of JEV envelope protein, cell-mediated immunity appears to be the major mechanism of protection. There is great hope that DNA vaccines can offer protection against serious infectious diseases. Most likely it is to become a tool to benefit mankind in the 21st century as such or in combination with recombinant/cell culture vaccines or as an adjunct to chemotherapy (Padmanaban, 1999).

In the case of malaria, DNA vaccines have distinct advantage, where plasmid DNA encoding different antigens and prepared by the same genetic procedure can be mixed and administered. A mixture of 4 plasmid DNA (pfCSP, pfSSP2, pfEXP-1 and pfLSA-1) has been injected into Rhesus monkeys and found to elicit multiple antigen-specific cytotoxic T lymphocytes.

It has been reported that the DNA vaccine coding for a mycobacterial heat-shock protein of molecular weight 65000 (HSP65) when administered in 4 doses to mice, 8 weeks after intravenous injection of virulent *M. tuberculosis* H37RV, leads to a dramatic decrease in number of live bacteria in spleen and lungs, 2 months and 5 months after the first dose of DNA. Certain other mycobacterial antigens and BCG did not have this effect.

ANTI-IDIOTYPIC VACCINES

When antibodies are generated against other antibodies that bind to a particular antigen, some of the anti-antibodies have the immunological properties of original antigen. Such antibodies are called anti-idiotypic antibodies. Thus, anti-idiotypic antibody is a surrogate antigen that can be formulated as vaccine because it

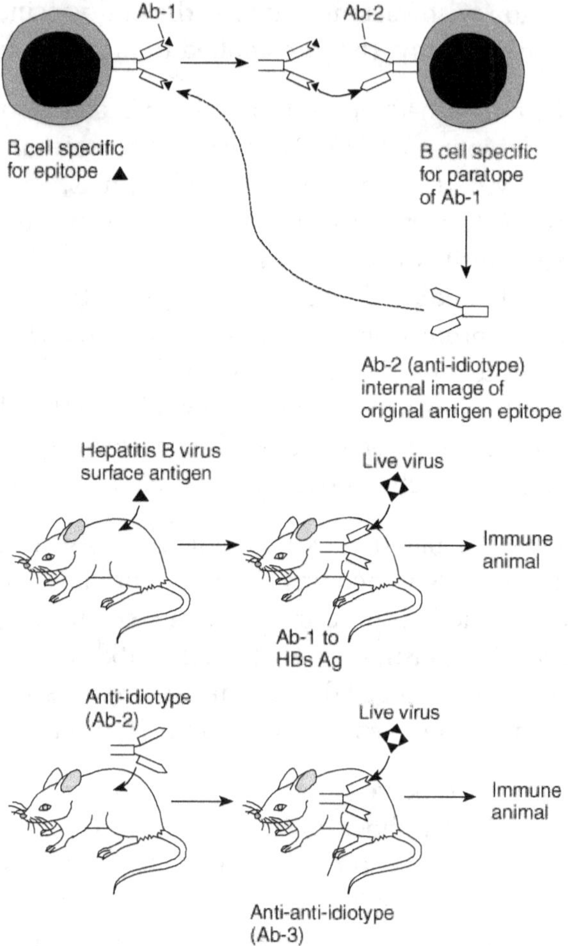

Figure 11.4 Use of anti-idiotype antibody as a vaccine. (a) The binding site on some anti-idiotype antibodies (Ab-2) resembles the structure of the eiptope on the original antigen. Such anti-paratope antibody can interact with B cells specific for the original antigen, thus inducing production of more antibody against the antigen. (b) Immunization with anti-idiotype antibody has been shown experimentally to protect mice against hepatitis B virus without exposing the animal to the virus. HBsAg = Hepatitis B surface antigen.

mimics a specific antigen (Figure 11.4). Anti-idiotypic antibodies could be used when a) antigenic material is not easily obtained; b) either the target molecule or the organism itself is only weakly antigenic; c) it is either dangerous or difficult to culture a pathogenic organism; d) a specific determinant of pathogen needs to be targeted for neutralization; and e) the antigen is not a protein.

Preliminary studies of anti-idiotypic antibodies as vaccines are carried out against *Trypanosoma rhodesiense* (the human sleeping sickness parasite) in mice and coccidiosis caused by the protozoan parasite *Eimeria tenella* in chickens. Anti-idiotypic antibodies that mimic features of certain viruses and bacteria effectively elicited neutralizing antibodies in host organisms.

Multivalent Subunit Vaccines

One of the limitations of synthetic peptide vaccines and recombinant protein vaccines is that these vaccines tend to be poorly immunogenic; in addition, they tend to induce a humoral antibody response but are less likely to induce a cell-mediated response. Now it becomes essential to construct a synthetic peptide vaccine that contains both immunodominant B cell and T cell epitopes. Furthermore, if a CTL response is desired, the vaccine must be delivered intracellularly so that the peptides can be processed and presented together with class I MHC molecules. A number of innovative techniques are being applied to develop multivalent vaccines that can present multiple copies of a given peptide or a mixture of peptides to the immune system.

One approach is to prepare solid matrix–antibody–antigen (SMAA) complexes by attaching monoclonal antibodies to particulate solid matrices and then saturating the antibody with the desired antigen. The resulting complexes are then used as vaccines. By attaching different monoclonal antibodies to the solid matrix, it is possible to bind a mixture of peptides or proteins, composing immunodominant epitopes of both T cells and B cells, to a solid matrix (Figure 11.5a). These multivalent complexes have been shown to induce vigorous humoral and cell-mediated immunity. Their particulate nature contributes to their increased immunogenicity by facilitating phagocytosis by phagocytic cells.

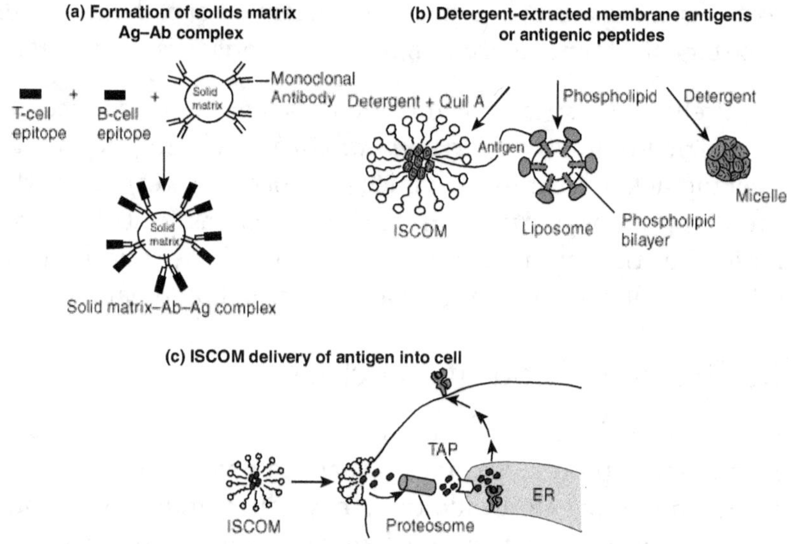

Figure 11.5   Multivalent subunit vaccines

Another means of producing a multivalent vaccine is to use detergent to incorporate protein antigens or synthetic antigenic peptides into protein micelles, into lipid vesicles (called liposomes), or into immunostimulating complexes (Figure 11.5b). Micelles are formed by mixing proteins in detergent and then removing the detergent. The individual proteins orient themselves with their hydrophilic residues towards the aqueous environment and the hydrophobic residues at the centre so as to exclude their interaction with the aqueous environment. Liposomes containing protein antigens are prepared by mixing the proteins with a suspension of phospholipids under conditions that form vesicles bounded by a bilayer. The proteins are incorporated into the bilayer with the hydrophilic residues exposed. Immunostimulating complexes (ISCOMs) are lipid carriers prepared by mixing protein or peptide antigens with detergent and a glycoside called Quil A.

Membrane proteins from various pathogens, including influenza virus, measles virus, hepatitis B virus, and HIV have been incorporated into micelles, liposomes, and ISCOMs and are currently being assessed as potential vaccines. In addition to their increased immunogenicity, liposomes and ISCOMs appear to fuse with the plasma membrane to deliver the antigen intracellularly, where the cytosolic pathway can process it and thus induce a cell-mediated response (Figure 11.5c).

GENETIC IMMUNIZATION

This is a variation of the recombinant vaccine strategy. A cloned gene encoding an antigen is delivered to cells in the ears of mice by biolistic transformation system. The gene gets incorporated into the chromosomal DNA and directs the synthesis of the protein antigen. This in turn activates the immune system of mouse to produce corresponding antibodies against target antigen. This procedure may be interesting because it surpasses costly and time-consuming procedures of purifying an antigen or creating recombinant vaccine. This approach is referred to as "genetic immunization" and may be useful for vaccination of domestic animals.

VACCINE EXPRESSION SYSTEMS

Once the specific protective proteins are identified, it is important to develop expression systems to produce large quantities of the specific protein in an economical fashion. At present, there are 4 different expression systems.

Bacterial expression   It is the viral protein produced in bacteria, which are often not folded properly for induction of the desired immune response. A considerable amount of activity is being directed towards using other viruses such as vaccinia, adenoviruses or herpes viruses to express specific viral proteins in mammalian systems.

Virus expression   Another very popular expression system is the application of baculovirus, an insect virus, to produce high quantities of animal virus proteins in insect cells. Vaccinia virus will be used as an example wherein a number of different viral proteins have been introduced into the vaccinia virus and are being used by a variety of different delivery systems to induce both local as well as systemic immunity. The advantage of vaccinia expression is that both humoral, as well as cellular immune response is elicited. Even more attractive is the large size of vaccinia genome and it is possible to delete large quantities of its genome and still maintain a viable virus. The non-essential vaccinia genes can be replaced with a number of genes coding for other proteins from other viruses. Expression of a number of genes in one virus would be much more economical than to culture each individual vaccine independently.

Vaccinia can replicate in a wide variety of hosts, making it attractive for controlling infectious diseases in veterinary medicine and in human medicine. Furthermore, its thermal stability and ability to replicate in a wide variety of hosts provide an opportunity to immunize wildlife against infectious diseases that can be transmitted to domestic livestock.

Eukaryotic cell system   This is the most natural method of producing non-infectious viral vaccines. The reason for the attractiveness of eukaryotic cells is that the proper level and degree of glycosylation and folding is more natural than in prokaryotic systems. At present, a number of viral genes have been successfully expressed in yeast, mammalian cells, and more recently in green algae and filamentous water fungi. Green algae and filamentous water fungi provide large quantities of cheap proteins with the correct post-translational modification required for proper recognition of the hosts' immune system.

Yeast (*Saccharomyces cerevisiae*) is used as an expression system because of our extensive experience with this organism, and animals already have antibodies to yeast. Moreover, yeasts do not have any oncogenes. This makes vaccines expressed in yeasts potentially safer than vaccines produced in mammalian cells. But sometimes yeast may over-glycosylate proteins, which may influence the immune response to that specific protein.

Mammalian cells Mammalian cells are used for the continuous production and secretion of viral proteins and glycoproteins. However, the level of expression is relatively low and the high cost of cell culture is a serious disadvantage in the use of mammalian cells for production of vaccines for veterinary use.

For mammalian cells to be an economically viable vehicle, we need Microcarrier, which produce large quantities of mammalian cells in a very concentrated environment, as well as strong promoters. Extensive progress is made in developing microcarriers to culture mammalian cells in a continuous fashion. In parallel, with the development of microcarrier systems are the requirement for new media and profusion of the bioreactors so that cells can grow continuously with minimal manipulations.

MONOCLONAL ANTIBODIES

Antibodies are special types of proteins, ordinarily glycosylated, called **immunoglobulins** (Ig) produced in response to antigens; an antibody specifically interacts with the antigen, which induced its production. The antigen–antibody interaction is highly specific, but some cases of **cross-reaction** (interaction of an antibody induced by one antigen with another antigen) are known possibly due to a similarity between the epitopes of the two antigens. **Epitope** or antigenic determinant is that portion of an antigen molecule, which is involved in antigen–antibody interaction. Interaction with an antibody renders the antigen molecule nonfunctional; this is the basis for protection or immunity produced by antibodies by an animal against their antigens. Therefore, production of antibodies by an animal in response to an antigen is known as **immune response**, and this activity of

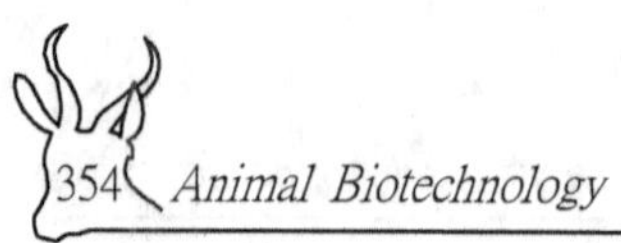

the antigen is called **immunogenicity**. In contrast, the ability of antigens to interact with antibodies is referred to as **antigenicity**.

In general, antigens are also immunogenic, and the same antigen epitope functions in both the phenomena. However, some small molecules show antigenicity but by themselves are unable to produce immune response; such molecules are called haptens. However, when a hapten is linked to a large molecule called carrier, it elicits an immunogenic response as well as induces antibody production against itself. The carrier molecule may itself be either non-immunogenic or immunogenic.

ANTIBODY STRUCTURE

An antibody molecule consists of two identical light chains (220 amino acids each) and two identical heavy chains (about 440–450 amino acids each) held together by disulphide bridges; this constitutes the monomeric form of an antibody. The enzyme papain cleaves each monomeric form into two fragments that bind to the antigen (designated as Fab or fragment with antigen binding) and one fragment, which does not bind to antigen but forms crystals and hence called crystal forming fragment Fc (Figure 11.6).

The amino-terminal ends of both light and heavy chains, which are about 100 amino acids long constitute their **variable region** denoted as V_L and V_H respectively; the amino acid sequence of this region varies among antibodies specific to different antigens. The remaining regions of the heavy and light chains are called constant region (designated as C_H and C_L respectively).

Each antibody molecule has two **antigen-binding sites** or **domains**, each domain being constituted by the variable regions of one light band one heavy chain. The constant regions of two heavy chains of an antibody molecule form its effector function domain, which determines its interaction with the other components of the immune system. The light chains are of two

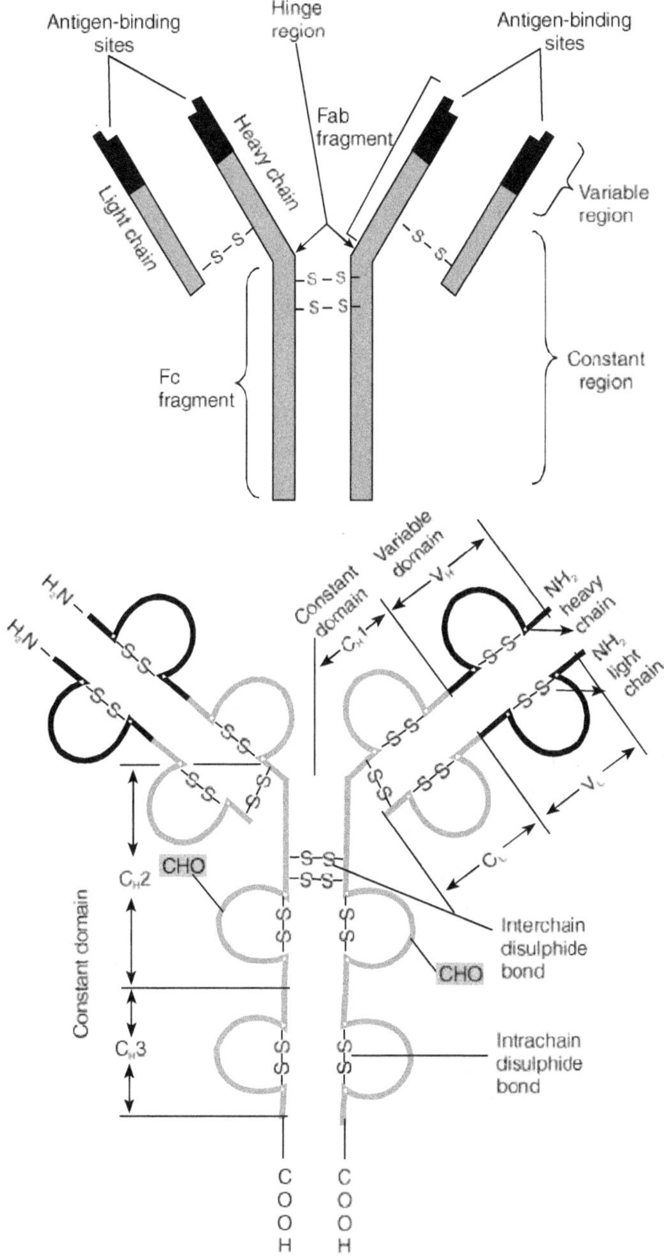

Figure 11.6 Diagrammatic representation of an immunoglobulin (antibody) structure

types: **Kappa** (κ) and **Lambda** (λ); the type of light chain is determined by its constant region. An antibody molecule contains either only two kappa or only two lambda light chains. Separate genes encode the kappa (located in human chromosome 2) and lambda (chromosome 22) light chains, and the heavy chain (chromosome 14).

The variable region of each chain contains 3 highly variable regions, called **hypervariable regions**, and denoted as CDR1, CDR2 and CDR3 (CDR = **complementarity-determining region**) separated by 4 invariant regions called **Framework regions** (designated as FR1, FR2, FR3, FR4). The constant region of each heavy chain has 3 homologous regions CH_1, CH_2 and CH_3) which most likely originated from a common parental gene (3 tandem repeats of the parental gene, followed by mutations).

CLASSES OF IMMUNOGLOBULINS

Immunoglobulins (antibodies) are grouped into five classes based on their unique antigenic properties, which reside in their heavy chains (in the constant region). The five classes are IgG, IgM, IgA, IgD and IgE. The heavy chains of an Ig class are designated by the small Greek letter corresponding to the class of Ig, i.e., γ heavy chain in IgG, μ in IgM, δ in IgA, ε in IgD and ε in IgE. An individual has antibodies for all the five classes of Ig.

The immunoglobulins are further divided into subclasses based on antigenic differences within each of the different classes.

IgG is a monomer, makes up 75% of the total human Ig, and is the only class of Ig which can cross the placenta and protect the newborn. IgM constitutes 10% of human Ig, is a pentamer of the monomeric units and has a fourth C_H domain ($C_H 4$). IgA is of 2 types: serum IgA and secretory IgA. IgD is very low in serum. Both IgD and IgM are present on the surface membranes of mature but virgin B lymphocytes. IgE is involved in hypersensitivity and allergy, and is present only in very low quantity, 0.00005 mg/ml, in the serum. The Fc portion of IgE binds strongly to a receptor

Lymphocytes are specialized cells which are responsible for the immune response of an animal. Lymphocytes originate from pluripotent stem cells present in bone marrow; these stem cells give rise to both lymphoid stem cells (giving rise to lymphocytes) and to those that produce the various blood cells (myeloid stem cells). The lymphoid stem cells, in turn give rise to lymphocyte progenitor cells, pre-B and pre-T cells, which are further processed in lymphoid-inducing environments of bone marrow and thymus. Bone marrow processed cells mature into **B lymphocytes** while those processed in thymus develop into **T lymphocytes** (Figure 11.7).

Bone marrow and thymus provide the sites for lymphocyte differentiation, and are called **primary lymphoid organs**. The mature lymphocytes migrate to spleen and lymph nodes, which constitute the secondary lymphoid tissues; the B- and T-lymphocytes are present in discrete areas of these organs.

B lymphocytes are antibody-producing cells, while T cells act as effector, helper or suppressor cells. The characteristic feature of B lymphocyte is that they possess cell surface immunoglobulin (sIg), while T lymphocytes express on their surface a differentiation antigen, **Thy 1**. The differentiation of pre-T cells is brought about by inducing factors secreted by thymus epithelial cells; the factors that have been identified are thymosin (induces expression of Thy 1), **thymopoietin** and **thymic humoral factor**. The factors associated with B-cell differentiation are not yet known.

Lymphocytes are the major type of cells of the lymph nodes, and they lie in a fine meshwork called reticulum. Lymphocytes circulate through the body, passing between blood and lymphatic

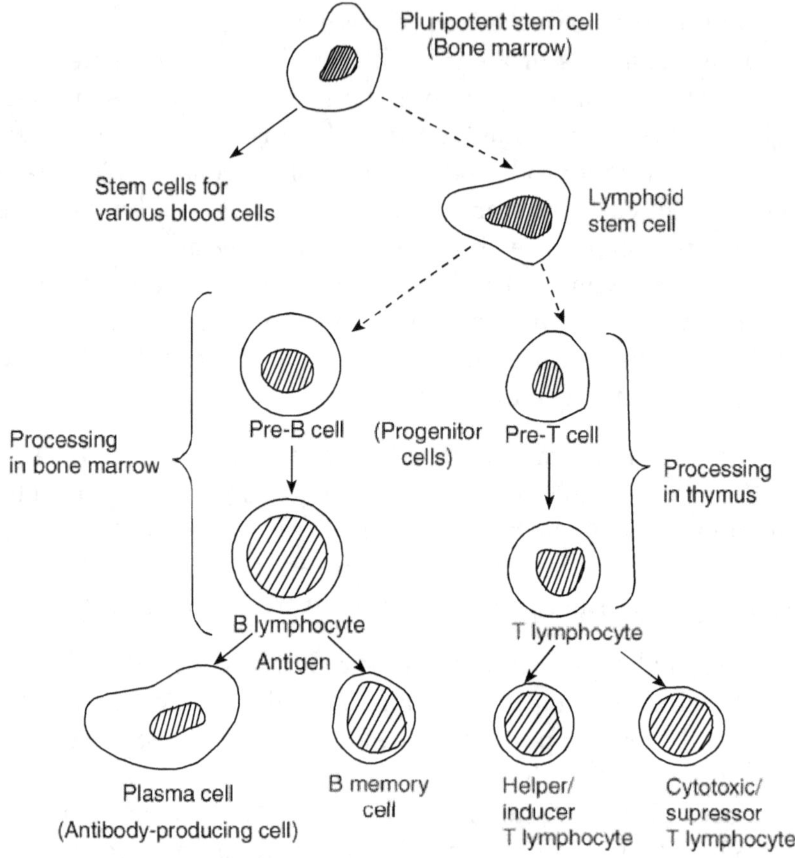

Figure 11.7 A schematic representation of lymphocyte differentiation

vessels at the lymph nodes. Some T cells act as **helper cells** to enable B cells to produce antibody (B cells are called **effector cells**). The helper T cells and effector B cells interact with different parts of the immunogen (the antigen inducing immune response). Thus, an immunogenic molecule must have one recognition site for helper T cells and another for effector B cells to be able to induce antibody production.

The surface immunoglobulin (sIg) present on a B cell has the same specificity as the antibody secreted by it. Thus the sIg molecules on a B cell surface act as antigen-specific receptors. However, the nature of T-cell receptor is not clearly understood, but the T cells react with the antigens in a specific manner. Binding of the antigen to sIg is necessary for the activation of B lymphocytes; however, it is also essential that the antigen must cause cross-linking of the receptor molecules for activation to occur. Once this has happened, proliferation and differentiation of the activated B cells is induced by Interleukin-4 (IL-4) produced by T lymphocytes. The differentiated B cells (plasma cells) produce and secrete the antibody.

MONOCLONAL AND POLYCLONAL ANTIBODIES

Antibodies are proteins synthesized in blood against specific antigens just to combat and give immunity to the body. They can be collected from blood serum of an animal. Such antibodies are heterogeneous and contain a mixture of antibodies, i.e., polyclonal antibodies. Hence, they do not exhibit specificity. If a specific lymphocyte, after isolation and culture *in vitro*, becomes capable of producing a single type of antibody, which bears specificity against a specific antigen, it is known as "**monoclonal antibody**.

HYBRIDOMA TECHNOLOGY

Hybridoma, a biotechnological tool, is the new path for achieving the goal of complete immunization of human body from infections. Hybridoma technique is a method of creating pure and uniform antibodies (immunoglobulins) against a specific target (antigen).

Hybridomas are the hybrid cells of myeloma (cancer) cells with antibody-producing cells (lymphocytes) from an immunized donor (animal). The hybrid cell or hybridoma resulting from this fusion has the ability to multiply rapidly and indefinitely *in vitro* and to produce an antibody of predetermined specificity, known as monoclonal antibody.

The technique of producing monoclonal antibodies in specialized cells is now popularly known as hybridoma technology. This technology was discovered in 1975 by two scientists, George Kohler of West Germany and Cesal Milstein of Argentina, who jointly with Niels Jerne of Denmark were awarded the 1984 Nobel prize for physiology and medicine.

Kohler (1974) successfully produced a hybridoma by fusing a P3 myeloma cell (resistant to azaguanine) and a lymphocyte (from the spleen of a mouse) immunized against sheep red blood cells. The experiment consisted of immunizing mice against red blood cells (the antibodies produced against this antigen are easily detected in the serum by means of an assay developed by Jerne and Nordin, 1963), then mixing mouse myeloma P3 cells with spleen cells from immunized mice in the presence of PEG (polyethylene glycol). These chimeric cells, called hybridomas, retained the property of immortality of the myeloma cells as well as that of secreting an antibody specific for an unknown antigen, resulting from the fusion of an antibody-secreting cell and a tumour cell.

Hybridoma cells are mass-cultured for the production of monoclonal antibodies (mAbs) either *in vivo* in the peritoneal cavity of mice or *in vitro* in large scale culture vessels.

Culture in Peritoneal Cavity

The hybridoma cells are transplanted into the peritoneal cavity of an appropriate strain of mice, and subsequently the ascitic fluid from the animals is harvested and the mAbs are purified. This method gives 50–500-fold higher antibody yields than *in vitro* culture of hybridomas. But, some of the disadvantages of this method are

1. The antibody preparations are of lower purity than those from cell cultures, particularly if serum-free media are used,

2. The procedure is labour-intensive and

3. Pathogen-free animals of specific genotypes are required.

Mass *In vitro* Culture

Large-scale culture of hybridoma cells has been done in stirred bioreactors, airlift fermenters and in vessels based on immobilized cells. The culture systems using immobilized cells enable the cultivation of cells at very high densities, which markedly enhance the antibody yields. The medium used for *in vitro* culture of hybridomas must serve two purposes:

1. Support adequate cell proliferation and

2. Give high antibody yields.

Most basal tissue culture media have been designed for cell proliferation or production. In general, four basal media (DMEM, IMDM, F12 and RPMI 1640) are used either singly or in combination, and are almost always supplemented with insulin, transferrin (stimulates cell proliferation), ethanolamine, and often with bovine serum albumin. Many hybridomas require additional hormones, and growth and differentiation factors. A 1:1:2 combination of DMEM, Ham's F12 and RPMI 1640 supplemented with amino acids, vitamins, insulin, transferrin, ethanolamine and selenite gave the best results for proliferation and IgM production from human hybridoma cells.

PRODUCTION OF MONOCLONAL ANTIBODIES

It is now possible to obtain consistently large number of hybrids and one can select the lines producing the antibody of choice. The critical component in the production of hybridomas is the preparation of cells for fusion. Where rodents are used as a source of cell for fusion, immunization protocols must be devised that optimize the proliferative response to antigen in the spleen.

Plasmablasts appear to be more suitable as fusion partners than mature plasma cells.

The production technique of monoclonal antibodies can be divided into three steps:

Step I: Fusion and Culture of Hybridomas

- Boost mice with an intravenous injection of antigen, 72 hours before use. On the day of fusion, collect the spleen and make a free cell suspension by injecting and flushing the spleen with sterile serum-free medium. Centrifuge the cells and resuspend in lysing solution. When sterile water is used, suspend in 1 ml of water and immediately dilute in 20 ml of medium. Any delay will result in the loss of plasmablasts and a consequent reduction in antibody-producing hybrids.

- Maintain plasma cells in cell culture and feed into fresh flasks and medium for 16–24 hours before fusion to ensure that they are in early phase of growth at the time of fusion. At the time of use, collect the cells, centrifuge, and resuspend in serum-free medium.

- Count both myeloma and spleen cells and then mix in the appropriate ratio. Depending on the properties of the tumour cells, the ratio may vary from 5 : 1 to 2 : 1.

- Then, the cells are centrifuged into a loose pellet by spinning at 1000 rpm for 10–15 minutes. Remove the supernatant and overlay the pellet with 1 ml of PEG. For 3 minutes, mix the PEG into the pellet. This helps in breaking the pellet into uniform small clumps.

- Following fusion, dilute the cells in 30 ml of serum-free medium with the first 10 ml medium being added and mixed at 1 ml per minute. Slow dilution reduces the risk of osmotic disruption of the fused cells. Centrifuge the cells and resuspend in complete medium containing HAT and then dispense into

96-well tissue culture plates, 10 plates for each 108 splenocytes used in the fusion. Add 106 thymocytes to each well to serve as feeder cells. The latter step has proven critical for optimizing the outgrowth of newly formed hybrids. Feeder cells and 2 mercaptoethanol in the medium exert a synergistic effect.

- Incubate the cultures for 3–4 days in CO_2 incubator with high humidity. Replace half the medium with fresh HAT medium every 4th day.

- Identify and mark the wells containing hybridomas on day 9 or 10 and then allow the colonies to grow to 500 or more cells. In rapidly growing cultures, supernatants can be collected 7 assayed for antibody activity by day 12–14. When appropriate, collect and test the supernatants from the largest colonies first; test the remainder 2–4 days later.

- After each test for antibody, transfer positive cell lines to 24 well plates. Add $3–5 \times 10^6$ feeder cells to each well to promote rapid cell growth. Maintain cells in static culture for a minimum of 2 weeks by removing 50–75% of the hybrid cells after 2–3 days interval. Select stable, rapidly growing, antibody-producing hybrids. Slow-growing hybrids and hybrids that cease to synthesize antibody are eliminated.

- After 2 weeks, make duplicate cultures and allow the cells to overgrow and die. Collect the supernatants and assay for the presence of antibody. Take cultures producing antibodies of the desired specificity from the master plate and expand into 6-well plates. Harvest the cells twice for preservation in liquid nitrogen. Replenish the cultures with fresh medium and allow the cells to overgrow and die. Collect the final supernatant for further analysis. The 6-well plates are essentially for minimizing labour and time during this phase of hybridoma production.

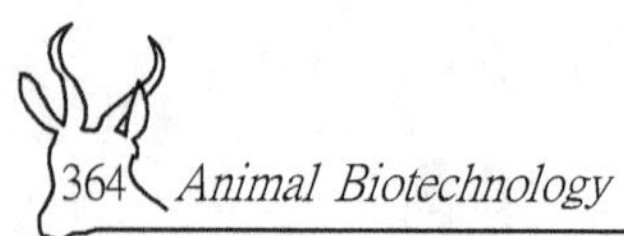

Step II: Cloning and Preservation of Hybridomas

- Following preliminary selection of hybrids, screen the final supernatants in detail to identify antibodies of immediate interest. Take the parent cell lines from the freezer and clone by limiting dilution. When viability is good, clone the cell lines immediately. Take the remainder of the excess cells and culture them in a T75 flask for 1–2 days. As soon as sufficient cells are present, harvest the cells and freeze one ampoule to replace the ampoule used for cloning. The culture should not be allowed to proliferate more than necessary to avoid change in composition of the cell line at this early stage of processing. As in the initial step, use feeder cells to promote growth.

- At 6–8 days, mark wells containing a single colony. At 12–14 days, assay supernatants from the marked wells. Transfer 24–48 positive cultures to 24 well plates and maintain in static culture.

- After identifying stable clone lines, expand 4–6 clones of each cell line into 6-well plates for cell preservation and production of antibody. Record the stable and unstable cell lines.

- Preserve 4–6 clones from each cell line in liquid nitrogen.

Step III: Production of Antibody

To produce antibody, culture the cloned cell lines *in vitro* or grow in ascites from the mice. Then assay the monoclonal antibodies in supernatants of hybridomas by ELISA, radioimmunoassay (RIA), immunofluorescence, cytotoxicity, flow cytometry, etc.

APPLICATIONS OF MONOCLONAL ANTIBODIES

Monoclonal Antibodies in Diagnosis, Screening and Therapy

Monoclonal antibodies are now an essential tool of much biomedical research and are of great commercial and medical value. For instance, ABO blood groups could be earlier identified with the help of human sera carrying antibodies of known specificity. In the U.K. these human sera have been replaced by monoclonal antibodies produced by hybridomas, for the identification of ABO blood groups. Besides the use of monoclonals in identification of blood groups in the U.K. (U.K. blood typing), three other uses of monoclonals namely diagnosis, immunopurification and therapy have been described. Only the first two of three make a definite market at present.

In diagnosis, pregnancy can be detected by assaying the hormone HCG in the whole blood, urine, plasma or serum in just 5 minutes sparing several days of culturing of cells earlier needed. Immunopurification involves separation of one substance from a mixture of very similar molecules. Monoclonal antibodies have proved to be useful in high-level purification and large-scale preparation of medically important substances. Immunoaffinity columns containing monoclonal antibodies coupled to cyanogen bromide-activated sepharose, when used for human leucocyte interferon, a product, which is 5,000 times richer in interferon, could be obtained from the original crude mixture in a single step. Prof. Bill Stimson stated that monoclonal antibodies can reduce rejections of organ transplants. Transplant patients produce leucocytes and killer lymphocytes. Leucocytes act as a barrier against bacterial and fungal attacks on the organ transplants, whereas killer lymphocytes tend to reject the transplant. The present drug therapy to curb transplant rejections such as anti-lymphocyte serum and corticosteroids not only attack the killer lymphocytes but also destroy leucocytes thus affecting the immune system. American scientists have now shown that monoclonal antibodies can be produced specifically for blocking killer lymphocytes without affecting leucocytes. Such monoclonal antibodies will then be used in injections of anti-lymphocyte serum.

Monoclonal antibodies will be useful for tissue typing and selection of the donor's organ, which is close to that of the recipient. This selection will be more accurate due to the high specificity involved in the use of monoclonal antibodies for typing. This will reduce the need for drugs, which are otherwise used to suppress rejection.

The quantity of radiolabelled antigen complexing with specific antibody (monoclonal) varies inversely with the amount of unlabelled antigen in test sample. A standard curve can be plotted using known concentration of antigen versus radioactivity and concentration of antigen in test sample obtained from the radioactivity measured.

Radioimmunoassays (RIA) using known radiolabelled antigen along with the test sample containing unknown concentrations of the same antigen brings about competition between them to combine with specific antibody and to thereby find out the unknown concentration. RIA, a popular technique at present is useful for the measurement of hormone level, which is normally low. Monoclonal antibodies are now being used in commercial RIA kits for the measurement of proteins and hormones in various diseases or disorder states.

Locating a particular antigen in tissue sections is already being done using a specific antibody and to combine it with a fluorescently labelled anti-antibody is called as **Sandwich technique** in immunohistochemistry. The sensitivity and efficiency of the method can be increased with the use of monoclonal antibody instead of just ordinary polyclonal antisera. An example of the current use of monoclonal antibodies in histopathological diagnosis of human disease is the examination of lymph node biopsies as an aid to classification of a particular type of lymphoid tumour (e.g. Hodgkin's disease, various lymphomas). A similar technique can be used for finding out the cell of origin of a tumour by the antigen it contains.

Diagnosis, prognosis and detection of the spread of tumour (metastases) are possible by identifying the tumour-derived markers using specific antibodies and RIA, fluorescent-microscopy-like techniques. Monoclonal antibodies will make these tests more specific and sensitive. Also monoclonal antibodies may help to identify new, more specific markers of tumours.

The antibodies were used for diagnosis nearly four decades prior to the introduction of precipitin-based methods. The introduction of radioimmunoassay (RIA) by Yalow and Berson revealed the extreme sensitivity possible with high-affinity antibodies. Since then, fast, sensitive immunoassays—RIAs, enzyme immunoassays (EIAs or ELISAs) and fluorescence immunoassays (FIA)—have been applied to a wide variety of molecules.

The literature teems with diagnostics improved, or made possible, by monoclonal antibodies (mABs). Monoclonals against unique microbial epitopes have made it possible to discriminate between closely related organisms—and even among subtypes. They can distinguish between closely related molecules: between morphine and heroin; between testosterone and related steroids; and even between enantiomers of the same molecule. Furthermore, pharmacokinetic studies may use mABs to monitor levels of drugs and their metabolites or to study the difference in the clearance of two enantiomers. Such selectivity is impossible with conventional polyclonal antisera.

Monoclonals do have some disadvantages, however. Cross-reactivity is one, a direct consequence of the reagent's monoclonality. For the same reasons, a monoclonal antibody may sometimes be too selective. Small changes in antigen structure due to genetic polymorphism, heterogeneous glycosylation, or slight denaturation may so change the epitope that the antibody cannot bind.

The growing understanding of monoclonal antibodies' distinctive features has obviously made it possible to design new types of high-performance assays. The two-site immunoassay, for example, uses a pair of antibodies, each recognizing a different epitope, one antibody binds the analyte to a matrix; the other, linked to a marker, reveals the analyte's presence. This assay performs well in comparison with standard RIA and EIA and has a definite advantage when one must use cross-reacting antibodies. Cross-reacting substances are unlikely to carry both antigenic determinants specifying the monoclonal pair.

So far, murine monoclonals have accounted for most applications. In general, the antibody's origin has no impact on the assay. Some human antigens (such as the Rh antigens), however, have stymied efforts to produce murine monoclonal antibodies with the desired specificities. In this case, human monoclonal antibodies have solved the problem. Human mAbs are now also available against other blood group antigens-including A, A1, Rh (G), Rh (c), Rh (E) and Kell. In the future, human monoclonals will likely replace polyclonal antisera for blood typing.

For therapeutic uses, monoclonal antibodies are so designed that they will neutralize the reaction or response by one defined antigen, but still preserve the reaction of all other antigens. Several antigens of T-cell receptor complex, including CD3, CD4 and CD8 have been the targets of specific antibodies for therapy. The most widely used monoclonal antibody is OKT3, for the treatment of acute renal allograft rejections. Monoclonal antibodies have also been used for patients with (i) malignant leukemic cells, (ii) B-cell lymphomas, and (iii) a variety of allograft rejections after transplantation. However, in several cases of therapy by monoclonal antibodies, there are undesirable side effects like high fever, vomiting, diarrhoea and respiratory distress; sometimes it leads to broad immunosuppression leading to increased incidence of infections. The effectiveness of murine antibodies is limited due to their short survival time in humans and relatively low cytocidal effects. Hence, humanized antibodies are being produced by genetic manipulations and in future these will be increasingly utilized for therapy. "Monoclonal antibodies–cytotoxic agent conjugates" called **immunotoxins** have also been designed as carriers of cytotoxic substances to the target cells.

Radiolabelled monoclonal antibodies have also been developed for delivery of a cytotoxic effector to target cells and for "radioimaging". The advantage of using radiolabelled antibodies is that they can kill cells from a distance and thus can also kill cells adjacent to antigen expressing cells.

Radioimmunodetection is also widely used. $(Fab')_2$ and Fab fragments are preferred for imaging, because both targeting and blood clearance are rapid. Radiolabelled antibodies can detect tumours as small as 0.5 cm, which are missed by other radiological methods or Fab fragments.

Antibody-mediated immunotherapy was first used over sixty years ago: haematopoietic tumours were treated with hyperimmune sera from rabbits. It was not, however, until the introduction of monoclonals that antibody-based therapy of different diseases could be systematically investigated. As analytical tools, monoclonals have helped researchers identify a number of tumour-associated antigens, virus and lymphokine-neutralizing epitopes, endotoxins and other important structures. The repertoire of cell-surface antigens has also been investigated on a number of human tumours, using murine monoclonal antibodies. No real tumour-specific antigens have yet been identified, except for the idiotypic T- and B-cell receptors in lymphoproliferative diseases.

This is not entirely surprising. The mouse immune system identifies foreign cells mainly by transplantation and blood group antigens. It might therefore overlook small antigenic changes specific for a human tumour cell. For example, mouse monoclonals exhibit none of the human antibody's fine-tuned specificity against the polymorphic structures of human histocompatibility antigens. Thus, the mouse immune system might ignore the small structural changes specific to human tumour-cell surfaces, which are conformations readily recognized by immunocompetent human cells.

Human mABs might thus be advantageous for treating human neoplasms. And, *in vivo*, human monoclonals might not elicit as strong an anti-immunoglobulin response as mouse immunoglobulin. Murine monoclonals have made it possible to detect a large variety of tumour-associated antigens expressed little—or not at all— by normal cells. Many experimental and clinical therapeutic systems have applied such antibodies (directed against melanomas, carcinomas, or sarcomas, for example) in different modalities and administration schedules, with varying degrees of success. Thus, unconjugated, complement-fixing, or antibody-dependent cellular cytotoxicity-mediating antibodies have been tested. Their main advantage is their relatively low systemic toxicity. Antibodies conjugated to toxins or radioactive isotopes have also been used but unspecific uptake in normal tissue remains a problem.

Patients sensitized to mouse proteins have suffered a number of side effects such as fever, rashes, vomiting, urticaria, bronchospasm, tachycardia, and dyspnoea. These side effects are normally transient and disappear as soon as the infusion of mouse antibodies stops.

Much more serious is the reduction of therapeutic effect as the patient mounts an anti-mouse response. The patient's immune response first targets the therapeutic monoclonal's constant regions but it focuses on an idiotypic response after only a few injections. These reactions can produce secondary allergic reactions due to immunocomplex deposition in various tissues: the kidneys, liver and lung. Therapeutic antibody dosage, the target antigen's tissue distribution and the antibody's reactivity all affect the anti-mouse immunoglobulin response. Human monoclonals should solve most of these problems though it remains to be seen just how these will be used.

At present, human monoclonals have been used only as native unconjugated molecules or as radiolabelled imaging agents. Current developments in immunoconjugates, especially those using small, highly toxic compounds, should be of utmost importance when applied to human antibodies.

Table 11.1 Clinical effects of human monoclonals against solid tumours

Tumour	No. of patients	Clinical effects	Comments
Malignant melanoma	8	2 complete remission 2 partial remission 2 objective response	Anti-GD2 (IgM) antibody No side effects
Glioma	1	Not available	Antibodies administered in a subcutaneously implanted culture chamber
Breast carcinoma	6	Localization	^{111}I – labelled IgM antibodies for tumour localization

Today, a few human monoclonals can be used to treat tumours, infectious diseases, autoimmune conditions and drug overdoses. Clinical studies are very scarce (Table 11.1) mostly because of the technical barriers to routine production. This is changing. Recent progress in *in vitro* immunization and immortalizing human B cells by Epstein–Barr virus (EBV) infection have made it possible to obtain antigen-specific human hybridomas at significantly higher frequencies.

Monoclonal Antibodies in Vaccine Production

Antibodies have also been used to immunize against certain diseases in humans and cattle. The most promising outcome is the prospect of developing antimalarial vaccine in the near future. Monoclonal antibodies that inhibit the *in vitro* multiplication of *Plasmodium*, and the antigametocyte antibodies that inactivate male gametes have been developed. Monoclonal antibodies that destroy merozoite infected RBCs have also been developed now. Such antibodies may prove useful as vaccines.

In 1981, two groups first proposed using antibodies as vaccines, a logical consequence of Jerne's idiotype–anti-idiotype network theory. Antibodies obviously bind to antigen epitopes. They can also serve as antigens themselves, to be recognized by still other antibodies, which bind to their variable regions (idiotypes). Some of these anti-idiotypic antibodies carry the internal image of the original antigen: they can therefore act as antigenic stand-ins, eliciting an antibody response.

Clearly, not all idiotypes can serve as vaccines: some will produce a protective immunity; some will not. There are at least two reasons for this:

First, only a few anti-idiotypes mimic the antigen and bind to the primary antibody's (Ab1) antigen-binding area (paratope). The other antibodies target parts of the Ab1 variable region that do not participate in antigen recognition and binding. (Note, however, that anti-idiotypes binding outside the paratope can sometimes elicit production of antibodies against the nominal antigen. Such anti-idiotypic antibodies may not make proper vaccines. They probably will not generate a relevant T-cell immunity.)

Second, some anti-idiotypes seem to induce suppressive rather than protective immunity in the vaccinated animal even though they elicit both B- and T-cell responses against the antigen.

Anti-idiotype vaccines could replace microbes and microbial toxins, both hazardous to the patient. Anti-idiotypes also offer peptide alternatives to some microbes' primarily glycan epitopes which are particularly important for vaccinating infants against bacterial polysaccharides. And modern hybridoma processes can produce these in almost limitless supply, a particularly significant consideration where carbohydrates make up the antigen's most important structures, putting them, at least for now, beyond the reach of recombinant DNA technology.

Anti-idiotypic antibodies have reportedly produced vaccines against microbial antigens such as parasites, bacteria and viruses.

Anti-idiotypic cancer therapy is obviously another area of great interest. Although Hollinshed has demonstrated the benefits of vaccinating lung tumour patients with autologous tumour extracts, immunization with tumour-associated antigens has generally been rather disappointing.

There are many reasons for this. Tumour-associated antigens are often unidentified, hard to purify, and "self". Anti-idiotypic antibodies have been tried in several experimental systems for vaccination against tumours (Table 11.2). Such idiotype-based antigens could potentially outperform conventional tumour-associated antigens especially since they are easy to mass-produce and are free of the tumour viruses that may be present in conventional tumour-extracted antigens. Furthermore, by

Table 11.2 Infective and other diseases treated with anti-idiotypic vaccines

Antigen source	Anti-idiotypic source	Species vaccinated
Streptococcus pneumoniae	Mouse monoclonal	Mouse
E. coli K13	Mouse monoclonal	Mouse
Trypanosoma cruzi	Rabbit serum	Mouse, rabbit, guinea pig
Hepatitis B surface antigen	Rabbit serum	Chimpanzee
Polio virus type II	Mouse monoclonal	Mouse
Rabies virus	Rabbit serum	Mouse
Reovirus	Mouse monoclonal	Mouse
Cytomegalovirus	Mouse monoclonal	Mouse
Human immunodeficiency virus	Rabbit serum	Mouse
Murine sarcoma	Mouse monoclonal	Mouse
Murine bladder tumour	Mouse monoclonal	Mouse
Murine B cell lymphoma	Mouse idiotypic IgM	Mouse
Human melanoma	Rabbit seruk	Mouse
Human T cell lymphoma	Mouse monoclonal	Mouse
Human colon carcinoma	Goat serum	Human
Human colon carcinoma	Human monoclonal	---

putting the epitope in a new molecular environment, such as an anti-idiotypic antibody coupled to keyhole limpet haemocyanin (KLH) or tetanus toxoid, it may mobilize T cell clones that would not otherwise participate in the anti-tumour antigen response.

Several laboratories have demonstrated immunity to tumour-associated antigens after vaccination with anti-idiotypic antibodies: melanoma, virally induced sarcomas, bladder tumour, B and T cell lymphomas, and colon carcinoma. The 17-1A murine monoclonal has had therapeutic effects in patients suffering from colorectal cancer. This may not be a direct effect; rather, the 17-1A may induce anti-idiotypic antibodies, immunizing the patient. More recently, others have obtained a set of human monoclonal anti-idiotypic antibodies, following WBV transformation of B cells from patients treated with 17-1A.

H. Kohler and his group have developed and characterized several anti-idiotypic antibodies, which induce T and B cell immune responses. Although these antibodies bind to the same paratope, some induce protective immunity, and some do not. The anti-idiotypic antibody that failed to induce protective immunity elicited suppression rather than protection. The reason is unclear: it may depend on the anti-idiotypic antibody's ability to activate T-effector cells via direct idiotype binding. Protective antibodies induced by vaccination with irradiated tumour cells reacted with the anti-idiotype giving protective immunity. On the other hand, serum antibodies (from individuals with growing tumours) bound to the anti-idiotype inducing suppression. This suggests an organic relationship between tumour development and the type of antibody response evoked. The results also caution us that anti-idiotypic vaccination can have complex effects on the immune system.

We can now produce enormous numbers of different antibody specificities as internal images, with a corresponding variety of biological functions. Thus, research has produced anti-idiotypic antibodies, containing internal images of such molecules

Table 11.3 Anti-idiotypic antibodies with receptor-binding ability

Antigen	Source of anti-idiotype	Activity	Receptor
Insulin	Rabbit serum	Agonist	Insulin
Angiotensin II	Rabbit serum	----	Angiotensin II
Adenosine	Mouse monoclonal	Agonist	Denosine I
Alprenolol	Rabbit serum	Agonist	Beta adrenalin
Nicotine	Mouse monoclonal	----	Rat brain nicotine
Morphine	Guinea pig serum	Agonist	opiate

as insulin, angiotensin II, and adenosine, and to adrenergic, nicotinic and opiate compounds. Such antibodies can bind to their respective receptors and some of them have demonstrated an agonistic effect (Table 11.3).

Monoclonal Antibodies as Enzymes (Abzymes)

The use of antibodies as enzymes is described as Abzymes. The antibodies may often bind specific ligands (haptens), but may not carry out chemical reactions. By modifying these ligands, antibodies may be generated that will catalyse specific reactions just like enzymes. Production of these abzymes is based on the following two principles:

1. Enzymes work by binding the transition state of a reactant better than the ground state;

2. Antibodies which bind to specific small molecules can be produced by coupling this small molecule to a protein carrier and using this protein for immunizing experimental

animals. If this small molecule is a transition state analogue, then the antibodies that were produced to bind to this molecule will function as enzyme towards the substrate of this reaction. Abzymes were shown to accelerate the reactions, which they were supposed to catalyse (Table 11.4).

Table 11.4 Antibody with enzymatic activity

Enzymatic activity of antibody	Substrate of reaction	Acceleration
Chorismate mutase	Chorismate	10^2
Esterase	Hydroxyester	1.7×10^2
Esterase	Carboxyl ester	9.6×10^2
		2.1×10^2
Esterase	Carbonate ester	7.7×10^3
Esterase	Coumarin ester	1.5×10^3
Esterase	Carbonate ester	8.1×10^2
Esterase	Carbonate ester	6.3×10^6
		1.2×10^5

Preparation (Purification) of Medically Important Products using Monoclonal Antibodies

Tiny amounts of enzymes and proteins of medical value in a crude preparation have to be extracted in maximum pure form in many cases. Monoclonal antibodies have proved to be useful in high-level purification and large-scale preparation of the medically important substances. Immunoaffinity columns containing monoclonal antibodies coupled to cyanogen bromide-activated sepharose, when used for human leucocyte interferon, a product which is 5,000 times richer in interferon could be obtained from the original crude mixture in a single step.

Treatment of Drug Overdose

Availability of highly specific monoclonal antibodies against drugs such as digoxin may be helpful in the treatment of drug overdose. Either monoclonal antibody can be injected into blood circulation or blood is allowed to circulate extracorporally through monoclonal antibodies bound to solid support. Application is still at research level.

Monoclonal Antibodies for Cytogenetic Analysis

The technique of ELISA, utilizing monoclonal antibodies has also been used for cytogenetic analysis in wheat. Monoclonal antibodies specific to chromosomes 1B, 6A and 6D are already available and those for other chromosomes will become available in future, due to proteins coded by chromosomes of all homologous groups.

Monoclonal antibodies can be used to enrich the number of aneuploids in a population, by rejecting seeds showing high level of reaction with monoclonal antibodies. The major advantage of using monoclonal antibodies is that a very large number of seeds can be examined non-destructively and the chromosome analysis can be greatly reduced, if not eliminated.

Other Uses of Monoclonal Antibodies

1. Dose determination of a medicine can be carried out by using the monoclonal antibodies of an animal immunized against this particular medicine.

2. They are used to detect allergies, to carry out hormone tests, to diagnose viral diseases, to detect certain types of cancers,

and to monitor the presence or appearance of malignant cells after surgical or radio-therapeutic treatments.

3. The use of these antibodies was also applied for the labelling and precise identification of specialized cells such as neurons in order to gain better knowledge of the way in which these cells associate and operate.

4. A monoclonal antibody is also of great value in the area of the structure of cell membrane as membrane proteins are hard to purify.

5. In the field of direct therapy, serotherapy can be made more effective with the administration of a monoclonal antibody.

6. Identification and classification of major histocompatibility antigens complex products useful in population.

7. For assays in forensic sciences monoclonal antibodies will replace conventional tissue section immunofluorescence assay.

8. Monoclonal antibody kits are also available to check mycotoxins in food.

DNA PROBES

Disease diagnosis is an important feature in animal health care. Diagnostic tests are currently being developed for the following purposes.

- to detect infectious diseases
- to detect specific forms of cancer
- to detect AIDS viruses in blood for transfusion
- to detect genetic disorder
- to detect high- and low-density lipoproteins which are indicative of a possible heart disease, etc.

Diagnostic tests are mainly of 2 types:

1.   those using monoclonal antibodies
2.   those using DNA probes

The role of monoclonal antibodies in disease diagnosis has been discussed in detail in the previous chapter.

Although DNA probes have direct competition from diagnostics using monoclonal antibodies, DNA probes will probably be more effective for detecting bacterial infections and genetic disorders. Further, while monoclonal antibodies are non-proprietary, DNA probes can be protected by patents and this offers the producers potentially larger shares of any market they penetrate.

WHAT ARE DNA PROBES?

A strand of nucleotides will bind to other strands of nucleotides only if there is complete matching of the base pairs in the sequence on two strands. This stringency of base-pairing, except for a very little mismatch allowed, is the basis of hybridization technique. Specific DNA fragments can be obtained with the help of restriction endonuclease enzyme of a target DNA to be analysed and then radiolabelled with P^{32} or if the exact amino acid sequence in proteins arising from a particular part of DNA is known, then a single-stranded DNA fragment can be chemically synthesized. Obtained by any method, it is single-stranded, radiolabelled DNA oligonucleotide, which can act as DNA probe for the detection of the particular DNA in question.

The specificity of nucleic acid hybridization makes possible the identification of particular DNA sequence in clinical material. Sequences as little as 13 bps should be sufficient in theory to detect unique genes or subfragments diagnostic of a particular condition or a specific infectious agent. DNA is stable to many harsh treatments, including formalization and so can be recovered even from long-dead tissues.

DNA probes can be used to identify specific plasmids or resistant genes in clinical isolates of bacteria. Such probes may be used for rapid determination of resistance properties of bacterial pathogens. This helps in deciding on the right antibiotic therapy. Specific probes of different resistance genes can also be important tools in studying epidemiology of drug resistance.

At present, the numbers of probes are limited. Probe technology will replace conventional antibiotic sensitivity testing in routine diagnosis one day.

PROBE TECHNOLOGY

- Double-stranded DNA is extracted from the sample and fixed to the support.

- Chemicals are added to separate the strands.

- Single strand of DNA complementary to the specific part of sample DNA are introduced.

- Probes attach themselves to sample DNA and any that remains free is washed away.

- Amount of labelled probe still fixed to the support via sample DNA, if any, is detected.

- If the sample DNA does not have a complementary DNA sequence, then no probe will get latched on to DNA on support and the test is negative indicating absence of DNA being searched for in the sample.

APPLICATIONS OF DNA PROBES

DNA probes can have applications like:

    a. Identification of viruses that may be difficult to cultivate in routine laboratory (10 viruses in 1000 l of water can be detected).

b. Rapid identification of fastidious or slow-growing bacteria such as *Mycobacterium tuberculosis* and *Mycobacterium leprae*.

c. Rapid identification of pathogens in tissues or body fluids (e.g. *Neisseria meningitis* in CSF).

d. Detection of *Salmonella* in foodstuff.

e. Diagnosis of genetic diseases.

Even a single base pair change in human gene can cause genetic diseases. Detection of such change may be useful in antenatal diagnosis of inherited diseases. Changes in genomic DNA sequences generate or abolish or alter the position of the recognition site for restriction enzymes, relative to other sites. This gives rise to restriction fragment length polymorphism (RFLPs) (Figure 11.8). These RFLPs can be detected by measuring the different sizes of restriction fragments that hybridize with specific DNA probes. Thus indirectly predictive screening of genetic defects in families is possible. An advantage of examining genetic defects by using RFLP is that a particular pathological condition is diagnosed even before the biochemical defect is known.

For prenatal diagnosis, now methods are developed to obtain foetal DNA in sufficient quantity in 6–10 weeks of pregnancy, amplify it in *in vitro* and conduct a study on it.

For using DNA probes in genetic disorder diagnosis, it is necessary that the structure of DNA and changes associated with different disorders are known. Today, of all genetic disorders, only the inherited haemoglobin disorders (sickle cell anaemia, thalassemia) are studied at the DNA levels.

With more than 4000 diseases known to arise because of genetic factors, the use of genetic probes to identify and quantify the risks of individual to certain diseases will change the practice of medicine. These DNA probes are claimed to be 1000 times more sensitive than the conventional diagnostic methods, which rely on a mixture of antibodies extracted from animal blood.

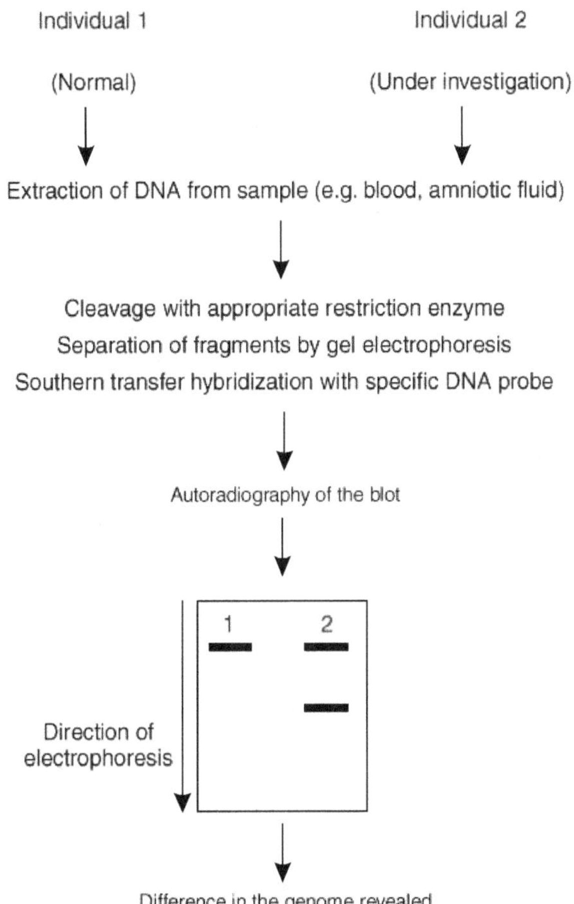

Figure 11.8 Examining genetic defects by using RFLP hybridizing with specific DNA probe

DNA probes are now developed even for the selection of organs for transplantation. The antigens which characterize an individual's organ and cells are encoded by a set of genes known as major histocompatibility complex (MHC). By comparing the abilities of DNA probes to bind to DNA from MHC of both donors and recipient cells, it will be possible to select a more matching donor appropriately.

DNA FINGERPRINTING

A DNA fingerprint is really just a Southern blot. It is widely used in forensic laboratories to identify individuals who have left blood or other DNA-containing tissues at the scenes of crimes. Such DNA typing has its roots in a discovery by Alec Jeffrey and his colleagues in 1985. These workers were investigating a DNA fragment from the gene for a human blood protein α-globin, when they discovered that this fragment contained a sequence of bases repeated several times. This kind of repeated DNA is called a "minisatellite". For example,

Let us consider the following DNA sequence.

5′-GACTGCCTGCTAAGATGACTGCCTGCTAAGATC
GACTGCCTGCTAAGATGACTGCCTGCTAAGATCT
GACTGCCTGCTAAGATGACTGCCTGCTAAGATAA
GACTGCCTGCTAAGATGACTGCCTGCTAAGATGA
TGACTGCCTGCTAAGAT-3′

Minisatellite sequence is composed of nine tandem repeats of the 16-base pair sequence GACTGCCTGCTAAGAT. Clusters of such sequences are widely dispersed in the human genome and the number of repeats at each locus ranges from 2 to more than 100. These loci are known as variable number tandem repeats.

Minisatellite

It includes the variable number tandem repeats (VNTRs). The repeating DNA sequences of VNTRs may be 15–100 bp long and is found within and between genes. Many such clusters are dispersed throughout the genome and they are referred to as minisatellites.

The number of repeats at a given locus is variable and each variation constitutes a VNTR allele. Many loci have dozens of alleles; as a result, heterozygosity is common.

More interestingly they found similar minisatellite sequences in other places in the human genome, again repeated several times. This simple finding turned out to have far-reaching consequences, because individuals differ in the pattern of repeats of the basic sequence. In fact they differ enough that two individuals have only a remote chance of having exactly the same pattern. That means that these patterns are like fingerprints. Indeed they are called DNA fingerprints.

The number of tandem copies of each specific sequence at each location varies in individuals, creating localized regions of 1000–5000 bp (1–5 kbp) in length. The variations in size (length) of these regions between individuals in humans was originally the basis for the forensic technique referred to as DNA fingerprinting.

MICROSATELLITES

Another group of tandemly repeated sequences consist of dinucleotides, also referred to as "microsatellites". Like VNTRs, they are dispersed throughout the genome and vary among individuals in the number of repeats present at any site. For example, in humans, the most common microsatellite is the dinucleotide $(CA)n$, where n equals the number of repeats. Most common n is between 5 and 50. These clusters are also used forensically and in addition have served as useful molecular markers during genome analysis.

METHODS OF DNA FINGERPRINTING

Investigators first cut the DNA under study with a restriction enzyme such as *Hae III*. Jeffrey chose this enzyme because the repeated sequence he had found did not contain a *Hae III* recognition site. That means that *Hae III* will cut on either side of the minisatellite regions but not inside. In this case, the DNA has three sets of repeated regions, containing four, three and two repeats respectively. Thus three fragments of different sizes bearing these repeated regions will be produced.

Next the fragments are electrophoresed, denatured and blotted. The blot is then probed with a labelled minisatellite DNA and the labelled bands are detected with X-ray film, or by phosphor imaging. In this case, three labelled bands occur, so three dark bands will appear on the film.

Real animals have a much more complex genome than the simple piece of DNA in this example, so they will have many more than three fragments that contain a minisatellite sequence that will react with the probe.

DNA FINGERPRINT ANALYSIS

DNA fingerprint analysis (fingerprinting or typing) is the process of preparing and interpreting barcode-like profiles of DNA segments for individual identification. Nanogram to microgram quantities of genomic DNA is isolated and polymorphic segments are directly analysed or amplified by means of the polymerase chain reaction (PCR) and analysed. With the exception of identical twins and clones, profiles of segments (usually separated and characterized according to length or by sequence differences) from humans, other animals, microorganisms, fungi and plants are unique to each individual.

The analysis of DNA is revolutionizing the field of identification. The need for only minute quantities of tissues, the stability of DNA and the high degree of assay accuracy and precision has contributed to universal application of DNA typing. The use of this procedure is rapidly altering the manner in which the genetic analyses are carried in human parentage, rape and homicide cases, in animal poaching, parentage, breeding and population studies, in medical analysis and in plant patent disputes, clone identity, investigations, parentage testing and gene bank management programs.

Principles

Specimen processing   DNA is the molecule of heredity. It is an integral component of all living matter with the exception of RNA viruses. Any living or non-living organic matter containing relatively intact DNA fragments can be used for DNA typing analysis. Common sources include human or other animal blood, semen, solid tissues, plant foliage and seeds.

Specimens are stored under clean, dry, cool conditions to reduce contamination by microorganisms and DNA degradation caused by cell lysis and DNase activity. Successful analysis has been carried out on DNA extracted from dried bloodstain, dried museum specimens and frozen tissue preserved for many decades. Indeed these studies have formed the basis of recent new theories in molecular evolution and anthropology.

A good yield of high molecular weight DNA devoid of organic and inorganic contaminants can be extracted from tissues. The enzyme proteinase K is generally used to assist with cell lysis and to digest proteins. The detergent SDS facilitates the separation of residual protein from DNA. Proteinaceous materials are removed by phenol extraction (or a non-organic NaCl or LiCl precipitation) and chloroform is added to remove traces of phenol. Most of the contaminating RNA can be eliminated provided the phenol is appropriately equilibrated. High molecular weight DNA is recovered in a relatively pure form by ethanol precipitation in the presence of salts. The extraction and purification steps can, under circumstances, be bypassed after cellular lysis. Fragments of DNA can be directly amplified by PCR and the amplified fraction used to construct a DNA fingerprint profile. If this approach is followed, possible inhibitors of the PCR reaction can be removed by means of a simple non-organic technique using an ion exchange resin such as Chelex 100 or by washing through a Centrican 100 filter.

The analyst may be presented with mixed specimens. It is possible to separate mixtures such as sperm and female cells often present in vaginal swabs from rape victims. The female cells are lysed in extraction buffer devoid of dithiothreitol (DTT) and the DNA is isolated from the supernatant. The sperm pellet is then lysed by adding extraction buffer containing DTT and increased detergent.

DNA quality can be evaluated by ethidium bromide staining or by Southern blotting and hybridizing with a species-specific probe. The degree of degradation is estimated by observing under UV-light, ethidium bromide-stained fragment lengths subjected to agarose gel electrophoresis. High molecular weight DNA appears as a single band, whereas partially degraded DNA forms a long smear of large to small fragments (>20,000 bp to a few hundred base pairs for human DNA). Ethidium staining cannot differentiate DNA from different organisms. Species-specific probes are used for this purpose.

Quantitation is the final step in preparing DNA for analysis. At least 3 different procedures are available. In the first technique, the absorbance of a DNA solution is measured in a spectrophotometer at 260 nm wavelength. Second the DNA and a series of standards of known concentration are subjected to agarose gel electrophoresis and stained with ethidium bromide. Last, an aliquot of the DNA is fixed on a nylon membrane and is hybridized with, for example, a labelled, highly repetitive, primate-specific α-satellite probe P^{17H8} that detects locus D1721.

Specimen analysis   The analysis objective is the isolation of DNA segment that will facilitate identity discrimination among individuals profile patterns from tissues, for the same individual should be identical.

Many polymorphic genetic markers (segments) consist of tandem repetitive sequences. Depending on the locus, each repeat can consist of a minimum of 2 to perhaps 80 bp and may contain single base pair polymorphic sites. The number of tandem repeats varies from one to hundreds.

A locus may, therefore, be polymorphic for the number of tandem repeats, referred to as variable number tandem reports VNTRs) and short (STRs) and for the base pair composition at specific sites within each repeat (minisatellite variant repeat or MVR). This variation, when analysed at one or more loci by RFLP, AMP-FLP or MVR-PCR techniques, is the basis of DNA profile construction. Polymorphic sites that may or may not be located in tandem repeat units provide another form of variation useful in identity analysis. The RAPD, AP-PCR or DAF system is based on the annealing of arbitrary sequences of PCR primers to template sequences sufficiently close together on complementary DNA strands. This facilitates the amplification system to differ in their specificity of the length of primers used, the amplification conditions and the resolution and visualization of the products. Amplification will not occur if the template sequences differ significantly from the primer sequence or perhaps if competition occurs between the primer and the hairpin loop structures in the template.

The final DNA pattern or profile can be visualized with colour dyes, fluorescent dyes, radioisotopes or other stains such as silver. There are 3 analytical approaches used for DNA fingerprinting.

1. VNTR analysis
2. Dot blot analysis of sequence polymorphisms
3. Direct sequence analysis

VNTR analysis The restriction-digested or amplified DNA fragments are separated, generally based on size, by electrophoresis in a sieving medium. The relative positions of the bands after electrophoretic separation are determined with reference to size standards.

Dot blot analysis   The target DNA sequence is amplified by the PCR and the amplified products are fixed to membranes. The DNA is then incubated with allele-specific probe. If the hybridization occurs, the dot will be marked. If the DNA does not contain the particular allele, the spot will remain blank. Alternatively, dot blots can be configured where the allele-specific probes are fixed to the membrane and the amplified DNA is allowed to hybridize to the fixed probes. This reserve dot blot approach is used for HLA-DQA typing.

Direct sequence analysis   The sequence of DNA bases can be read directly.

A number of analytical approaches can, therefore be used to isolate and identify profile patterns. Isolated and purified DNA or non-extracted DNA may be used. DNA from each of a number of loci may be analysed separately and the results combined to form a profile. Otherwise DNA consisting of similar base pair sequence from multiple loci may be analysed, coincidentally. The DNA may be analysed directly or it may be amplified and then analysed.

Data processing   Genetic profiles are presented by one of the following ways.

1. The presence or absence of dots

2. The position or size of DNA fragments separated by an electrophoretic approach or

3. Direct sequence information
    (e.g. AATCGTACCTGATCC)

The results produced by any of these methods can be evaluated visually. However as data becomes more abundant and complex, computer-assisted analysis is needed to aid in pattern interpretation, data storage and manipulation.

For dot blot profiles, most interpretations are performed by the "eye" and the data are generally transcribed manually into a computer storage bank. However, since most dots are coloured or visible on X-ray film, densitometric scanning systems can be used automatically to store the data for future analysis.

For most VNTR analyses, which are based on size or position, the first generation of data acquisition used to determine the relative size of unknown DNA fragments was processed by measurement with a ruler and comparisons with known size standards (on autoradiograms). With the use of fluorescent UV detection system, labelled DNA fragments can be detected in real time analysis and automatically transcribed to a computer. Both VNTR typing and sequencing can be automated using this approach.

The main advantages of non-manual data acquisition are the reduction of human transcriptional errors and hopefully reduction in labour. Once the data has been stored in the computer, various simple and sophisticated statistical analyses can be performed for determination, such as likelihood of occurrence of DNA profiles for identity purposes and for linkage analysis.

Quality assurance Laboratory quality assurance is a documented verification that proper procedures have been carried out by skilled and highly trained personnel to ensure that valid and reliable results can be obtained. The validity of DNA typing results centres on correctly identifying non-matches and matches from a prospective DNA fingerprint profile. Test reliability measures reproducibility under defined condition of use and should transcend different laboratories and practitioners. Quality assurance must encompass all significant aspects of DNA typing process which include personnel education, training, documentation of records, data analysis, quality control of reagents and equipments, technical controls, proficiency testing reporting of results and auditing of the laboratory procedures. Quality assurance and appropriate standards for DNA typing evolved initially from the experience of clinical laboratories more recently; however, they have been carefully defined through consensus from forensic laboratories. It is important to note that quality assurance guidelines in forensic science must retain enough flexibility to accommodate the unique nature of forensic samples as well as future advances in recombinant DNA technology and molecular biology. Good quality control management is paramount for obtaining quality results. It is also a key factor in establishing uniform reliability between laboratories.

FORENSIC USES OF DNA FINGERPRINTING

A valuable feature of DNA fingerprinting is the fact that although almost all individuals have different patterns, parts of the pattern (set of bands) are inherited in a Mendelian fashion. Thus fingerprints can be used to establish parentage.

An immigration case in England illustrates the power of DNA fingerprinting. A Ghanaian boy born in England had moved to Ghana to live with his father. When he wanted to return to England to be with his mother, British authorities questioned whether he was a son or nephew of the woman. Informations from blood group genes were equivocal, but DNA fingerprinting of the body demonstrated that he was indeed her son.

In addition to testing parentage, DNA fingerprinting has the potential to identify criminals. This is because a person's DNA fingerprint is in principle unique, just like a traditional fingerprint. Thus if a criminal leaves some of his cells (blood, semen, or hair, for example) at the scene of a crime, the DNA from these cells can identify him. But most DNA fingerprints are very complex. They contain dozen of bands, some of which smear together, which make them hard to interpret.

To solve this problem, forensic scientists have developed probes that hybridize to a single DNA locus that varies from one individual to another, rather than to a whole set of DNA loci as in a classical DNA fingerprint. Each probe now gives much simpler patterns, containing only one or a few bands. This is an example of a restriction fragment length polymorphism (RFLP). RFLP's occur because the pattern of restriction fragment sizes at a given locus varies from one person to another. Of course, each probe by itself is not as powerful an identification tool as a whole DNA fingerprint with its multitude of bands but a panel of four or five probes can give enough different bands to be definitive.

One advantage of DNA fingerprinting is its extreme sensitivity. Only a few drops of blood or semen are sufficient to perform a test. However, sometimes forensic scientists have even less to go on—a hair pulled out by the victim. Although the hair by itself may not be enough for DNA typing, it can be useful if it is accompanied by hair follicle cells. Selected segments of DNA from these cells can be amplified by PCR and typed.

In spite of its potential accuracy, DNA typing has been effectively challenged in court, most famously in the O.J. Simpson trial in Los Angeles in 1995. Defence lawyers have focused on two problems with DNA typing. First, it is tricky and must be performed very carefully to give meaningful results. Second, there has been controversy about the statistics used in analysing the data. This second question revolves around the use of the product rule in deciding whether the DNA typing result uniquely identifies a suspect. Let us say that a given probe detects a given allele (a set of bands in this case) in one in a hundred people in the general population. Thus the chance of a match with a given person with this probe is one in a hundred or 10^{-2}. If we use five probes, and all five alleles match the suspect, we might conclude that the chances of a match are the product of the chances of a match with each individual probe or $(10^{-2})5$ or 10^{-10}. Because fewer than 10^{-10} (10 billion) people are now on earth, this would mean that DNA typing would statistically eliminate everyone but the suspect. Prosecutors have used a more conservative estimate that takes into account the fact that members of some ethnic groups have higher probabilities of matches with certain probes. Still, probabilities greater than one in a million are frequently achieved and they can be quite persuasive in court.

Applications

DNA fingerprinting is applicable to any identification problem for which a minimal quantity of relatively intact DNA is available. The requirement for intact DNA depends on the method of analysis. Typing by means of RFLP generally requires high molecular weight DNA, whereas most genetic marker typing based on PCR can make use of much smaller, even somewhat degraded fragments.

Human   Suspect or victim tissue found at the scene of a rape or a homicide provides a source of forensic evidence for determining whether a suspect is the source of the crime scene specimens. Accident or homicide victims unidentifiable from their physical features can be identified provided typable DNA is isolated from the victims remains and is matched with DNA, perhaps from hair roots, obtained from his or her hair brush. If the victim's putative parents or other relatives are available, parentage (or other biological relationship) testing also may be carried out.

Parentage can be determined for child custody and immigration cases and for counselling about genetic diseases. The bands in the offspring's DNA profile must have been inherited from the biological parents. Barring mutations, the presence of bands not found in either parental profile is indicative of non-parentage. Applications in medicine include genetic counselling, tracing the parentage of donor versus recipient cells in bone marrow transplants, determination of possible "contamination" of foetal chorionic villi sample (CVS) tissue with maternal tissue, tissue culture, cell line identification and the confirmation of twin zygosity.

Identification monitors are feasible for security purposes in the armed forces and for identification in mass disasters or war. For these applications, DNA profiles are recorded for future match comparisons.

Animal Parentage, poaching, identification in cases of theft or loss, population studies, detection of trait markers and breeding programs are areas of application in animal husbandry.

Proof of animal parentage may be important for pedigree registration and for establishing sale value. A DNA profile from animal remains can be compared with a profile from a frozen steak or from a trophy specimen in a poaching case.

The evolution of population can be traced, for example, by analysing DNA from dried museum specimens or from insects housed in amber for millions of years. Base pair sequences and profile bands can be compared with those of the modern relatives.

DNA profiling is used in breeding programs for populations of endangered species with small gene pools. Animals exhibiting the greatest number of DNA profile differences and therefore probably, the greatest genetic diversity are chosen for breeding. Also biological relationship can be confined in artificial insemination and embryo transfer programs for domesticated animals.

Microorganisms DNA fingerprinting provides a tool for microorganism identification. Different bacterial strains for example have unique profiles. Profiling also provides a basis for issuing patents for microorganisms such as yeast used in the brewery.

Plants The field of plant identification for patent, parentage, theft and trait marker purposes has recently received considerable impetus because of DNA analysis, especially using the RAPD-type systems.

The infringement of breeders' rights relative to the origin of cloned material such as cuttings, parentage of plants with desirable traits, the theft of expensive trees and gene bank management for identification and the measurement of variations are examples of application.

DNA FINGERPRINTING VERSUS OTHER IDENTIFICATION TECHNIQUES

Many of the conventional biochemical identification techniques such as isoenzyme electrophoresis are being replaced by DNA typing, especially where more information is required. Although conventional methods such as blood typing are rapid, simple and definitive for exclusions, they are much less powerful than DNA analysis for the probability of inclusion. If a mother and a putative father both have blood type O, their offspring cannot be type A, B or AB. The putative father is readily excluded as the biological father in this sample. If the offspring is type O, possible exclusion of the putative father is considerably more difficult using conventional methods.

Additional key factors in choosing DNA for analysis include the availability of the hereditary material, as well as its highly polymorphic nature, continuity within the same organism, stability relative to that of enzymes, and ability to amplify segments *in vitro* when a minimal quantity of tissue is available or the DNA is moderately degraded.

FUTURE DIRECTIONS

Automation is a primary objective. Current methodology is labour-intensive because of a number of isolation steps including combinations of extraction, amplification, endonuclease digestion, electrophoresis, labelling, hybridization, blot scanning, profile comparison matching, data reduction and probability calculation. Automation is especially desirable for direct sequencing of highly polymorphic regions of the genome, because it will result in the elimination of most of the tedious current techniques.

Although fingerprinting approaches using RAPD (AP PCR) systems circumvent the need for probes, a limitation arises in the calculation of probabilities. It is extremely difficult to determine population frequencies of the amplified fragments. Without these data, the calculation of the probability of profile matches is not readily feasible. Probes (together with their population allele frequencies) specific to each species are, therefore, urgently needed.

DNA fingerprinting began in 1985, with multilocus analysis of sperm and blood in a double rape–homicide case in Britain. In less than a decade, applications have reached into every taxonomic kingdom—Monera, Protista, Fungi, Animalia and Plantae. It appears that the potential uses of molecular biology techniques to identity testing have only begun to be realized.

DNA Fingerprinting to Identify WTC Victims of 9/11

During the second half of September 2001, Myriad Genetics Inc., a biotechnology company in Salt Lake City that normally provides breast cancer tests, was flooded with a very different type of medical sample–frozen DNA from people whom had presumably perished in the terrorist attack on the WTC on the 11th September 2001. The laboratory also received cheek-brush scrapings from relatives of the missing, amassed at DNA collection centers set up through out New York City in the days after the disaster and tissue from the victim's toothbrushes, razors, and hairbrushes. The workers used the PCR to determine the number of copies of 4-base sequences of DNA, called short tandem repeats or STRs, at 13 locations in the genome. They also determined the sex chromosome constitution. The chance that any two individuals have

(Contd.)

the same 13 markers by chance is 1 in 250 trillion, so that the STR pattern of a sample from the crime scene matched a sample from a victim's toothbrush, identification was fairly certain. Mitochondrial DNA fingerprinting was attempted on tissue that had been too degraded to yield to the more accurate STR typing.

DNA fingerprinting can counter the uncertainties of traditional forensic techniques such as dental patterns, scars, fingerprints and clues such as jewellery, wallets and rolls of film found with the victims.

Summary

- Vaccine is a preparation containing a pathogen either in killed or attenuated state.

- Vaccines should be tested and standardized before use.

- The vaccines based on recombinant proteins are also called subunit vaccines.

- Birth control vaccine developed in India is proved to be safe and devoid of any side effects.

- A second generation of HB vaccine has been developed by employing the techniques of genetic engineering.

- Vaccinia virus has been widely employed as a vector vaccine.

- DNA vaccines induce both humoral and cell-mediated immunity.

- Antibodies generated against other antibodies that bind to a particular antigen, have the immunological properties of original antigen and are called anti-idiotypic antibodies.

- The particulate nature of multivalent subunit vaccines induce vigorous humoral and cell-mediated immunity and contribute to their increased immunogenicity by facilitating phagocytosis.

- It is important to develop expression systems to produce large quantities of the specific protein in an economical fashion.

- A specific lymphocyte, after isolation and culture *in vitro*, becomes capable of producing a single type of antibody, which bears specificity against specific antigen and is known as "Monoclonal Antibody".

- The production of monoclonal antibodies in specialized cells through a technique is called as Hybridoma Technology.

- Hybridoma cells are mass cultured either *in vivo* in the peritoneal cavity of mice or *in vitro* in large scale culture vessels.

- Monoclonal antibodies can reduce rejections of organ transplants.

- Monoclonal antibodies will be useful for tissue typing and selection of the donor's organ.

- Immunotoxins have been designed as carriers of cytotoxic substances to the target cells.

- Disease diagnosis is an important feature in animal health care.

- Monoclonal antibodies and DNA probes are mainly used for diagnostic tests.

- DNA probes are more effective in detecting bacterial infections and genetic disorders.

- Single-stranded, radiolabelled DNA oligonucleotide, act as DNA probe for the detection of the particular DNA in question.

- Probe technology will replace conventional antibiotic sensitivity testing in routine diagnosis.

- A DNA fingerprint is really just a Southern Blot.

- A DNA fragment containing a sequence of bases repeated several times is called a Minisatellite.

- Group of tandemly repeated sequences consisting of dinucleotides are referred to as Microsatellites.

- DNA fingerprinting is the process of preparing and interpreting barcode-like profiles of DNA segments for individual identification.

- DNA fingerprinting is applicable to any identification problem for which a minimal quantity of relatively intact DNA is available.

REVIEW QUESTIONS

1. Define vaccine.

2. State the features of an ideal vaccine.

3. Summarize the different types of vaccines.

4. Briefly explain the purified antigen vaccines.

5. Write a note on recombinant vaccines.

6. Distinguish recombinant polypeptide vaccine and recombinant vector vaccine.

7. Give a detailed account of DNA vaccines.

8. Describe in detail multivalent subunit vaccines.

9. What are anti-idiotypic vaccines?

10. Give a brief note on vaccine expression systems.

11. Define:

    i. epitope

    ii. hapten

12. Brief on the structure of antibody with diagrammatic illustration.

13. Explain the mass culturing of monoclonal antibodies.

14. Describe the three steps involved in the production of monoclonal antibodies.

15. Discuss the role of monoclonal antibodies in diagnosis, screening and therapy.

16. What are abzymes?

17. Monoclonal antibodies can be used for cytogenetic analysis. Comment.

18. What is DNA probe?

19. State the applications of diagnostic tests in animal health care.

20. Brief on probe technology.

21. List out the applications of DNA probes.

22. Define DNA fingerprinting.

23. Define minisatellite and microsatellite.

24. Write short note on the methods of DNA fingerprinting.

25. Give a detailed account on DNA fingerprint analysis.

26. Elaborate on the forensic uses of DNA fingerprinting.

ANIMAL FEED ADDITIVES

Feed additive is an ingredient or combination of ingredients added to the basic feed mix or parts thereof incorporated in an animal feed for the purpose of improving rate of gain, feed efficiency, or for preventing and controlling disease. It is used in microquantities.

A variety of feed additives are used in animal feeding. Some are approved to be used as implants or in an injectable form. Feed additives are broadly classified into the following types.

1. Nutrient feed additives, e.g. amino acids, minerals and vitamins.

2. Non-nutrient feed additives, e.g. antibiotics, hormones, immunomodulators, enzymes, probiotics, antioxidants, surfactants, sweetening agents, etc.

ANTIBIOTIC FEED ADDITIVES

Antibiotics are groups of soluble organic substances produced from microorganisms, which in small concentration have the capacity of inhibiting the growth of other microorganisms and even of destroying them. The quantity of antibiotics to be added as an additive is much less than that used for therapeutic purpose.

Factors such as age of the animal, kind of production, nutritional status of the animal, level of hygiene in the farm, stresses of the animals, etc. are to be considered before deciding the quantity.

Antibiotic feed additives can be categorized into **ionophore antibiotics** and **non-ionophore antibiotics**.

Ionophores are low molecular-weight molecules that bind ions of various minerals and modulate their movement across cell membranes. These are produced by several strains of *Streptomyces* spp. Some examples of ionophore antibiotics are monensin, lasalocid, salinomycin and lysocellin.

Monensin   It is a polyether ionophore antibiotic sold under the trade name Rumensin. It is produced by *Streptomyces cinnamonensis*. It is useful as an anticoccidial agent for broilers and lambs. It is approved for feed efficiency but not for growth promotion in cattle.

Comparative slaughter data revealed a trend of higher protein in monensin-fed goats compared with controls. Magnitude of improvement tended to be greater in low dietary protein treatments than in high protein treatments, suggesting protein-sparing effect of monensin.

Lasalocid   It is produced by the fungi *Streptomyces lasoliensis* and is more potent than monensin. Lasalocid acts specifically against hydrogen-producing bacteria and results in higher propionate production. Mechanism of action of lasalocid is similar to that of monensin.

Lysocellin   It is a divalent polyether antibiotic obtained from *Streptomyces cacaoci* var. *asoensis*.

Non-ionophore antibiotics appear to have a complex mode of action. Chlortetracycline, oxytetracycline, bacitracin, flavomycin, etc. are some of the non-ionophore antibiotics. They have been suggested to have a metabolic effect, in which the drug affects various enzyme systems such as some oxidative phosphorylation reactions. They also exhibit nutrient-sparing effect, which may be due to stimulation of microorganisms in the gastrointestinal tract, suppression of organisms which compete for critical nutrients, improved nutrient absorption from the gastrointestinal tract, reduction of vitamin D requirements for normal bone calcification and lower manganese requirements for growth and prevention of perosis. Non-ionophore antibiotics have been proved to control the multiplication of toxin-producing microorganisms. Also, they suppress the sub-clinical level of infection and increase feed efficiency and promote growth.

Both the ionophore and non-ionophore antibiotics have been used in nonruminants and pre-ruminants, while only the ionophores have been successfully used in adult ruminants.

HORMONES

Hormones are the active principles secreted by the endocrine glands into the blood for transportation to target organs and tissues. They are of endogenous origin and are broadly classified into two types.

Anabolic hormones   These include somatotropin, thyroxine, and androgens. Somatotropin stimulates growth of endochondrial bones and epiphysis of long bones while in protein metabolism it aids nitrogen retention and overall protein synthesis. Thyroxine also stimulates growth of long bones as well as protein synthesis. Testosterone is a potent androgen and at low doses it increases the epiphyseal diameter and promotes muscle growth by augmenting nitrogen retention.

Catabolic hormones   These include oestrogen and glucocorticoids. Oestrogen inhibits skeletal growth although in ruminants, it increases nitrogen retention. Glucocorticoids decrease growth of epiphysis and also aid in degrading protein and amino acids and thereby inhibit protein synthesis in extrahepatic tissues.

Some oestrogenic activity is present in some clovers, soybean, sesbania, etc. Some are synthesized chemically. There are exogenous sources of hormones. They are administered orally as feed additives or through parenteral injections.

Hormones can be used as carcass modifiers because they enhance nitrogen retention in the body by way of significantly decreasing urea in blood and urine and urinary nitrogen and result in the production of leaner carcasses.

Anabolic steroids like Oestradiol, Trenbolone acetate (TBA) and zeranol are also used as animal feed. TBA interferes with catabolic action of glucocorticoids on muscle protein thereby enhancing the rate of protein synthesis, an 82% increase in daily carcass protein.

Steroid hormones are orally active and they are not species-specific. Somatotropin is a polypeptide and therefore not orally active, and it is species-specific. Hence the little, if any, somatotropin present in animal products gets degraded in the gastrointestinal tract and thus has no influence on human health.

Growth hormone (GH) secreted by the anterior pituitary and somatotropin influence growth. The secretion of these hormones is enhanced following feeding of protein. The activity of somatotropins is manifested in increased nitrogen retention in the body. It stimulates various biosynthetic processes and increases the growth of skeletal tissue as well. Growth hormone is homeorhetic because it manifests its actions by chronic influence on metabolism and involves partitioning of nutrients for selected processes such as growth and milk production. Growth hormone is a more powerful lactogenic stimulant than an anabolic agent.

Treatment of growing ruminants with exogenous GH increases nitrogen retention, which is associated with as increase in the proportion of protein to increase the fractional synthetic rate of muscle tissues. But this effect is mediated by age, and adult animals are unaffected by exogenous GH or the duration of administration of GH. This is in agreement with the concept that there is a substantial influence of the developmental age on protein turnover and growth.

Thyroid-active materials are used to stimulate growth of body tissue, wool and milk secretion by creating mild hyperthyroidal state. However, they seem to have little practical importance in livestock feeding.

β-adrenergic agonists (β-agonists) are structural analogues of the naturally occurring catecholamines, adrenaline (epinephrine) and noradrenaline. These are orally active and are used to enhance the lean content and reduce the fat content of animals. Thus they increase feed efficiency and growth promotion.

β-agonists react with special cell receptors and increase concentration of cyclic AMP which in turn results in reduced lipogenesis and increased lipolysis, e.g. Clenbuterol, Cimaterol, Ractopamine.

Growth hormone-releasing factor (GHRF) is the main endogenous stimulator of somatotropin secretion. Hence as an alternative to direct administration of growth hormone (GH), stimulation of endogenous GH secretion, using GHRF is highly effective in increasing lean and reducing fat of animals.

Immunomodulators are compounds obtained from organisms or synthesized chemically which are capable of enhancing the defence mechanism of animals, including fish and shrimps. These are of two types namely natural immunomodulators and synthetic immunomodulators. The use of the immune response potentially has a lot to offer the animal production industry as a method of growth promotion and manipulation of carcass composition. Cell wall preparations, vitamin C, vitamin E, Levamisole and quaternary ammonium compounds (QAC) are some of the widely used immunomodulators.

ENZYME FEED ADDITIVES

Maize and soybean meal are the most preferred feed for poultry, swine, etc. The high cost of these ingredients spurred the pursuit of alternative ingredients such as wheat, oats, barley, etc., and sunflower seed meal, rapeseed meal, etc. These alternate feedstuff have deleterious pictures in the utilization of nutrients. Feed enzyme additives act as biocatalysts to assist the digestion process and support utilization of nutrients that otherwise go unused.

The cellulolytic or hemicellulolytic enzymes (e.g. β-glucanase, xylanase, mannase, pectinase, β-galactosidase, 1–4, β-galactanase) degrade gel-forming viscous and structural polysaccharides like cellulose, arabinoxylans and mixed-linked β-glucans.

Feed enzymes such as β-glucanase and xylanases have enabled barley or wheat in poultry diets up to 50–60%. A combination of endo-1-4 β-galactanase and β-galactosidase could be used to improve the nutritive value of soybean meal. Feed enzyme complexes containing arabinases, xylanases and pectinases break down the arabinoxylans and pectins present in sunflower seed meal, rapeseed meal, lupin seed meal, etc. and release the protein and other nutrients. This enzyme complex can convert sunflower meal into soybean meal in terms of nutritive value.

A multi-enzyme preparation with cellulolytic and proteolytic activity (cellulases, β-glucanases and protease) can degrade the structural polysaccharides and proteins. Phytase enzyme helps to increase the utilization of phytin phosphorus by swine, poultry, etc. Enzymes which can withstand high temperature (95°C) employed during pelleting are now available. Feed enzyme additives combine both economic and ecological benefits.

"Fibrozyme" (Alltech) is the first feed-grade enzyme that is rumen-stable. It significantly increases dry matter digestibility and carbohydrate utilization in cows which are fed diets containing high amounts of fibre. Glycosylation process protects the fibrozyme enzyme in the rumen, and also during pelleting. Feeds containing fibrozyme can be pelleted with only a slight loss in enzyme activity. Enzymes that assist in digestion in piglets of 3 weeks of age are protease, xylanase, pectinase and hemicellulase.

Some of the commercial feed enzymes are Avizyme series, Natugrain, Natuphos, Vevozyme, Kenzyme, Grindazym, Flavodan.

PROBIOTICS

Over the last several years, considerable attention has been given to the use of probiotics, yeast cultures, and acidifiers in pit, and poultry feeds primarily. Much of this interest has been generated because of increased public awareness and objection to the use of antibiotics as growth promotant feed additives.

The term "Probiotic" was coined by Parker in 1974 who defined it as "organisms and substances, which contribute to intestinal microbial balance." The term Probiotic means "for life". Probiotics are advocated as an alternative to antibiotics for growth promotion. Probiotics are live cultures of non-pathogenic organisms, which are administered orally. Later Fueller (1989) redefined probiotics as live microbial feed supplements which beneficially affect the host animal by improving its intestinal microbial balance. Probiotic products are available in the form of oral pastes, water-dispersible powders or liquids or directly fed feed additives and include microbial cells, microbial cultures and microbial metabolites. Most probiotics get destroyed by up to 80% in the presence of antibiotics or when mixed with anti-mycoplasma drugs in the feed. Sometimes the form "pronutrients" is used in the place of probiotics. The US Food and Drug Administration used the term direct fed microbials (DFM) instead of probiotic. Viability of the microbes is assured in the pelleted feeds. Some probiotics supply viable bacterial spores of selected *Bacillus* strains, which are heat-resistant. The recovery rate of organisms after pelleting is 95%. Micro-granulated probiotics are available and facilitate reduced dust, improved flowability, and homogeneous mixing even at very low inclusion levels, with improved stability during storage and pelleting, e.g. Paciflor Microgranulated.

Microorganisms used as probiotics are *Lactobacillus acidophilus, L. bifidus, L. bulgaricus, L. casei, L. lactis, Aspergillus oryzae, Streptococcus faecium, S. thermophilus, Saccharomyces cerevisiae and Torulopsis.*

L. acidophilus produces lactic acid and the enzyme amylase. *L. casei* compliments the growth of *L. acidophilus*. *Bifidobacterium bifidum* is found commonly in mother's milk and the intestine of humans and animals. *A. oryzae* produces the enzyme cellulase. *Torulopsis* is the mother culture of yeast. The enzyme lipase is exhibited by *Torulopsis*.

Characteristics of a Good Probiotic

1. The culture should exert a positive effect on the host. It should be gram-positive, acid-resistant and bile-resistant, and contain a minimum of 30×10^9 CFU (colony forming unit) per gram.

2. The culture should possess high survival rate and multiply faster in the digestive tract. It should be strain-specific.

3. The culture microorganisms should neither be pathogenic nor toxic to the host.

4. The adhesive capability of microorganisms must be faster and more firm.

5. It should be durable enough to withstand the stress of commercial manufacturing, processing and distribution so that the product can be delivered alive to the intestine.

Mode of action may be competitive with the harmful enteric microorganisms, stimulatory for increasing growth rate and thus may be productive and nutrient-sparing or a combination of these effects, e.g. the main metabolites of lactobacilli are lactic acid and H_2O_2. The former is responsible for preventing the growth of coliform organisms by reducing the pH while the latter has bactericidal effect. Finally, an acid environment is conducive to increased enzymatic activity within the digestive system. Probiotics at the specific concentration stimulate the immune system.

Pigs and Probiotics

Probiotics have shown the greatest potential in very young and growing pigs. Probiotics prevent and control the incidence of diarrhoea in pigs. Feeding probiotic (*L. acidophilus and S. faecium*) improved weight gain and feed efficiency. Growing pigs receiving the probiotic in their feed showed either equal or superior daily gain, feed intake and feed efficiency compared to pigs fed with the antibiotic. The probiotic species *Bacillus subtilis* reduce the number of *E. coli* and thus reduce the incidence of *E. coli* scours in piglets.

Poultry and Probiotics

The role of the intestinal microflora in health and digestion is more extensively understood in poultry than in mammalian species. There is a delicate balance between beneficial and pathogenic bacteria. A pH of 4–5 favours *Lactobacillus* sp. and a high pH of 6 to 7 is optimal for *E. coli*. It was reported that lactobacilli could be found in crop epithelial tissue as early as day one of life. The crop of fowl normally contains a microbial population in which lactobacilli predominate.

These microbes attach to the crop epithelium and colonize the surface. Attached bacteria are able to inoculate newly ingested food. Lactic-acid-producing bacteria moving with digesta to the intestine serve to influence that microbial community. The result ensures dominance of lactic acid species for the suppression of *E. coli*.

Shortly after hatching, the chick or poult has a nearly sterile digestive tract with a pH range of between 5.5 and 6.0. In addition, the young bird lacks sufficient gastric acid secretion to maintain the ideal acidic pH. It was reported that the poults were born with high numbers of coliforms. So supplementation of a lactobacillus product in the water or feed along with an acidifying agent would be effective in controlling the coliform proliferation.

In addition to the production of lactic acid, the inhibitory effects of *Lactobacillus* on other bacterial species have been attributed to hydrogen peroxide formation and production of an inhibitory substance formed called "acidolin". Lactobacillus probiotic and Zinc bacitracin had similar effects in stimulating weight gain and feed efficiency of broilers. The probiotic caused a distinct change in microbial flora of the caeca and small intestine in that by nine days of age, enterococci had essentially disappeared.

YEAST CULTURE

Hungate (1966) observed that rumen bacterial concentrations increased when fermentation products such as yeast cultures were added to the diet. According to him, the stimulatory effect was due to metabolites in the fermentation products, which served as nutrients for the bacteria. By-products of fermentation include dried brewer's yeast, dried distillers solubles and dried bacterial press cakes.

Yeast cells are destroyed by pelletizing the feed, storage conditions and strong mineral mixtures. Yeast is sensitive to heat and does not survive conditions used in pelleting commercial feed. Hence use of live yeast culture is recommended.

The feeding of live yeast culture, *Saccharomyces cerevisiae,* has attracted the attention of animal nutritionists to improve the microbiological balance of the host animal and thereby extract the nutrients to the maximum extent and get them deposited in the end products (milk, meat and eggs) for human use which simultaneously reduces environmental pollution. Live yeast cells are highly probiotic.

Live yeast culture has the following effects:

Effect on animal physiology Reduces the temperature in heat-stressed animals. The greatest benefit occurs during the hotter months. Fungal cultures like *Aspergillus oryzae, Armillaria heimii,* etc. (3 to 5 g/day) in the diet decreased body temperatures and respiration rates in hot, but not in cool weather. However, the mechanism is not clear.

Effect on the rumen   *Saccharomyces cerevisiae* could act by production of growth-stimulating factors in the rumen, stabilization of rumen pH, and reduction of lactic acid production in the rumen. The probiotic activity might be due to the metabolites in the yeast culture. The increased population of cellulolytic bacteria and total bacteria increases the rate of forage degradation which eventually enhances feed intake and the supply of microbial protein from the rumen (Figure 12.1).

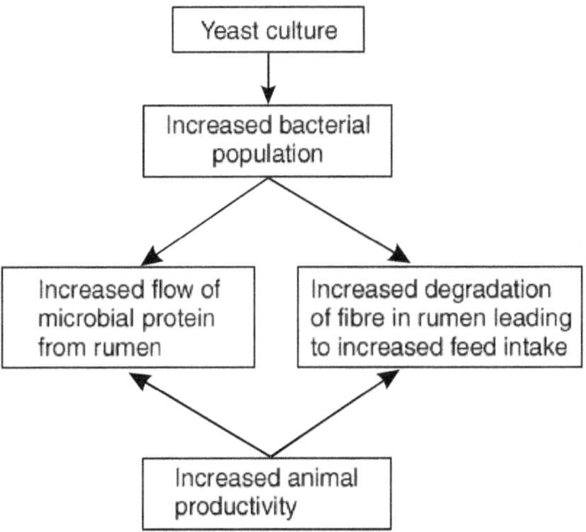

Figure 12.1   Action of yeast culture in ruminants

Effect on milk production   An increase of 7.8% in milk yield was observed using yeast culture as probiotic. Responses are greater during early lactation. The protein and energy output in the form of milk was increased using yeast cultures.

Live Yeast Culture for Poultry

Yeast culture is considered a probiotic by many when that culture supplies live yeast cells. However, it is better known for its ability to enhance feed utilization. Hence live yeast culture with lactic-acid-producing bacteria (*L. acidophilus* and *S. faecium*) was supplemented in broilers (1 kg/ton) and the results showed that weight gain and feed conversion were improved.

Combined use of *Lactobacillus* and yeast culture in the feed and water has been shown to be effective in reducing morbidity and mortality, and improving growth performance and production characteristics.

Mannan oligosaccharide (MOS) MOS, a complex carbohydrate extracted from yeast cell wall, improves the health and performance of monogastric animals. MOS blocks attachment of pathogenic bacteria to the animal's intestine and prevents colonization that can result in disease. In addition, MOS may stimulate the animal's immune system, thereby reducing the risk of disease, further. MOS increases the release of cytokines, which coordinate activity among different cells of the immune system. MOS also enhances interleukin-2 concentration. The immune function requires interleukin-2 for T-cell proliferation and differentiations.

Fructo-oligosaccharides (FOS) Short chain fructo-oligosaccharides encourage the growth of beneficial bacteria in the gut, such as *Lactobacillus* sp., *Bifidobacterium* sp. and *Bacteroides* sp. Feeding FOS helps proliferation of these probiotic bacteria, which inhibit growth of more harmful bacteria and reduction of flatulence (since FOS are not digested by host intestinal enzymes) in animals. So these (FOS, MOS) are termed as "prebiotics".

Research findings strongly favour dietary supplementation of live yeast culture as a stress reliever during the hot season of the year and of intensive rearing of livestock and as a probiotic to increase the rate of fibre degradation and thus increased feed intake and eventually animal productivity.

Apart from living organisms, chemical substances like acidifiers, antioxidants, pigments, and buffers are also added as feed additives. It is reported that there is synergy between formic acid and phytase to improve the growth performance and feed conversion rates of young pigs in the 22–47-kg weight range.

In diets for laying hens, the feed additive, calcium formate (1.3%), which is converted to formic acid in the crop, offers the following potential benefits beyond reducing *E. coli* and *Salmonella* populations.

- Mould inhibition effect and general preservative effects.
- Improved consistency of dropping, resulting in fewer dirty eggs.
- Better calcium absorption and egg shell quality.
- Chemically stable in storage, safe in food and the environment and without destructive effects for other feed supplements, such as vitamins and yolk pigments.

Calcium formate additive is non-corrosive, odour-free and non-caustic. It is a relatively safe acidifier for use in feed manufacturing process. It is approved by European Commission (EC) for use as a food and feed preservative.

Antioxidants Oxidation of feed fats causes rancidity and spoils the taste and flavour of the feeds through a process known as lipid peroxidation or auto-oxidation. These rancid fat-containing diets impart undesirable off flavour to milk and milk products. Oxidation also causes much loss to carotenes, vitamin A and vitamin D. The use of antioxidants limits this oxidation spoilage.

Yolk colour is improved by the addition of alfalfa leaf meal at 2–3%, which is the source of carotenoids.

Feeding high-grain (low fibre) diets to meet the energy requirements of high-yielding (over 35 kg milk/day) cows to minimize the energy crisis during early lactation leads to changes in rumen pH and rumen fermentation pattern. Buffers like sodium bicarbonate, magnesium oxide, calcium carbonate, etc. are used to correct these changes.

ADVANTAGES OF FEED ADDITIVES

1. *Increase feed quality and feed palatability*   Emulsifiers and pelleting agents are used to meet the demands of feed manufacturers, while antioxidants, fungistatic agents and fermentation inhibitors ensure proper shelf life of feed.

2. *Improve animal performance*   Feed additives are mixed with feeds in nontherapeutic quantities for the purpose of promoting animal growth, lowering feed consumption and protecting the animal against all sorts of harmful environmental influences (stresses).

3. *Improve the final product*   Addition of antioxidants to diets produce grades of meat in which the fat does not become rancid or does so more slowly. The use of additives also makes end products more homogeneous and improves their quality.

4. *Economize the cost of animal protein*   Low levels of additives, mainly of antibiotics or other growth promoters and related compounds in animal feed contribute to increased production of animal proteins for human consumption. Feeding antibiotics and other additives lower the cost of meat, milk and egg production.

DISADVANTAGES OF FEED ADDITIVES

Feed additives suffer from the following setback:

- Use of hormones and antibiotics leave their residues is meat, milk and eggs, which is objectionable.

- Feeding of low concentration of antibiotics may favour the proliferation of antibiotic-resistant microorganisms, which could have serious consequences for disease control in humans or domestic animals.

However, it is difficult to envisage and develop intensive animal breeding without antibiotic feed additives. As animal concentrations increase in intensive farms, the use of antibiotic feed additives become even more indispensable.

To conclude, feed additives, though less expensive and safe from public health point of view, need more exploration about the active ingredients and the mode of their action so that their potential can be explained in terms of animal production.

Summary

- ✔ Any chemical incorporated in an animal feed for the purpose of improving rate of gain, feed efficiency, or preventing and controlling disease is called feed additive.

- ✔ Feed additives are broadly classified into nutrient feed additives and non-nutrient feed additives.

- ✔ Ionophore and non-ionophore antibiotic feed additives have been used in non-ruminants and pre-ruminants.

- ✔ Hormones, either anabolic or catabolic, can be used as carcass modifiers.

- ✔ Immunomodulators are used to enhance the defence mechanism of animals, including fish and shrimps.

- ✔ A multi-enzyme preparation with cellulolytic and proteolytic activity can degrade the structural polysaccharides and proteins.

- ✔ Probiotics are defined as organisms and substances which contribute to intestinal microbial balance.

- Live yeast cells are highly probiotics. The probiotic activity might be due to the metabolites in the yeast culture.

- Apart from living organisms, chemical substances like acidifiers, antioxidants, pigments, and buffers are also added as feed additives.

- Antioxidants are used to limit the oxidation spoilage.

REVIEW QUESTIONS

1. Define feed additive.

2. Discuss the advantages and disadvantages of feed additives.

3. Brief on antibiotic feed additives.

4. Explain the role of hormones as feed additives.

5. Write a short note on feed enzyme additives.

6. Define probiotics. State the characteristics of a good probiotic.

7. Briefly explain:

    i.   pigs and probiotics

    ii. poultry and probiotics

8. Discuss the effect of yeast culture on the rumen.

9. Write short notes on MOS and FOS.

10. Antioxidants limit oxidation spoilage. Justify.

13

STEM CELL
TECHNOLOGY

Cloning and Stem Cell Technology

Cloning is the creation of a genetic replica of an individual. The technique transfers a nucleus from a somatic cell into a cell whose nucleus has been removed, and then a new individual develops from the manipulated cell. There are two goals of cloning. Reproductive cloning seeks to clone a baby using the nucleus from the cell of a particular individual who will then, supposedly, be duplicated. In contrast, the goal of therapeutic cloning is to use very early embryos as sources of stem cells. Specifically, a nucleus from a person's somatic cell is transferred to an enucleated oocyte. The embryo develops until the inner mass forms, and these cells are then used to establish cultures of stem cells that are genetically identical to the donor of the somatic cell nucleus. The idea is that the stem cells match cells of the person, so the immune system will not reject them. Therapeutic cloning holds much promise to treat spinal cord injuries as well as neurodegenerative disorders such as Parkinson disease and Alzheimer disease, both prevalent in our aging people.

The bioethical issues that reproductive and therapeutic cloning raise are complex. Some people argue that therapeutic cloning violates rights of early-stage embryos. Proponents of the research argue that not to allow this research violates the rights of the people who might benefit from stem cell therapy. They suggest that "leftover" fertilized ova currently deep-frozen in many reproductive health clinics could be used in therapeutic cloning. As the debate rages, research in some nations has been put on hold, as politicians and the public argue the pros and cons of the research.

INTRODUCTION

Research on stem cells is advancing knowledge about how an organism develops from a single cell and how healthy cells replace damaged cells in adult organisms. This promising area of science is also leading scientists to investigate the possibility of cell-based therapies to treat disease, which is often referred to as **regenerative** or **reparative medicine**.

Study of stem cells is one of the most fascinating areas of biology today. But like many expanding fields of scientific inquiry, research on stem cells raises scientific questions as rapidly as it generates new discoveries.

Stem cells have two important characteristics that distinguish them from other types of cells. First, they are unspecialized cells that renew themselves for long periods through cell division. The second is that under certain physiological or experimental conditions, they can be induced to become cells with special functions such as the beating cells of the heart muscle or the insulin-producing cells of the pancreas.

Scientists primarily work with two kinds of stem cells from animals and humans: **embryonic stem cells** and **adult stem cells**, which have different functions and characteristics that will be explained in this chapter. Scientists discovered ways to obtain or derive stem cells from early mouse embryos more than 20 years ago. Many years of detailed study of the biology of mouse stem cells led to the discovery, in 1998, of how to isolate stem cells from human embryos and grow the cells in the laboratory. These are called **human embryonic stem cells**. The embryos used in these studies were created for infertility purposes through *in vitro* **fertilization** procedures and when they were no longer needed for that purpose, they were donated for research, with the consent of the donor.

Stem cells are important for living organisms for many reasons. In the 3- to 5-day-old embryo, called a **blastocyst**, stem cells in developing tissues give rise to the multiple specialized cell types that make up the heart, lung, skin, and other tissues. In some adult tissues, such as bone marrow, muscle, and brain, discrete populations of adult stem cells generate replacements for cells that are lost through normal wear and tear, injury, or disease. The blastocyst includes three structures: the trophoblast, which is the layer of cells that surrounds the blastocyst; the blastocoel, which is the hollow cavity inside the blastocyst; and the inner cell mass, which is a group of approximately 30 cells at one end of the blastocoel.

It has been hypothesized by scientists that stem cells may, at some point in the future, become the basis for treating diseases such as Parkinson's disease, diabetes, and heart disease.

Scientists want to study stem cells in the laboratory so they can learn about their essential properties and what makes them different from specialized cell types. As scientists learn more about stem cells, it may become possible to use the cells not just in **cell-based therapies**, but also for screening new drugs and toxins and understanding birth defects. However, as mentioned above, human embryonic stem cells have been studied only since 1998. Therefore, in order to develop such treatments scientists are intensively studying the fundamental properties of stem cells, which include:

1. determining precisely how stem cells remain unspecialized and self-renewing for many years; and

2. identifying the **signals** that cause stem cells to become specialized cells.

UNIQUE PROPERTIES OF STEM CELLS

Stem cells differ from other kinds of cells in the body. All stem cells, regardless of their source, have three general properties:

they are capable of dividing and renewing themselves for long periods; they are unspecialized; and they can give rise to specialized cell types.

Scientists are trying to understand two fundamental properties of stem cells that relate to their **long-term self-renewal**:

1. Why can **embryonic stem cells** proliferate for a year or more in the laboratory without differentiating, but most **adult stem cells** cannot?

2. What are the factors in living organisms that normally regulate stem cell **proliferation** and self-renewal?

Discovering the answers to these questions may make it possible to understand how cell proliferation is regulated during normal embryonic development or during the abnormal cell division that leads to cancer.

Stem cells are unspecialized   One of the fundamental properties of a stem cell is that it does not have any tissue-specific structures that allow it to perform specialized functions. A stem cell cannot work with its neighbours to pump blood through the body (like a heart muscle cell); it cannot carry molecules of oxygen through the bloodstream (like a red blood cell); and it cannot fire electrochemical signals to other cells that allow the body to move or speak (like a nerve cell). However, unspecialized stem cells can give rise to specialized cells, including heart muscle cells, blood cells, or nerve cells.

Stem cells are capable of dividing and renewing themselves for long periods   Unlike muscle cells, blood cells, or nerve cells which do not normally replicate themselves, stem cells may replicate many times. When cells replicate themselves many times over it is called proliferation. A starting population of stem cells that proliferates for many months in the laboratory can yield millions of cells. If the resulting cells continue to be unspecialized like the parent stem cells, the cells are said to be capable of long-term self-renewal.

The specific factors and conditions that allow stem cells to remain unspecialized are of great interest to scientists. It has taken scientists many years of trial and error to learn to grow stem cells in the laboratory without them spontaneously differentiating into specific cell types. For example, it took 20 years to learn how to grow **human embryonic stem cells** in the laboratory following the development of conditions for growing mouse stem cells. Therefore, an important area of research is understanding the signals in a mature organism that cause a stem cell population to proliferate and remain unspecialized until the cells are needed for repair of a specific tissue. Such information is critical for scientists to be able to grow large numbers of unspecialized stem cells in the laboratory for further experimentation.

Stem cells can give rise to specialized cells   When unspecialized stem cells give rise to specialized cells, the process is called **differentiation**. Scientists are just beginning to understand the signals inside and outside cells that trigger stem cell differentiation. The internal signals are controlled by a cell's **genes**, which are interspersed across long strands of DNA, and carry coded instructions for all the structures and functions of a cell. The external signals for cell differentiation include chemicals secreted by other cells, physical contact with neighbouring cells, and certain molecules in the **microenvironment**.

Adult stem cells typically generate the cell types of the tissue in which they reside. A blood-forming adult stem cell in the bone marrow, for example, normally gives rise to the many types of blood cells such as red blood cells, white blood cells and platelets. Until recently, it had been thought that a blood-forming cell in the bone marrow which is called a **haematopoietic stem cell** could not give rise to the cells of a very different tissue, such as nerve cells in the brain. However, a number of experiments over the last several years have raised the possibility that stem cells from one tissue may be able to give rise to cell types of a completely different tissue, a phenomenon known as plasticity.

Examples of such **plasticity** include blood cells becoming **neurons**, liver cells that can be made to produce insulin, and haematopoietic stem cells that can develop into heart muscle. Therefore, exploring the possibility of using adult stem cells for cell-based therapies has become a very active area of investigation by researchers.

GROWING EMBRYONIC STEM CELLS IN THE LABORATORY

Growing cells in the laboratory is known as **cell culture**. Human embryonic stem cells are isolated by transferring the **inner cell mass** into a plastic laboratory culture dish that contains a nutrient broth known as **culture medium**. The cells divide and spread over the surface of the dish. The inner surface of the culture dish is typically coated with mouse embryonic skin cells that have been treated so they will not divide. This coating layer of cells is called a **feeder layer**. The reason for having the mouse cells in the bottom of the culture dish is to give the cells of the inner cell mass a sticky surface to which they can attach. Also, the feeder cells release nutrients into the culture medium. Recently, scientists have begun to devise ways of growing embryonic stem cells without the mouse feeder cells. This is a significant scientific advancement because of the risk that viruses or other macromolecules in the mouse cells may be transmitted to the human cells.

Over the course of several days, the cells of the inner cell mass proliferate and begin to crowd the culture dish. When this occurs, they are removed gently and plated into several fresh culture dishes. The process of replating the cells is repeated many times and for many months, and is called **subculturing**. Each cycle of subculturing the cells is referred to as a **passage**. After six months or more, the original 30 cells of the inner cell mass yield millions of embryonic stem cells. Embryonic stem cells that have proliferated in cell culture for six or more months without differentiating, are **pluripotent**, appear genetically normal, and are referred to as an **embryonic stem cell line**.

Once cell lines are established, or even before that stage, batches of them can be frozen and shipped to other laboratories for further culture and experimentation.

IDENTIFICATION OF EMBRYONIC STEM CELLS

At various points during the process of generating embryonic stem cell lines, scientists test the cells to see whether they exhibit the fundamental properties that make them embryonic stem cells. This process is called characterization.

As yet, scientists who study human embryonic stem cells have not agreed on a standard battery of tests that measure the cells' fundamental properties. Also, scientists acknowledge that many of the tests they do use may not be good indicators of the cells' most important biological properties and functions. Nevertheless, laboratories that grow human embryonic stem cell lines use several kinds of tests. These tests include:

- *Growing and subculturing the stem cells for many months* This ensures that the cells are capable of long-term self-renewal. Scientists inspect the cultures through a microscope to see that the cells look healthy and remain **undifferentiated**.

- *Using specific techniques to determine the presence of surface markers that are found only on undifferentiated cells* Another important test is for the presence of a protein called Oct-4 which is typically made by undifferentiated cells. Oct-4 is a transcription factor, meaning that it helps turn **genes** on and off at the right time, which is an important part of the processes of cell **differentiation** and embryonic development.

- ☯ *Examining the chromosomes under a microscope*   This is a method to assess whether the chromosomes are damaged or if the number of chromosomes has changed. It does not detect genetic mutations in the cells.

- ☯ *Determining whether the cells can be subcultured*   This is done after freezing, thawing, and replating.

- ☯ *Testing whether the human embryonic stem cells are pluripotent*   This is tested by 1) allowing the cells to differentiate spontaneously in cell culture; 2) manipulating the cells so that they will differentiate to form specific cell types; or 3) injecting the cells into an immunosuppressed mouse to test for the formation of a benign tumour called a **teratoma**. Teratomas typically contain a mixture of many differentiated or partly differentiated cell types—an indication that the embryonic stem cells are capable of differentiating into multiple cell types.

STIMULATION OF EMBRYONIC STEM CELLS TO DIFFERENTIATE

As long as the embryonic stem cells in culture are grown under certain conditions, they can remain undifferentiated (unspecialized). But if cells are allowed to clump together to form **embryoid bodies**, they begin to differentiate spontaneously. They can form muscle cells, nerve cells, and many other cell types. Although differentiating embryoid cells are healthy, it is not an efficient way to produce cultures of specific cell types.

So, to generate cultures of specific types of differentiated cells—heart muscle cells, blood cells, or nerve cells, for example—scientists try to control the differentiation of embryonic stem cells. They change the chemical composition of the culture medium, alter the surface of the culture dish, or modify the cells by inserting specific genes. Through years of experimentation scientists have established some basic protocols or "recipes" for the **directed**

differentiation of embryonic stem cells into some specific cell types (Figure 13.1).

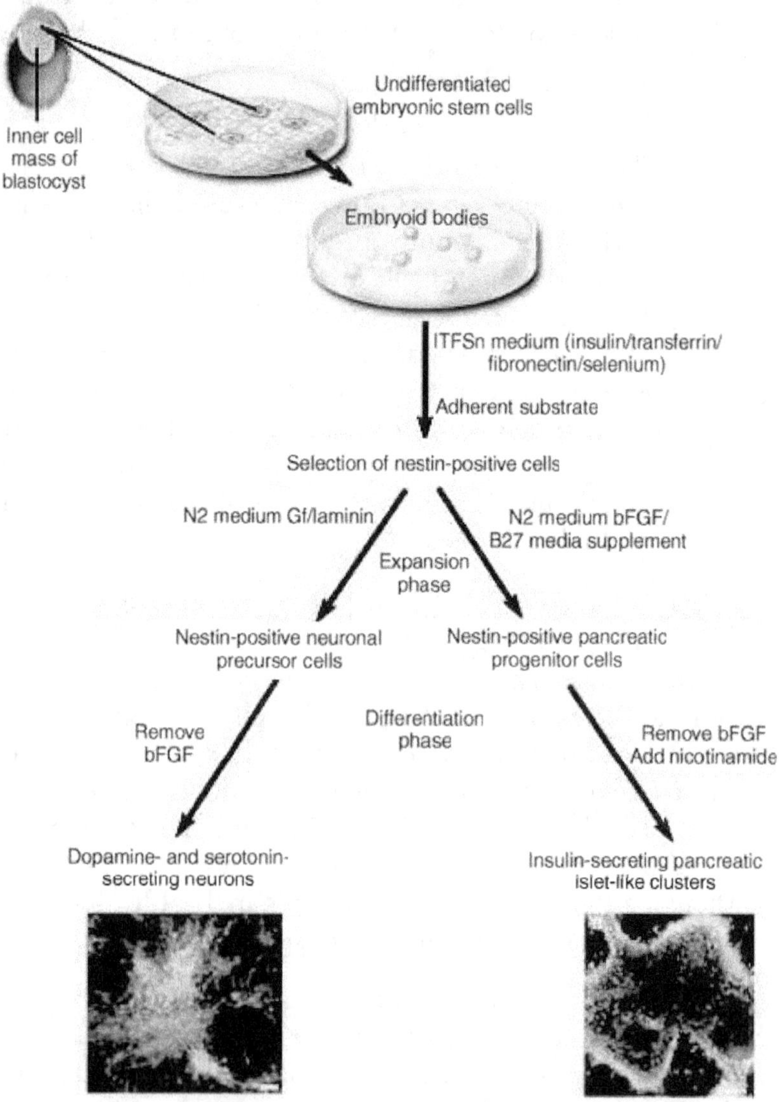

Figure 13.1 Directed differentiation of mouse embryonic stem cells

If scientists can reliably direct the differentiation of embryonic stem cells into specific cell types, they may be able to use the resulting, differentiated cells to treat certain diseases at some point in the future. Diseases that might be treated by transplanting cells generated from human embryonic stem cells include Parkinson's disease, diabetes, traumatic spinal cord injury, Purkinje cell degeneration, Duchenne's muscular dystrophy, heart disease, and vision and hearing loss.

ADULT STEM CELLS

An adult stem cell is an **undifferentiated** cell found among differentiated cells in a tissue or organ, can renew itself, and can differentiate to yield the major specialized cell types of the tissue or organ. The primary role of **adult stem cells** in a living organism is to maintain and repair the tissue in which they are found. Some scientists now use the term **somatic stem cell** instead of adult stem cell. Unlike **embryonic stem cells** which are defined by their origin (the **inner cell mass** of the **blastocyst**), the origin of adult stem cells in mature tissues is unknown.

Research on adult stem cells has recently generated a great deal of excitement. Scientists have found adult stem cells in many more tissues than they once thought possible. This finding has led scientists to ask whether adult stem cells could be used for transplants. In fact, adult blood-forming stem cells from bone marrow have been used in transplants for 30 years. Certain kinds of adult stem cells seem to have the ability to differentiate into a number of different cell types, given the right conditions. If this differentiation of adult stem cells can be controlled in the laboratory, these cells may become the basis of therapies for many serious common diseases.

The history of research on adult stem cells began about 40 years ago. In the 1960s, researchers discovered that the bone marrow contains at least two kinds of stem cells. One population, called **haematopoietic stem cells**, forms all the types of blood cells in the body. A second population, called **bone marrow stromal cells**, was discovered a few years later. **Stromal cells** are a mixed cell population that generates bone, cartilage, fat, and fibrous connective tissue.

Also in the 1960s, scientists who were studying rats discovered two regions of the brain that contained dividing cells, which become nerve cells. Despite these reports, most scientists believed that new nerve cells could not be generated in the adult brain. It was not until the 1990s that scientists agreed that the adult brain does contain stem cells that are able to generate the brain's three major cell types—astrocytes and oligodendrocytes, which are non-neuronal cells, and neurons, or nerve cells.

Occurrence and Functions of Adult Stem Cells

Adult stem cells have been identified in many organs and tissues. One important point to understand about adult stem cells is that there are a very small number of stem cells in each tissue. Stem cells are thought to reside in a specific area of each tissue where they may remain quiescent (non-dividing) for many years until they are activated by disease or tissue injury. The adult tissues reported to contain stem cells include brain, bone marrow, peripheral blood, blood vessels, skeletal muscle, skin and liver.

Scientists in many laboratories are trying to find ways to grow adult stem cells in **cell culture** and manipulate them to generate specific cell types so that they can be used to treat injury or disease. Some examples of potential treatments include replacing the dopamine-producing cells in the brains of Parkinson's patients, developing insulin-producing cells for type I diabetes and repairing damaged heart muscle, following a heart attack, with cardiac muscle cells.

Tests for Identifying Adult Stem Cells

Scientists do not agree on the criteria that should be used to identify and test adult stem cells. However, they often use one or more of the following three methods:

1. labelling the cells in a living tissue with molecular markers and then determining the specialized cell types they generate;

2. removing the cells from a living animal, labelling them in cell culture, and transplanting them back into another animal to determine whether the cells repopulate their tissue of origin; and

3. isolating the cells, growing them in cell culture, and manipulating them, often by adding growth factors or introducing new genes, to determine what differentiated cell types they can become.

Also, a single adult stem cell should be able to generate a line of genetically identical cells known as a **clone** which then gives rise to all the appropriate differentiated cell types of the tissue. Scientists tend to show either that a stem cell can give rise to a clone of cells in cell culture, or that a purified population of candidate stem cells can repopulate the tissue after transplant into an animal. Recently, by infecting adult stem cells with a virus that gives a unique identifier to each individual cell, scientists have been able to demonstrate that individual adult stem cell clones have the ability to repopulate injured tissues in a living animal.

Adult Stem Cell Differentiation

As indicated above, scientists have reported that adult stem cells occur in many tissues and that they enter normal **differentiation** pathways to form the specialized cell types of the tissue in which they reside. Adult stem cells may also exhibit the ability to form specialized cell types of other tissues, which is known as **transdifferentiation** or **plasticity**.

Normal differentiation pathways of adult stem cells   In a living animal, adult stem cells can divide for a long period and can give rise to mature cell types that have characteristic shapes and specialized structures and functions of a particular tissue. The following are examples of differentiation pathways of adult stem cells (Figure 13.2).

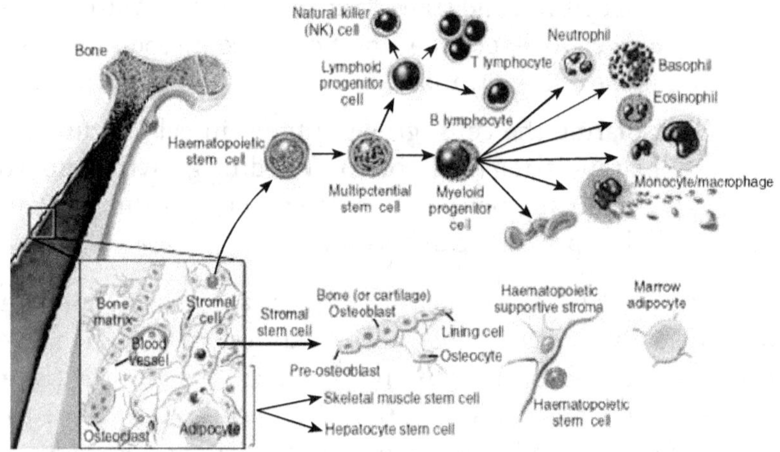

Figure 13.2   Haematopoietic and stromal stem cell differentiation

- Haematopoietic stem cells give rise to all the types of blood cells: red blood cells, B lymphocytes, T lymphocytes, natural killer cells, neutrophils, basophils, eosinophils, monocytes, macrophages, and platelets.

- Bone marrow stromal cells (**mesenchymal stem cells**) give rise to a variety of cell types: bone cells (osteocytes), cartilage cells (chondrocytes), fat cells (adipocytes), and other kinds of connective tissue cells such as those in tendons.

- **Neural stem cells** in the brain give rise to its three major cell types: nerve cells (neurons) and two categories of non-neuronal cells—astrocytes and oligodendrocytes.

- Epithelial stem cells in the lining of the digestive tract occur in deep crypts and give rise to several cell types: absorptive cells, goblet cells, Paneth cells, and enteroendocrine cells.

- Skin stem cells occur in the basal layer of the epidermis and at the base of hair follicles. The epidermal stem cells give rise to keratinocytes, which migrate to the surface of the skin and form a protective layer. The follicular stem cells can give rise to both the hair follicle and to the epidermis.

Adult Stem Cell Plasticity and Transdifferentiation

A number of experiments have suggested that certain adult stem cell types are **pluripotent**. This ability to differentiate into multiple cell types is called plasticity or transdifferentiation. The following list offers examples of adult stem cell plasticity that has been reported during the past few years.

- Haematopoietic stem cells may differentiate into: three major types of brain cells (neurons, oligodendrocytes, and astrocytes); skeletal muscle cells; cardiac muscle cells; and liver cells.

- Bone marrow stromal cells may differentiate into: cardiac muscle cells and skeletal muscle cells.

- Brain stem cells may differentiate into: blood cells and skeletal muscle cells.

Current research is aimed at determining the mechanisms that underlie adult stem cell plasticity. If such mechanisms can be identified and controlled, existing stem cells from a healthy tissue might be induced to repopulate and repair a diseased tissue (Figure 13.3).

EMBRYONIC VERSUS ADULT STEM CELLS

Human embryonic and **adult stem cells** each have advantages and disadvantages regarding potential use for cell-based regenerative therapies. Of course, adult and embryonic stem cells differ in the number and type of differentiated cell types they can become. **Embryonic stem cells** can become all cell types of the body because they are **pluripotent**. Adult stem cells are generally limited to differentiating into the cell types of their tissue of origin. However, some evidence suggests that adult stem cell **plasticity** (Figure 13.3) may exist, increasing the number of cell types a given adult stem cell can become.

Large numbers of embryonic stem cells can be relatively easily grown in culture, while adult stem cells are rare in mature tissues and methods for expanding their numbers in **cell culture** have not yet been worked out. This is an important distinction, as large numbers of cells are needed for stem cell replacement therapies.

A potential advantage of using stem cells from an adult is that the patient's own cells could be expanded in culture and then reintroduced into the patient. The use of the patient's own adult stem cells would mean that the cells would not be rejected by the immune system. This represents a significant advantage as immune rejection is a difficult problem that can be circumvented only with immunosuppressive drugs.

Embryonic stem cells from a donor introduced into a patient could cause transplant rejection. However, whether the recipient would reject donor embryonic stem cells has not been determined in human experiments.

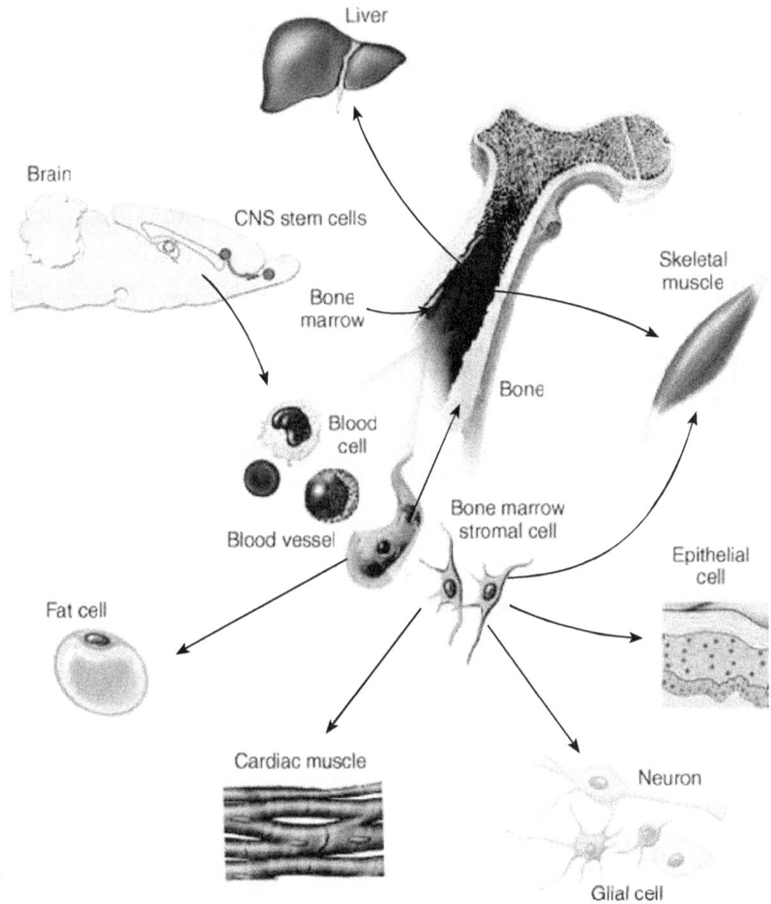

Figure 13.3   Plasticity of adult stem cells

ADVANTAGES AND DISADVANTAGES OF STEM CELL TECHNOLOGY

There are many ways in which human stem cells can be used in basic research and in clinical research. However, there are many technical hurdles between the promise of stem cells and the realization of these uses, which will only be overcome by continued intensive stem cell research.

Studies of **human embryonic stem cells** may yield information about the complex events that occur during human development. The primary goal of this work is to identify how **undifferentiated** stem cells become differentiated. Scientists know that turning genes on and off is central to this process. Some of the most serious medical conditions, such as cancer and birth defects, are due to abnormal **cell division** and **differentiation**. A better understanding of the genetic and molecular controls of these processes may yield information about how such diseases arise and suggest new strategies for therapy. A significant hurdle to this use and most uses of stem cells is that scientists do not yet fully understand the **signals** that turn specific genes on and off to influence the differentiation of the stem cell.

Human stem cells could also be used to test new drugs. For example, new medications could be tested for safety on differentiated cells generated from human **pluripotent** cell lines. Other kinds of cell lines are already used in this way. Cancer cell lines, for example, are used to screen potential anti-tumour drugs. But, the availability of pluripotent stem cells would allow drug testing in a wider range of cell types. However, to screen drugs effectively, the conditions must be identical when comparing different drugs. Therefore, scientists will have to be able to precisely control the differentiation of stem cells into the specific cell type on which drugs will be tested. Current knowledge of the signals controlling differentiation fall well short of being able to mimic these conditions precisely to consistently have identical differentiated cells for each drug being tested.

Perhaps the most important potential application of human stem cells is the generation of cells and tissues that could be used for **cell-based therapies**. Today, donated organs and tissues are often used to replace ailing or destroyed tissue, but the need for transplantable tissues and organs far outweighs the available supply. Stem cells, directed to differentiate into specific cell types, offer the possibility of a renewable source of replacement cells and tissues to treat diseases including Parkinson's and Alzheimer's diseases, spinal cord injury, stroke, burns, heart disease, diabetes, osteoarthritis, and rheumatoid arthritis.

For example, it may become possible to generate healthy heart muscle cells in the laboratory and then transplant those cells into patients with chronic heart disease. Preliminary research in mice and other animals indicates that bone marrow stem cells, transplanted into a damaged heart, can generate heart muscle cells and successfully repopulate the heart tissue. Other recent studies in **cell culture** systems indicate that it may be possible to direct the **differentiation** of embryonic stem cells or adult bone marrow cells into heart muscle cells (Figure 13.4).

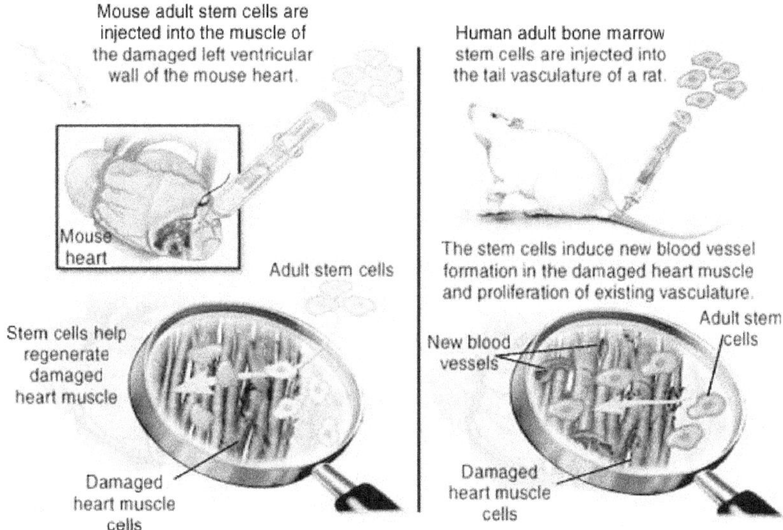

Figure 13.4   Heart muscle repair with adult stem cells

In people who suffer from type I diabetes, the cells of the pancreas that normally produce insulin are destroyed by the patient's own immune system. New studies indicate that it may be possible to direct the differentiation of human embryonic stem cells in cell culture to form insulin-producing cells that eventually could be used in transplantation therapy for diabetics.

To realize the promise of novel cell-based therapies for such pervasive and debilitating diseases, scientists must be able to easily and reproducibly manipulate stem cells so that they possess the necessary characteristics for successful differentiation, transplantation and engraftment. The following is a list of steps in successful cell-based treatments that scientists will have to learn to precisely control to bring such treatments to the clinic. To be useful for transplant purposes, stem cells must be reproducibly made to:

- Proliferate extensively and generate sufficient quantities of tissue.

- Differentiate into the desired cell type(s).

- Survive in the recipient after transplant.

- Integrate into the surrounding tissue after transplant.

- Function appropriately for the duration of the recipient's life.

- Avoid harming the recipient in any way.

Also, to avoid the problem of immune rejection, scientists are experimenting with different research strategies to generate tissues that will not be rejected.

To summarize, the promise of stem cell therapies is an exciting one, but significant technical hurdles remain that will only be overcome through years of intensive research.

Summary

- Stem cell technology is a cell-based therapy to treat disease, which is often referred to as regenerative or reparative medicine.

- Stem cells are unspecialized cells that renew themselves for long periods through cell division.

- ✔ Under certain physiological or experimental conditions, stem cells can be induced to become cells with special functions.

- ✔ Stem cells are categorized into embryonic stem cells and adult stem cells.

- ✔ Embryonic stem cells clump together to form embryoid bodies and begin to differentiate spontaneously.

- ✔ Adult stem cells exhibit plasticity.

- ✔ Human stem cells could be used to test new drugs, for e.g. cancer cell lines are used to screen potential anti-tumour drugs.

REVIEW QUESTIONS

1. Differentiate stem cells from other types of cells.

2. State the importance of stem cells.

3. Explain in detail the unique properties of stem cells.

4. Discuss the process of culturing ES cells in the laboratory.

5. Give a detailed account of characterization of ES cells.

6. Explain directed differentiation of mouse ES cells.

7. What are adult stem cells?

8. Explain the tests performed to identify adult stem cells.

9. Give an account of adult stem cell differentiation.

10. Compare and contrast embryonic and adult stem cells.

14

ETHICS, MORALITY AND ANIMAL BIOTECHNOLGOY

In several parts of the world, projects are underway to record genetic and other types of health information on citizens.

Many population-based studies of gene variants are in the planning stages. Bioethicists have suggested several strategies to ensure fair use of genetic information. They include:

- Preserving choice in seeking genetic tests

- Protecting the privacy of individuals by legally restricting access to genome information

- Tailoring tests to those genes that are most relevant to an individual

- Refusing to screen for trivial traits in offspring, such as eyes or hair colour, or traits that have a large and controllable environmental component, such as intelligence

- Educating the public to make informed decisions concerning genetic information, including evaluating the risks and benefits of medical tests, judging the accuracy of forensic data, or eating genetically modified foods

If these goals are met, human genome information will reveal the workings of the human body at the molecular level, and add an unprecedented precision and personalization to health care.

In this chapter we focus mainly on ethical, moral and social issues surrounding relatively recent developments that involve genetic modification, i.e., the direct manipulation of an animal's genetic make-up. Genetic modification of animals was first achieved with mice in 1980, and of cattle, sheep and pigs by about 1985.

We will also consider issues raised by the new technology called nuclear transfer. Here, whole nuclei and the genes which they carry are transferred. This is the process that was used to produce the sheep Dolly, and subsequent cloned animals. It does not involve altering the genes, it rather involves copying them. In this respect it resembles what gardeners do when they "take cuttings" of a plant to propagate it.

A TALE OF TWO SHEEP

The ewe on the left is genetically modified: she carries a copy of a human gene for an enzyme inhibitor called alpha-1-antitrypsin which can be used to treat some lung disorders. Because the sheep carries a copy of a human gene she may be said to be "transgenic". The ewe on the right is Dolly—the world's first mammal cloned from an adult cell. She is neither genetically modified nor transgenic. Dolly has the same genes as the ewe from which an udder cell was taken and fused with an "empty" egg to produce her.

WHY ARE GENETICALLY MODIFIED ANIMALS PRODUCED?

There are five reasons why genetically modified animals are produced:

1. *To help scientists to identify, isolate and characterize genes in order to understand more about their function and regulation.*

   Genetic modification can be used to knock out the activity of a particular gene. By correlating loss of function with this "knock out" it is possible to gain information about the role of the gene and the product for which it codes.

2. *To provide research models of human diseases, to help develop new drugs and new strategies for repairing defective genes ("gene therapy").*

   Animal models of diseases have been used for many years by exploiting naturally occurring mutations in genes, and inbreeding laboratory strains of animals carrying the mutation. An example is a mouse model of Duchenne Muscular Dystrophy. Genetic modification has been used to produce animal models of many diseases including mice with predisposition to cancers, and mice with cystic fibrosis. These models may be made by "knocking out" the activity of genes, as described above, or by inserting defective genes.

   By inserting additional copies of a gene into laboratory mice and observing the effects, scientists have recently confirmed the role of this gene in a disorder of human babies that is associated with increased susceptibility to childhood cancers. This will aid the design of new medical treatments.

"NUDE" MOUSE

A particular strain of mouse that is hairless and has a very small number, if any, of a particular type of cell of the immune system. The mice are frequently used to maintain human tumours, which can then be studied more easily. Due to their genetic make-up, the mice do not reject the tumours.

3. *To provide organs and tissues for use in human transplant surgery.*

There is a shortage of human donors of hearts, kidneys and other organs for transplantation. Animals can be genetically modified so that they carry copies of the human genes that code for proteins that inhibit the immune response to foreign tissue. This means that organs from such animals might escape rejection and so could possibly be used for transplantation into human patients.

4. *To produce milk which contains therapeutic proteins; or to alter the composition of the milk to improve its nutritional value for human infants.*

Sheep, goats and cattle have been produced that make medically important proteins only in their milk. This has been achieved by inserting copies of human genes for these proteins and attaching to them regulatory genes that make sure that the inserted gene works only in the mammary gland. Two products from transgenic animals are already in phase-two clinical trials.

5. *To enhance livestock improvement programmes.*

In future, genetic modification might enable breeders both to accelerate the rate of improvement in livestock performance, and to take advantage of genes not accessible through conventional selective breeding. Targets for improvement include enhanced disease resistance, for example, to develop chickens that resist infection by *Salmonella*.

Genetic modification of some cows could enable replacement of one or two "cow" proteins by "human-identical" proteins, so that milk from these cows could be used for premature human infants who cannot tolerate ordinary cow's milk.

WHY ARE ANIMALS USED INSTEAD OF GENETICALLY MODIFIED MICROBES OR PLANTS?

Scientists' ability to move genes from, say, a plant into a microbe, or from an animal into a microbe, arises from the fact that all living things share the same genetic code. This means that a gene generally codes for the same sequence of amino acids (the building blocks of proteins) whether it is working in an animal, a plant or a microbe. It might be argued, therefore, that for some applications, transgenic plants or microbes might replace transgenic animals, so why are animals preferred? There are two main reasons.

First, animals may be preferred because of their closer biochemical similarity to humans. This is important for making therapeutic molecules. Many animal proteins need to be modified before they can carry out their function. Usually, the enzymes that are needed to do this only exist in animal cells. For example, a gene that produces the protein alpha-1-antitrypsin can be inserted into a plant, but plants lack the mechanism to attach carbohydrate groups to this protein. Without such groups the protein is removed from the human body 50 times faster than the natural product. So to make a useful therapeutic form of the protein, production must be in animal cells.

Another reason why animals are sometimes preferred for some genetic modification is because they can make large amounts of product. For example, extracting large quantities of a therapeutic protein from animal milk is technically more straightforward than purifying it from the fermentation broth of large-scale fermentation chambers of cultured plant or microbial cells. For some therapeutic proteins, which are required in large quantities, because patients need to take large doses, transgenic animals may be the only economically feasible way of making the proteins. Production in plants is relatively inexpensive and is an appropriate route for some products but not for others.

MORAL AND ETHICAL CONCERNS

We have now summarized the scientific practicalities of animal biotechnology, but the issues raised by this technology are not only scientific ones. Modern biotechnology has the potential to throw up a wide range of what are often referred to as "moral and ethical concerns" about which it seems difficult, if not impossible, to reach any substantial degree of consensus. This is certainly true of animal biotechnology, probably because of fundamental disagreements about what our attitudes and behaviour towards animals should be.

To call something a **moral** concern does not necessarily mean that it is of much **ethical** significance. A number of surveys have shown that, if asked, people will express moral concern about modern biotechnology, but this does not tell us whether they have done any **ethical** thinking about the issues. According to this suggested distinction moral concerns are felt about what it is right or wrong to do, while ethical concerns are about the reasons and justifications for judging those things to be right or wrong.

How can Moral and Ethical Concerns be Evaluated?

No conclusive, prescriptive **answers** will be offered about the rightness or wrongness of animal biotechnology, for ethics cannot provide final "proof" of this kind derived from factual data. One cannot prove that animals ought or ought not to be used in cancer research in the same way that one can prove that cancer kills animals. Ethical judgments may be argued for or against, and shown to be more or less rational and informed, but their rightness or wrongness can never be comprehensively established.

ANIMAL ETHICS

Before focusing on the moral and ethical issues concerning animal biotechnology, we need to consider briefly how animals and their treatment can raise any such issues at all. We use plants, minerals and all kinds of other natural materials for our own benefit and pleasure. Animals also can be extremely useful to us in many ways. So what, if anything, is wrong with using a pig or a monkey or a rat for our own ends, as we might use a tree or a rock?

Many people in the past have seen nothing at all wrong in doing this. Bernard, a famous French physiologist, asserted that the proper attitude for scientists was to disregard totally the pain suffered by unanaesthetized animals in experiments, and went so far as to practise what he preached upon the family dog, whose fate was shared by numerous animals used for experimental and teaching purposes.

This "instrumental" view of animals is also evident today, despite the gradual change, which seems to have taken place this century in our attitudes towards animals, particularly in western countries. Most people are not averse to using animals as a major source of food; large numbers of animals are still used in medical and other forms of scientific research and testing; and cases are regularly reported of domestic animals being used as children's toys or for their owners' temporary amusement, and then being abandoned when their period of usefulness has expired.

However, it is difficult and dangerous to generalize about "instrumental" attitudes towards animals. Probably much depends upon the context and the particular set of values, which individuals hold. Some meat-eaters, for example, while accepting animal slaughter for food purposes, may be opposed to fox hunting or to zoos; some vegetarians may not object to the killing of rats and mice for public health reasons being happy to benefit from the use of animals, directly or indirectly, and has always done so. The mere use of animals by human beings, then, has not in itself normally been seen as a matter for moral concern; indeed, some types or species of animal might well cease to exist if we did not use them. It is the particular **kinds** of use, which have come more and more under the ethical spotlight.

Animal Welfare and the Moral Community

The issue, which has increasingly come to be seen as ethically significant, is not the use of animals but their **welfare**. According to this view, the crucial distinction between animals and other natural materials, which we may use, is that animals can be said to fare well or badly; they can be treated in ways which either enhance or diminish their well-being; they can have experiences which are pleasant or unpleasant.

This concern for animal welfare is sometimes portrayed and dismissed purely as a matter of emotional response, particularly if the animals in question happen to have furry coats and soulful eyes. Philosophers, however, have long debated the possible ethical status of animals and explored the rational basis of arguments and claims about how human beings ought to behave towards them.

Sentiency, or the capacity to experience pain and pleasure, has increasingly come to be seen as an overriding qualification for membership of the moral community, and thereby for the possession of certain rights. The capacity of sentient beings to experience pleasure and pain, satisfaction and dissatisfaction, comfort and discomfort, means that they may be said to have such things as wants, needs, interests and quality of life, though the sentiency of many animals probably differs from that of human beings in certain respects, for example, in terms of self-conscious awareness and the ability to imagine and anticipate painful or pleasurable experiences.

Speciesism

If we admit that animals can suffer and that, that capacity is ethically significant, how might this affect our attitudes towards animals? Peter Singer, an Australian philosopher whose work on animal ethics has been highly influential, claims that equal consideration should be given to all beings capable of suffering: "if a being suffers, there can be no moral justification for refusing to take that suffering into account." Such a refusal, involving preferential consideration for human beings over animals, has been labelled "speciesism." Just as racism or sexism involve preferring the interests of a particular race or sex simply because one is a member of that race or sex, so speciesism implies that the mere fact of belonging to a particular species (e.g. *Homo sapiens*) is ethically significant in determining one's membership of the moral community and the nonmembership of other species. Critics of speciesism, then, claim that membership of a particular species is as irrelevant as membership of a particular sex or a particular race in deciding who deserves moral respect and considerate treatment.

INTRINSIC CONCERNS ABOUT ANIMAL BIOTECHNOLOGY

Animal biotechnology is a morally sensitive issue because many people have concerns not only about the treatment of animals but also about the nature of modern biotechnology itself. Some feel that all forms of genetic modification may be wrong in themselves, regardless of what they are being applied to and what consequences may result. Animal biotechnology, then, may for a variety of reasons be thought to be either intrinsically wrong **in itself** or extrinsically wrong **because of its consequences**. This important distinction can be applied to a large number of moral issues and can often help in identifying the precise grounds of any moral concern.

As intrinsic concerns are in many respects more fundamental than extrinsic ones they are best explored first.

RELIGIOUS CONCERNS

It is possible to hold religious views to the effect that modern biotechnology is blasphemous. These views may rest upon the belief that God has created a perfect, natural order; for people to attempt to "improve" that order by manipulating DNA, the basic ingredient of all life, and in some cases crossing species boundaries instituted by God, is not merely presumptuous but sinful. Some religions place great importance on the "integrity" of species, and object to any attempts to change them by genetic modification.

The essence of this concern, then, is that modern biotechnology is trying to "displace the first Creator", or to "play God", but in assessing such claims, the following points need to be noted.

- By no means **all** religious believers would make these claims. Different religions have different perspectives upon the nature of God and his creation. Even among Christians, for example, there is no unanimous condemnation of modern biotechnology *per se*. There is, for example, scriptural support for the view that humanity has been given by God an approved, privileged position of "dominion" over Nature. Some modern theologians even see biotechnology as a challenging, positive opportunity for us to work with God as "co-creators".

Animal biotechnology may move genes from one species to another, but religious believers do not necessarily hold that the boundaries between species are sacred and immutable, nor indeed that they are so regarded by God. For many religious believers evolutionary theory may suggest a view of species as provisional and fluid collections of individuals, each species playing its part in a developing process, initiated by God, of which we ourselves are a fairly recent product.

INSULIN

It is a case history that touches on many aspects of the current debate about the use of animals in research, and genetic modification technology.

In 1921, Frederick Banting and Charles Best in Canada showed that an animal pancreatic extract could control blood sugar levels in a diabetic dog. After purification processes and further animal tests, it became possible to test the extract on humans. The first use of insulin to treat a critically ill diabetic boy occurred in the following year. Subsequently people with diabetes around the world were treated with insulin which was derived from the pancreatic cells of cattle or pigs.

> Within the last 20 years, advances in molecular biology and gene technology have enabled scientists to make "human insulins" which are less likely to cause allergic reactions and which have other potential clinical advantages. Such insulins can be made either by converting the pig insulin into the human form, or by using genetic modification to insert a synthetic copy of the human gene for insulin into bacteria or yeast cells. In the latter case, the yeast and bacteria cells are cloned and used as "mini-factories" to produce large amounts of the human insulin.
>
> There are an estimated 60 million people with diabetes worldwide. As well as treating human diabetes, insulin is also used to treat the condition in cats and dogs.

EXTRINSIC CONCERNS ABOUT ANIMAL BIOTECHNOLOGY

Extrinsic concerns must carry weight only in proportion to the likelihood of the predicted consequences actually occurring. So to appraise the validity of these concerns becomes in part a technical matter of trying to establish what really is most likely to happen.

Ethical questions, however, can still be directed at these extrinsic concerns despite their technical nature. Indeed, it is essential that they should, for the following three reasons:

1. Even if agreement is reached about likely consequences (which is rare) this does not automatically answer the moral and ethical questions. We still have to ask what is good or bad, right or wrong, **about** those consequences and examine the moral claims and

assumptions surrounding them (e.g. of what ethical significance is animal welfare?).

2. There is never just one consequence to any activity but a whole set of consequences, often occurring at different times. The consequences of any activity, therefore, cannot simply be morally approved or condemned *en bloc,* for they will often produce conflicting advantages and is advantages (e.g. human beings may benefit from animal suffering).

3. Consequences, then, have to be **weighed** and **compared** against each other, and this cannot be a matter of purely factual assessment. Attempts to estimate the likely costs and benefits of an activity can of course be made on a straightforward financial basis, but this does nothing to address the moral issues. (Presumably a financial assessment of this kind was made in deciding the method of extinction to be used in Nazi concentration camps). Ethical judgements have still to be made about the **value** or **priority** to be placed upon different possible costs and benefits produced by different possible consequences.

ANIMAL WELFARE

Critics of animal biotechnology claim that no benefits have resulted yet for the animals themselves from the new technologies and that there is evidence of actual harm being caused. An early example of such harm which is often quoted was provided by the "Beltsville" pigs (named after the US Department of Agriculture Research Station where they were born). Growth genes were inserted into these animals to produce faster growth and leaner meat, but the animals also suffered from a number of serious and disabling disabilities, as have sheep treated in a similar way. The unpredictable nature of such experiments clearly has considerable potential to cause animal suffering, particularly in the early stages of a new technology, though more recent work in this particular area is claimed to have overcome the problems.

Other grounds for claiming that animal biotechnology may be detrimental to considerations of animal welfare include:

1. Disease resistance is aimed mainly at diseases, which are endemic to intensive farming methods, where animals and birds are highly vulnerable to infection. Increasing the productivity of food animals by genetic modification might also increase stress and performance-related diseases.

2. The production of pharmaceutical proteins in animal milk might increase pressure to extend the length of lactation or the frequency of milking.

3. Providing organs for transplants to humans will involve similar confinement and isolation in a clinical environment. The Government Advisory Report on the Ethics of Xenotransplantation (1997) accepted that pigs used for this purpose may be exposed to harm and suffering.

4. Changes to the growth rate of the embryo may cause reproductive stress for the mother.

On the other hand it can be argued that animals may benefit from biotechnology. A wide range of diseases and disorders could be tackled in this way, and while some of these may be exacerbated by ethically questionable farming practices, animal welfare might still be said to have improved if the overall level of disease is reduced. Other undesirable practices might be reduced, such as the destruction of day-old male chicks, castration and taildocking. Many genetic disorders (for example, in dogs) might also be prevented. Those who do not object to changing animals' telos if they are thereby caused less suffering or discomfort would also argue that it could benefit animals to be genetically modified so as to fit them for living conditions for which they are not "naturally" suited. Finally, there is the possibility of using genetic technology to save endangered species from extinction. Disagreement exists then, over the possible impact of genetic technology on animal welfare, but concerns about the potential for animal suffering are by no means the sole province of philosophers and animal welfare groups. In 1995, for example, the Banner Committee in the UK produced a report for Government Ministers on the ethical implications of emerging technologies in the breeding of farm animals, which recognized that genetic modification might result in welfare problems, some of which might not be immediately apparent. This Committee produced three general principles for future practice:

1. Harms of a certain degree and kind ought under no circumstances to be inflicted on an animal.

2. Any harm to an animal, even if not absolutely impermissible, nonetheless requires justification and must be outweighed by the good which is realistically sought in so treating it.

3. Any harm, which is not absolutely prohibited by the first principle, and is considered justified in the light of the second, ought, however, to be minimized as far as is reasonably possible.

Most animal biotechnology procedures at present, however, are practised not on the farm for food purposes, but in the laboratory for research purposes, and genetically modified laboratory animals raise particular and difficult problems in terms of animal welfare. Such animals may be used in various ways, and these are controlled by regulations intended to minimize suffering, but it is as models of human diseases that they create especially thorny ethical dilemmas. Animals have in fact been used as disease models for many years, but the production of genetically modified animals for this purpose has recently attracted particular attention and debate.

The best-known example of such an animal is the so-called oncomouse, the first creature ever to be (controversially) patented, which is genetically designed to develop cancer. The creation of this animal poses the question of whether an ethical distinction can and should be drawn between **inducing** cancer in a laboratory animal and genetically **engineering** an animal to develop cancer.

While there seems little difference in terms of the animal's welfare (or lack of it), some have argued that it is ethically objectionable to alter an animal's genetic structure with the clear purpose of causing it to be defective and to develop a painful, lethal condition.

Even more problematic than the oncomouse, however, is the use of animals to model other human diseases. At least in the case of cancer research, regulations require procedures to minimize the suffering of laboratory animals by employing euthanasia at a certain stage of the tumour's development. However, it has been suggested that research into a number of devastating human genetic diseases (e.g. Lesch–Nyhan disease and HPRT deficiency), where animals are likely to be increasingly used as models, will need to keep those animals alive for as long as possible in order to study the full course of the disease. As Bernard Rollin points out, there is no way to study many of these diseases in acute, terminal or short-term experiments, and as anaesthesia cannot be maintained for long periods there appears to be the potential here for a considerable amount of animal suffering.

The issue of animal welfare and the distinction between genetically modified food animals and laboratory animals throw into sharp relief the ethical problems, which have been the focus of this chapter.

The task of ethics here is to examine the justifiability of the alternative courses of action and the general principles on which they might be based. Ultimately, however, these principles are likely to reflect fundamental beliefs and value judgments, which cannot simply be "proved" right or wrong, about what is to count as human development and progress and what our relationships and obligations should be towards other species.

The examples of moral and ethical issues examined in this chapter are not intended as an exhaustive list, though they probably encompass the main areas of current concern. One of the challenges presented by our ever-developing range of knowledge and skills, however, is that it continues to throw up novel and more complex questions about what is right and wrong to do, and these questions cannot be answered by referring back to some previously agreed moral rule-book, partly because no such rule-book exists and partly because, even if it did, it would be inadequate to deal with fresh and often unforeseen developments. Science and ethics, then, need to proceed hand in hand in exploring new territories such as animal biotechnology, to ensure that the wide-ranging implications for human and animal welfare are kept under constant review.

Summary

- Genetic modification of animals was first achieved with mice in 1980.

- Modern concerns are those regarding what is right or wrong to do, while ethical concerns are about the reasons and justifications for judging those things to be right or wrong.

- The mere use of animals of human beings has not in itself normally been seen as a matter for moral concern.

- Sentiency is the capacity to experience pain and pleasure.

- ☑ Speciesism is ethically significant in determining one's membership of a particular species and the non-membership of the other species.

- ☑ Animal biotechnology may be thought to be either intrinsically wrong in itself or extrinsically wrong because of its consequences.

- ☑ Oncomouse is the first creature ever to be patented, which is genetically designed to develop cancer.

REVIEW QUESTIONS

1. Explain in detail the reasons for producing genetically modified animals.

2. What is "Nude" mouse?

3. State the reasons for using animals instead of genetically modified microbes or plants.

4. Elaborate on the intrinsic and extrinsic concerns about animals biotechnology.

5. Discuss the beneficial and detrimental aspects of biotechnology on animal welfare.

Appendix 1

ELECTROPHORESIS

Electrophoresis is used to separate molecules by size. It works on the principle that charge molecules, such as proteins or DNA will be drawn through a slab of gel when a current is passed across it. A number of different gel types can be used such as hydrolysed starch, cellulose acetate, agarose or polyacrylamide. Polyacrylamide gels are normally oriented vertically while other gel types are usually positioned horizontally. Polyacrylamide gel electrophoresis is sometimes known by the acronym PAGE. In vertical polyacrylamide gels, samples are placed at the top of the gel and are separated from one another by a comb-like structure, or by spaces. In horizontal gel systems, samples are inserted into slots in the gel close to, or at, one end. An example of a horizontal gel system is the starch gel apparatus. The strength of gels can be adjusted to make the pore size similar to the size of the molecules being separated so that some sieving effect can take place in addition to the electrical charge dragging the molecules through the gel.

Once the current has been run for sufficient time to separate the faster-migrating from the slower-migrating molecules, electrophoresis is stopped and the gels are prepared for visualization of the resulting bands of protein or DNA. For proteins, a general non-specific protein stain can be used, though in the case of enzymes the positions of the different bands (allozymes) on the gel are identified using substrate-specific stains. High concentrations of DNA are usually stained with ethidium bromide, which fluoresces under UV light, but lower concentrations require the more sensitive staining method. In radiolabelling, a radioactive isotope of an element, such as sulphur-35 (^{35}S) or potassium-32 (^{32}P), is incorporated into the DNA before electrophoresis, then the gel is dried and placed adjacent to a sheet of film (the autoradiograph negative) and the radioactive decay of the element exposes the negative at the point of the signal. Automated DNA analysis machines use chemo-luminescent stains that can be read by a laser. This removes the risks of working with radioisotopes and, by virtue of different-coloured stains, enables more DNA sequences to be obtained from a single gel. DNA can be radiolabelled after electrophoresis, but this requires it to be transferred from the fragile gel to a more robust membrane by a technique known as Southern blotting before hybridization with single-stranded DNA complementary to the sequence of interest.

Various standards can be run on gels alongside samples for comparison. Dyes and samples from individuals of known genotypes are run on protein and enzyme gels, while DNA bands of known sizes (in base pairs (bp) or kilobases (kb)) are used as molecular size standards in DNA electrophoresis. The size of DNA fragments run on sequencing gels are found by comparison with a known sequence that is run alongside.

When we have lots of high-quality copies of the target DNA in a pure solution, we can use a standard sequencing method to identify the precise sequence of the bases (A,C,G and T) along the DNA. Comparison of the sequence between individuals, between populations, between species or between higher order systematic divisions, provides information about the relatedness between these categories. Of course, different classes of DNA are needed to address these different levels of relatedness. Although we can generally assume that the chances of a point mutation occurring are the same anywhere along the DNA molecules that make up the genome of a particular species, the important question is what the consequences might be of such a point mutation. Sequencing of DNA is now a highly automated and, in some cases, roboticized procedure. All well-provisioned large genetic laboratories will have their own sequencers and there are a number of commercial companies that provide a relatively cheap sequencing service to institutions such as marine stations or aquaculture institutions where genetic study is usually only a small part of their activities.

DNA sequencing uses DNA polymerase enzymes, such as those used in PCR, to copy the DNA strand but with two added twists. The first is that, one of the dNTPs is fluorescently radiolabelled, so that the copies can be visualized. The second trick is that we sabotage the copying process. We do this by introducing a small proportion of dideoxynucleotides (ddNTPs) along with the dNTPs. Like dNTPs, the polymerase joins ddNTPs to the new DNA strand, but unlike dNTPs they lack the bond, which would enable another dNTP to be, joined after them, so they stop the copying process. Sequencing one piece of DNA involves carrying out four separate reactions for adenine, cytosine, guanine and thymine using ddATP, ddCTP, ddGTP and ddTTP respectively. The proportion of ddNTP to dNTP is balanced so that copy strands of many different lengths are produced from those that only extend a few bases from the sequencing primer to strands hundreds of bases long. But each will end in a ddNTP. So when we run the four reactions out on a sequencing gel, the ddATP reaction will produce bands of many lengths, but we will know that each band shows the length of a DNA fragment which ends with the nucleotide adenine. Imagine that the sequence has adenine occurring at positions 2nd, 5th, 6th, 9th, 12th, 13th, etc., after a 20-base sequencing primer. In that case the A series will contain molecules of 22, 25, 26, 29, 32, 33, etc. bases long. Similarly, the ddCTP, ddGTP and ddTTP reactions will consist of molecules of lengths specific to the positions of the bases cytosine, guanine and thymine, respectively, along the DNA. These four series of molecules are run in four lanes, side by side, down a high-resolution polyacrylamide gel. The sequence of the DNA then can be read from these four ACGT lanes from the bottom of the gel upwards.

Sequence data have now been obtained from a great range of organisms and this information is collected together in DNA databases such as the one at EMBL in Europe and GenBank in the USA. Scientists have free access to these databases and powerful computer programs are available to analyse new sequences and to compare them with all other available sequences on the databases. This field of bioinformatics is rapidly expanding.

Gel Electrophoresis

Gel electrophoresis is a commonly used technique for resolving proteins or nucleic acids. In general, a sample of one particular macromolecule (protein, DNA or RNA) is placed in a well at or near the end of a gel matrix (gel). The composition of an electrophoresis gel is a semi-solid, open meshwork of interlinked linear strands. A gel is cast as a thin slab with a number of sample wells. After the wells of a gel are loaded with sample, an electric field is applied across the gel, and charged macromolecules of the same size are driven together in the direction of the anode through the gel as discrete invisible bands. The speed with which the sample moves into a gel depends on the mass of its macromolecules and the size of the openings (pore size) of the gel. The smaller macromolecules travel further than the larger ones.

The progress of gel electrophoresis is monitored by observing the migration of a visible dye (tracking dye) through the gel. The tracking dye is a charged, low-molecular-weight compound that is loaded into each sample well at the start of a run. When the tracking dye reaches the end of the gel, the run is terminated. The bands, which are aligned in a lane under each well, are visualized by staining the gel with a dye that is specific for protein, DNA, or RNA. Discrete bands are observed when there is enough material present in a band to bind the dye to make the band visible and when the individual macromolecules of a sample have distinctly different sizes. Otherwise, a band is not detected. If there is little or no difference among the sizes of the macromolecules in a concentrated sample, a smear of stained material is observed. The intensity of a stained band reflects the frequency of occurrence of a macromolecule in a sample.

The molecular mass (molecular weight) of a gel-fractionated macromolecule (band) is determined from a standard curve that is based on a set of macromolecules of known molecular mass (size markers) that covers the separation range of the gel system and is run in one or both of the outside lanes (calibrator lanes) for the same gel as the samples. The logarithm of the molecular mass of a size marker is related to its relative mobility (R_f) through a gel. The value of R_f is defined as the distance travelled by a band divided by the distance travelled by the tracking dye (ion front). The relationship between the logarithm of the molecular mass of each size marker and its R_f value is plotted. Then, with this standard curve, a molecular mass is calculated for each band in a lane. The units of molecular mass for proteins and double-stranded and single-stranded nucleic acids are daltons, base pairs, and bases, respectively. The size markers are included in the same gel as the samples because the extent of mobility of a macromolecule(s) varies from one electrophoretic run to the next.

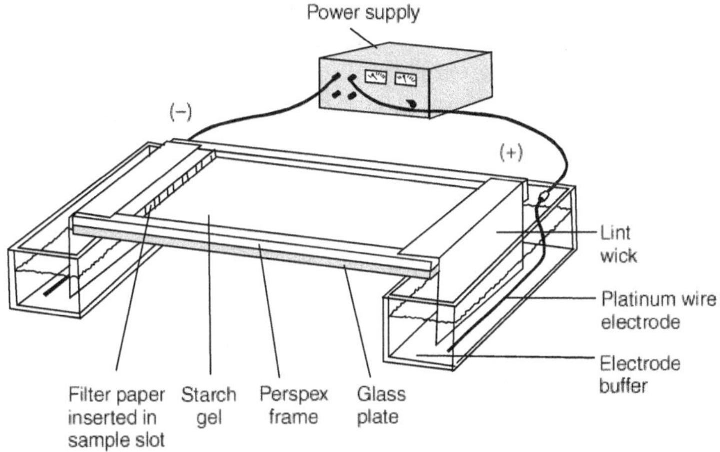

Figure A1 Apparatus for the separation of allozymes by horizontal starch gel electrophoresis

Polyacrylamide is the preferred gel system for separating proteins. Copolymerization of monomeric acrylamide and the cross-linked bisacrylamide forms a lattice of cross-linked, linear polyacrylamide strands. The pore size of a polyacrylamide gel is determined by the concentration of acrylamide and the ratio of acrylamide to bisacrylamide. For many applications, a protein sample is treated with the anionic detergent sodium dodecyl sulfate (SDS) before electrophoresis. The SDS binds to proteins and dissociates most multichain proteins. Each SDS-coated protein chain has a similar charge-to-mass ratio. Consequently, during electrophoresis, the separation of the SDS-protein chains is based primarily on size, and the effect of conformation is eliminated. SDS-polyacrylamide gel electrophoresis with a 10% polyacrylamide gel resolves proteins that range from 20 to 200 kilodaltons (kDa).

Agarose, which is a polysaccharide from seaweed, is used routinely as the gel matrix for the electrophoretic separation of medium-sized nucleic acid molecules. A 1.0% agarose gel will resolve duplex DNA chains that range from about 600 to 10,000 bp. Specialized agarose gel electrophoresis systems are available for fractionating DNA molecules with millions of base pairs, denatured DNA, and denatured RNA. In addition, for specific purposes, polyacrylamide gels are used for separating DNA molecules. For example, DNA chains that are as small as six bases and differ from each other by one nucleotide can be resolved with a 20% polyacrylamide gel.

AUTORADIOGRAPHY

Autoradiography is used to detect the site of a radiolabelled entity in a cell or sample of fractionated macromolecules. In principle, autoradiography consists of placing a radioactive source next to a radiosensitive photographic film that contains silver bromide. The energy from the decay of the radioisotope hits the photographic emulsion and produces electrons that are trapped by specks of silver bromide crystals in the emulsion. The negatively charged specks attract silver ions, and metallic silver is formed. The grains of metallic silver are visualized by developing the photographic film. Thus, an exposed dark region on a developed film indicates that the underlying material was radiolabelled. Parenthetically, fluorography is the term used for the exposure of light-sensitive photographic film to molecules that directly or indirectly generate light as the source of energy that reduces silver in the photographic emulsions.

Proteins and nucleic acids that are radiolabelled and separated by gel electrophoresis can be visualized by placing an X-ray film on a dried gel and developing the film after a suitable exposure time. All autoradiographic steps are carried out in the dark to avoid inadvertent exposure of the X-ray film by light. A number of autographic techniques have been devised for the quantitative and qualitative analysis of proteins and nucleic acids.

One of the major applications of autoradiography is the detection of the hybridization of a radiolabelled DNA probe to a DNA molecule that has been electrophoretically fractionated. However, DNA molecules in a gel are not accessible to hybridization with a DNA probe. Consequently, the DNA molecules in the gel are transferred by blotting or electrotransfer technique to a nitrocellulose or nylon membrane. The transfer process retains the same positions on the membrane as the DNA molecules had in the gel. The DNA molecules that are transferred to a membrane are denatured, bound to the membrane, and hybridized with a radiolabelled DNA probe. Autoradiography of the membrane reveals whether the probe is hybridized to a particular DNA band(s).

The transfer of DNA from a gel to a membrane is called Southern blotting (Southern DNA blotting) after Edwin Southern, who devised the original DNA blotting strategy. Northern blotting and western blotting are methods for the transfer of RNA and protein, respectively, from a gel to a membrane. The terms "northern" and "western" have nothing to do with direction and were coined by molecular biologists both to give Edwin Southern further credit for developing the notion of blotting macromolecules from a gel to a membrane and to distinguish which molecules were transferred. The designations "northern" and "western" are also examples of molecular biology humour.

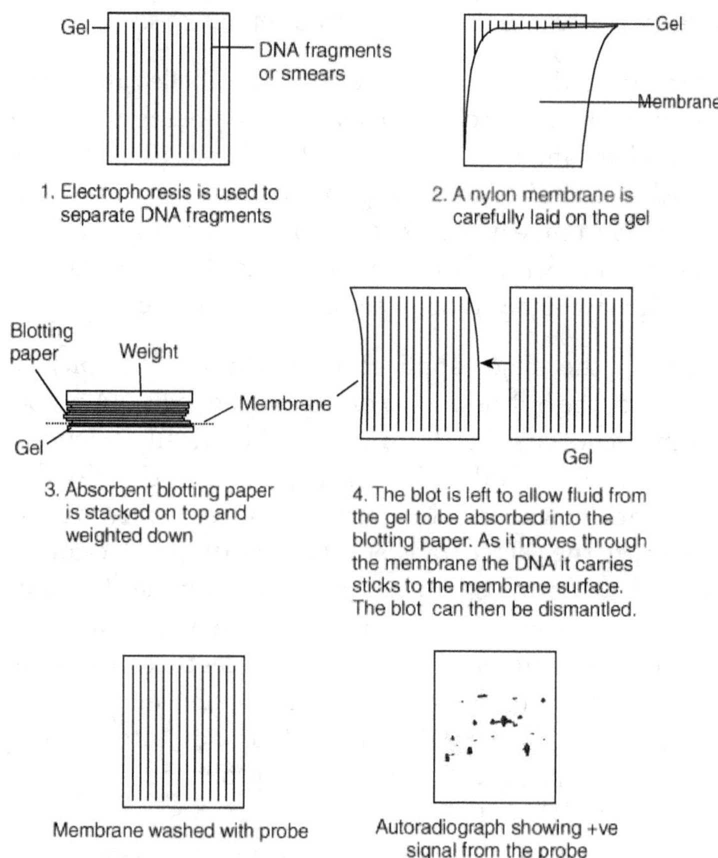

Figure A2 The Southern blotting method of transferring DNA from a gel to a membrane

POLYMERASE CHAIN REACTION

The PCR technique makes millions of copies of a particular target DNA sequence. The whole amplification process takes place in microtubes or in microwells in plastic plates in a small thermal-cycling machine on the bench. Each microtube contains a number of ingredients together with the template DNA that is to be copied from. Million of copies of a pair of primers-short single-stranded sequences of DNA each complementary to one end of the target DNA sequence-are included. A thermostable DNA polymerase enzyme, e.g. *Taq* polymerase, derived from the bacterium *Thermus aquaticus*, a resident of hot springs, is present together with the four deoxynucleotide triphosphates (dATP, dCTP, dGTP and dTTP, collectively dNTPs) in a buffer. Using *Taq* polymerase, the maximum length of the target DNA is effectively around 3–4 kb, because longer fragments cannot be successfully amplified and inaccuracies begin to accumulate to unacceptable levels. There are special DNA polymerases available for those who need long and accurate PCR replication.

Figure A3 A thermal cycler in which PCR is carried out. During PCR, the heated lid is closed over the microtubes which are positioned in the heating block.

There are three stages to PCR—denaturation, primer annealing and polymerization—each one lasting only about a minute and each operating at a different temperature.

- *Denaturation step*   The contents of the microtubes are heated to above 90°C to separate the two strands of the template DNA.

- *Primer annealing step*   The temperature is decreased rapidly to a predetermined annealing temperature, usually around 55°C, to allow the primers to 'sit down', that is, to become annealed to their complementary sequences on the template DNA.

- *Polymerization step*   The temperature is increased to 72°C, the temperature at which the *Taq* polymerase is most active, to enable the synthesis of new DNA in the 3′ direction away from the primers.

These steps are then repeated:

- *Denaturation step 2*   The mixture is again heated to about 94°C to denature all the newly built molecules, and any other parts of the template DNA which have become annealed by chance.

- *Primer annealing step 2*   Primers anneal as in the first cycle, but this time some will anneal to the newly manufactured strands of DNA.

- *Polymerization step 2*   New synthesis of molecules takes place and results in some molecules which have one strand of the precise length defined by the primer sequence at each end.

- *Denaturation step 3*   The strands of DNA are again separated ready for the annealing step.

From this point on, the number of newly synthesized molecules of the precise length specified by the two primers increases exponentially at each new cycle. Usually 20 to 40 cycles are

Start with one strand of the template DNA on which are the forward and reverse primer sites:

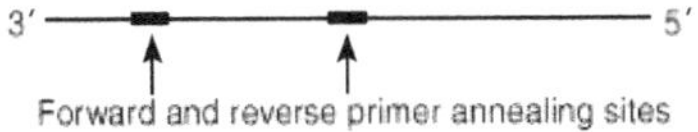

CYCLE 1

The forward primer anneals to one end of the target stretch and is elongated in the 5′ to 3′ direction by DNA polymerase:

The result of the first cycle is the original template DNA and an "overextended fragment":

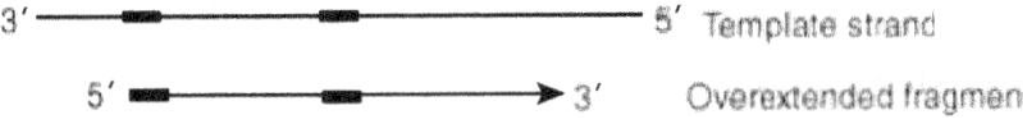

For each cycle one overextended fragment is produced. If the PCR is run for 35 cycles, there will be 35 such overextended fragments per original template strand.

CYCLE 2

The reverse primer anneals to the overextended fragments and are elongated by DNA polymerase, but only as far as the forward primer site, where the fragment ends. This produces the desired fragment, bound by the forward and reverse primers.

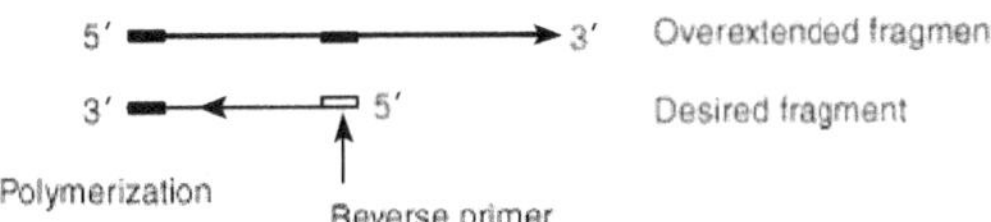

Each overextended fragment produced from the template DNA will produce perfect fragments in each remaining cycle and thus the number increases.

CYCLE 3–35

The perfect fragment makes a copy of itself; then in the next cycle, both these copies copy themselves and in the following cycle all four of those copy themselves, and so on. This is the "chain reaction" element of PCR.

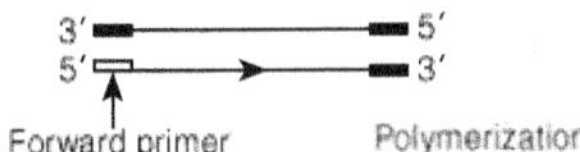

The first perfect fragment, produced in cycle 2, will double in each of the remaining 33 cycles, giving 2^{33} or 8,589,934,592 perfect fragments. Since all subsequent perfect fragments produced from each overextended fragment each PCR cycle. After the first two cycles the number of perfect fragments grows super-exponentially, to give a total of $(1 \times 2^{33}) + (2 \times 2^{32}) + (3 \times 2^{31}) \ldots + (33 \times 2^{1}) + (34 + 2^{0}) = 34359738332$ perfect fragments. It helps to understand these numbers if we realise that the total number of fragments doubles each cycle, so that after n cycles we have 2^{n} pieces of DNA, of which one will be the original strand and n will be overextended fragments.

Thus, in theory, if we start with one DNA template strand, by the end of 35 cycles we have:

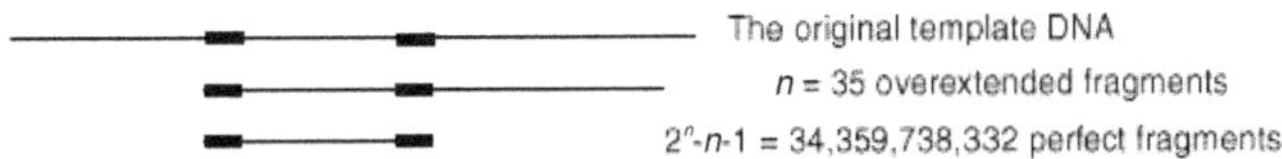

In practice these numbers are not achieved because dNTPs and primers run out and the DNA polymerase is not 100% efficient, but they serve to illustrate the DNA amplifying power of the polymerase chain reaction.

Figure A4   Polymerase chain reaction (PCR)

used and the resulting PCR product should consist of a very high copy number of the target sequence together with a small amount of original and fragmented DNA.

The above describes the theory. In practice, the PCR method is extremely sensitive to small variables. For each pair of primers, there is an optimum annealing temperature and optimum Mg^{2+} concentration. The slightest contamination of the template DNA with proteins or other material can often inhibit PCR amplification. *Taq* polymerase and buffers from different manufacturers have slightly different characteristics and may require re-optimization. And, of course, while temperatures are changing between steps, all the ingredients are free to interact in the most unpredictable way. In spite of these considerations, the PCR method has become routine in the laboratory and provides a simple and effective means of producing high copy numbers of specific DNA sequences.

The sizes of pieces of DNA produced from cloning or PCR can be determined by subjecting the DNA to electrophoresis alongside known size-standards. Since electrophoresis separates fragments based on their sizes, it can be used to purify DNA fragments. For example, the results of a PCR reaction can be run (electrophoresed) on an agarose gel. The DNA is stained during or after electrophoresis with ethidium bromide, which fluoresces under UV light. Hopefully there will be a nice bright band of the right size, our desired PCR product, which can then be cut out and the DNA extracted from the gel. We thus have the desired PCR product without leftover components of the PCR reaction, such as primers, which might have interfered with later DNA sequencing.

RESTRICTION FRAGMENT LENGTH POLYMORPHISM (RFLP)

The fact that restriction enzymes will only cut DNA at specific sequences presents us with a simple way of identifying genetic variation caused by point mutations. Let's say we have a 2-kb length of DNA from an individual animal, which can be amplified to a high copy number, and we then incubate this amplified DNA in a microtube with a suite of restriction enzymes. The restriction enzymes will cut the DNA into a number of fragments that can then be easily size-separated on agarose gel and stained with ethidium bromide. In other individuals of the same species, point mutations will have altered the sequence at one or more restriction enzyme cut sites, or may have produced a cut site where one was not present before. This will result in different individuals producing variation in the size and number of fragments when their DNA is incubated with this suite of restriction enzymes. Genetic variation identified in this way is called restriction fragment length polymorphism (RFLP).

RFLP data from a sample of a population can be analysed in two ways. First, all the different fragment patterns detected on the gels are counted and the frequencies of each determined. These RFLP frequency data can then be compared between populations. Secondly, RFLP data can be analysed on the basis of the proportion of nucleotides that differ between individuals. Of course, the number of nucleotides actually sampled is limited by the number of restriction enzymes used and the number of bases each enzyme has in its cut site. Nevertheless, such data are of value in establishing relationships between populations, species or higher taxa.

Primers designed for use with PCR are usually 20–25 bp in length to ensure that they will be specific to the particular DNA sequence being targeted. However, if a much shorter primer is used, say of 10 bp length, it is likely to anneal to many regions of the genome during the annealing step of PCR. If, by chance, the primer anneals to opposite strands of the DNA within a region up to about 3–4 kb in length, then a PCR product spanning the annealing sites will be produced. Some 10-bp RAPD primer sequences will produce no PCR product, while others may produce a number of fragments of different size. The PCR products are run on agarose gel electrophoresis to identify the fragments according to their size.

How does this Identify Genetic Polymorphisms?

Consider the situation where a point mutation is present in the sequence at one of a pair of RAPD primer sites in some individuals in a population. PCR of the DNA from these individuals will not produce the fragment present in other individuals and therefore that band on the gel will be absent. Thus, presence or absence of particular bands can be scored as markers of genetic variation. This information of presence (+) or absence (–) of bands is called 'dominant' data. It is not possible to distinguish individuals that are homozygous (+/+) from those which are heterozygous (+/–) because both genotypes will produce a band on the gel. Allele frequencies, however, can be calculated based on the assumption that the locus is in agreement with the Hardy–Weinberg (H–W) model. The proportion of individuals scored for absence of the band (that is, –/– homozygotes) should equal the square of the (–) allele frequency.

When first developed, the RAPD technique seemed be a relatively quick and cheap method to obtain molecular genetic data that required no prior knowledge of DNA sequences. However, agreement with the H–W model has to be assumed and may not hold. Also, considerable care has to be taken to ensure that RAPD data are repeatable. Banding patterns are easily altered by subtle changes in the PCR conditions and some bands have been found which violate Mendelian inheritance. There is also the problem that it is difficult to identify bands that have been amplified from contaminating DNA (from parasites, gut contents, epibiota, etc.). For these reasons, the RAPD technique has fallen out of favour with many geneticists though it is still widely used by those who work on clonal organisms as a way to identify particular clones.

AMPLIFIED FRAGMENT LENGTH POLYMORPHISM (AFLP)

Like RAPDs, the AFLP method provides size fragment-markers from DNA of unknown sequence. Unlike RAPDs, AFLP markers are extremely reliable and reproducible. DNA extracted from an organism is first cut using a pair of restriction enzymes, which leave cohesive ends-one a frequent (four-base) cutter, the other a rare (six-base) cutter. The majority of fragments produced will have two four-base-cut ends, the minority will have two six-base-cut ends and the remainder will have one end with a six-base-cut, and the other with a four-base cut. Two oligonucleotide 'adapters' designed to attach to the cohesive cut ends of both the six-base and four-base fragment terminals are then ligated to the resulting fragments. These adapters are used as the basis for polymerase chain reaction (PCR) primers that are also designed to overlap with the DNA fragment by one, two or three specified bases. The primer for the less frequent (six-base) cut adapter is radiolabelled before PCR amplification and polyacrylamide gel electrophoresis, so only those fragments, which were cut by the six-base restriction endonucleases, are visualized by autoradiography. By varying the number of specified bases on the primers the number of fragments amplified can be controlled. Typically 50–100 restriction fragments can be identified and separated in a single run of the AFLP method.

Appendix II

BIOTECH FIRMS

Amersham pharmacia

Biotech Asia Pacific Ltd.,
I-Shanti Niwas, Janpath Lane,
Connaught Place,
New Delhi – 110001.
pbiotech@de12. vsnl.net.in &
ranga@del13.vsnl.net.in

Amity Institute of Biotechnology

E-27, Defence Colony,
New Delhi – 112224.
ssaran@akcgroup.com

AV Biotech

The Sheveroy Estates Ltd.
(A V Thomas Group)
Cochin – 682030.
shevero@vsnl.com
www.avtbiotech.com

Avestha Gengraine Technologies Pvt. Ltd.

Plant Genome Biology
Laboratory, Discoverer,
9th floor, Unit 3,
International Tech Park,
Whitefield Road,
Bangalore – 560001.
www.avesthagen.com

Bangalore Genei Pvt. Ltd.

Regd. Office : No. 6th Main, BDA
Industrial Suburb. Near SRS Road,
Peenya, Bangalore – 560058.
gebeuvkr@vur,vsnl.net.in

Bharat Biotech International Ltd.

762, Road No. 3, Banjara Hills,
Hyderabad – 500034.
ellas@bharatbiotech.com
www.bharabiotech.com

Bharat Immunologicals & Biologicals Corporation Ltd. (BIBCOL)

Village Cholla,
Bulandsher – 203203.

Bio-Rad (India) Pvt. Ltd.

C-248, Defence Colony,
New Delhi – 110024.
biorad.delhi@sm4.sprintrpg.emsvsnl.net.in

Biotech India

B.29/10, Nadigram,
Lanka, Varanasi – 221005.
biotech@lwl.vsnl.net.in
www.biotechindia.com

Biotech Consortium (BCIL) India Ltd.

5th Floor, Anuvrat Bhawan,
Deendayai Upadhyay Marg,
New Delhi.
biotech@giasd101.vsnl.net.in

Cadila Health Care/ Pharmaceuticals Ltd.

IRM House,
CG Road,
Navrangpura,
Ahmedabad – 380008.

Care Biomedicals

Head Office : CB House,
Shrinath 1, Jayadeep Nagar,
Bandup (East),
Mumbai – 400042.
sales@carebio.com

Care Well Biotech Pvt. Ltd.

M2-3, Guru Arjun Dev Bhawan,
Ranjit Nagar
Commercial Complex,
New Delhi – 110008.

Chemibiotek Research International

Block GP, Plot G, Sector 5,
Kolkata – 700091.
aghosh@chembiotek.com

Crop Health Product Ltd.

R-12/50 Raj Nagar,
Ghaziabad – 201002.

Dabur Research Foundation

22, Site IV, Sahibabad,
Ghaziabad – 201010.
badrin@dabur.com

Devender Pal & Associates

Technology & Management
Consultants, 71-Dharam Jyot.
AB Nair Road, Juhu,
Mumbai – 400049.
dpals@bom3.vsnl.net.in

Dr.Reddy's Laboratories

7-1-27, Ameerpet,
Hyderabad – 500016
hr@drreddys.com
www.drreddys.com

Dupont Agricultural Product

E.I. Dupont India Ltd. DLF Plaza
Tower, 8th Floor, DLF Qutab
Enclave, Phase-I,
Gurgaon – 122002.

EID Parrys (India) Ltd.

Parrys Corner, Chennai – 600001.

E merck (India) Ltd.

424-A Bhandari Road,
Asaf Ali Road,
New Delhi – 110002.

East India Pharmaceutical Works Ltd.

6, Little Russell Street,
Calcutta – 700071.
eipwl@cal.vsnl.net.in

Ernst & Young

Corporate Advisory Services,
SRB House, 20 Community Centre
Madangir,
New Delhi 110062.
veena.raman@in. eyi.com
www.cyindia.com

Genetix

C-80, Kirti Nagar,
New Delhi – 110015.

Genome Engineering Laboratory Ltd.

Mevalurkuppam village on
Bangalore Highway,
Sriperumbudur Taluk,
Kancheepuram Dist.
Tamil Nadu – 602105.

Regd. Off : 18-5-288, Aliyabad,
Hyderabad – 500053.
venkatesh@genome-labs.com
mvgenome-labs.com

Hysel India

41, DDA Shopping Centre,
Sukdev Vihar,
Okhla-Mathura Road,
New Delhi – 110025.

Hi Media Laboratories

23, Vandna Industrial Estate,
LBS Marg, Mumbai – 400086.

Imperial Biomedic Private Limited

B-5, Ashoka Chambers,
Flat No. 3-A Rajindra Park,
Pusa Road, New Delhi – 110060.

Indian Vaccines Corporation

National Institute of Immunology
(NII) Campus,
Aurna Asaf Ali Marg,
New Delhi – 10067

Indo Biogene Pvt. Ltd.

Suite-14, Allenby Court,
Allenby Road, Kolkata – 700020.
Delhi Office
Ground Floor B-63,
Dayanand Colony,
Lajpat Nagar IV,
New Delhi – 110024.

Indo-American Hybrid Seeds (India) Pvt. Ltd.

P.O. Box No. 099, 17th Cross.
2nd "A" Main, K.R.Road, BSK 2nd
Stage, Bangalore – 560070.

Indian Immunologicals

Unit of National Dairy
Development Board,
Hyderabad – 500021.

In vitro Biotek

28-1-18/A, 9th Lane,
Shanti Nagar,
Eluru – 534007.

ITC Ltd.

Chowringee Lane,
Kolkata – 700071.

J Mitra & Co. Ltd.

A-180, Okhla Industrial Area,
Phase I,
New Delhi – 110020.
delhi,jmitra@axcess.net.in

Jain Biologicals Private Ltd.

Regd. Off. S.C.O.371-372,
II Floor, Swastik Vihar Mansa
Devi Complex,
Panchkuta – 134109.
puneera@vsnl.dot.in

Lab India Instruments Pvt. Ltd.

201. Nand Cha LBS Marg,
Near Vandna Cinema,
Thane – 400602.

Life Technologies

GIBCO BRL, INDIA Pvt. Ltd,
4F,G,H, 4th Floor,
Gopala Towers,
25 Rajindra Place,
New Delhi – 110088.
gibcobri.infotech@sm4,srrintripg.sprint.com

Maharashtra Hybrid Seeds Co. Ltd.

Mahyco Life Sciences Research Centre, Dawalwadi, P.O.Box. 76, Jalna – 421203.
ksravi@mahyco.com
Bombay Office
Resham Bhawan, 4th Floor,
78 Veer Nariman Road,
Mumbai – 400020.
Delhi Office
Ashok Centre, 4E/15, 3rd Floor,
Jhandewalan Extn. New Delhi.

MAHYCO Monsanto

Biotec. (India) Pvt. Ltd., Resham Bhawan, 4th Floor, 78,
Veer Nariman Road,
Mumbai – 400020.
mahyco@giasbm01.vsnl.net.in

Methalix

Methalix Life Science Pvt.Ltd.,
35/11 Sri Vasan Temple Road,
VV Puram, Bangalore – 560004.
k.k.narayanan@meta-helix.com
www.meta-helix.com

Nicholas Priamal

Nicholas Products Division, 16/36 Old Rajindra Nagar,
New Delhi – 110060.
Mumbai Office : 100 Centre Point, Ground Floor,
Dr. Ambedkar Road, Parel,
Mumbai – 400012.

Novartis India Ltd.

141 Jamshedji Tata Road,
Mumbai – 400020.

Nunhems Proagro Seeds Pvt. Ltd.

Corporate Office : Dhumaspur Road, Badshahpur,
Gurgaon – 122001.
ppgs@ndf.vsnl.net.in

Panacea Biotec. Ltd.

B-1, Ext./A-27,
Mohan Co-op Indl. Estate,
Mathera Road,
New Delhi – 110044.
panbio.panbio@rme.sprintrpg.cms.vsnl.net.in
www.panacea-biotec.com

Pfizer Ltd.

S V Road, Jogeswari West,
Mumbai – 400102.

Professional Biotech Pvt. Ltd.

G/167-B, Airtel Tower Building,
Hari Nagar,
New Delhi – 110058.
P.O.Box 9663,
New Delhi – 110058.
www.professional.biotech.com
dharma@vsnl.com

Rallis India Ltd.

Rallis Research Centre 21 & 22,
Peenya Industrial Area,
Phase II, Bangalore – 560058.

Rama Biotechnologies Pvt. Ltd.

302, Seshagiri Apts.
6-1-106 & 107,
Padmara Nagar,
Secunderabad – 500025.
rbtlalit@amccs.uunet.in

Ranbaxy Eili–Lilly Ltd.

8, Balaji Estate,
Guru Ravi Das Marg, Kalkaji,
New Delhi – 110019.
khera-anurg@lilly.com

Reliance Group of Industrial/ Central Recruitment Cell

Post Box No. : 11717, 222 Maker
Chamber IV, Nariman Point,
Mumbai – 400021.
crc-rtc@ril.com
http://www.ril.com

Sigma-Aldrich Corporation

India (Liaison Office), B-4/158
Safdarjung Enclave
New Delhi – 110029.

Serum Institute of India

1401, Ansal Tower,
38 Nehru Place,
New Delhi – 110019.
serum@nda.vsnl.net.in

Smithkline Beecham Pharmaceuticals (India) Ltd.

Bharat Yuvak Bhawan-1,
Jai Singh Road,
New Delhi – 110001.
http://www.sb.com

SPAN Diagnostics Ltd.

173-B, New Industrial Estate,
Udhna, Surat – 394210.

Shriram Biotech. Ltd.

H.No.7-1-29, United Avenue
(North end), G4, "B" Block,
Ameerpet, Hyderabad – 570016.

shriramkodi@eth.net;
dbssrs@yahoo.com

Seemax Industries Ltd. (Biotech Division)

Seemax House, F-33/6, Okhla
Phase II, New Delhi – 110020.
seemax.ind@elnet.esm.vsn.net.in

Shanta Biotech

Serene Chambers, IIIrd Floor,
Road, 7, Banjara Hills,
Hyderabad – 500034.
shanta@hdl.vsn.net.in

Spinco Biotech Pvt. Ltd.

105, A3-A4,
Guru Ram Das Bhawan,
Ranjit Nagar Commercial
Complex, New Delhi – 10008.
Grams : BIOSPIN
Regd & H.O. : 35,
Vaidyaram Street, T. Nagar,
Chennai – 600017.

Saver Biotech Ltd.

1442, Wazir Nagar,
Kotla Mubaraqpur,
New Delhi – 110003.

Southern Petrochemical Industries (SPIC)

SPIC Centre,
97 Mount Road,
Chennai – 600032.
joethom@md3.vsnl.net.in
http://www.spicgroup.com

Sumka Son Instrumentation

137-B, Thadagam Road,
RS Puram,

Coimbatore – 461002.
sukma@vsnl.com

Thapar Centre for Industrial Research & Development
Thapar Technology Campus,
Bhadson Road,
P.B. No. 68,
Patiala – 147001.

dg@tcrdcpt.re.nic.in
www.thapartech.org.

Transgene Biotech Ltd.
C-44, Madhura Nagar,
SR Nagar,
Hyderabad.
transgene_biotek_ltd@usa.net
kolliparakrao@usa.net

Glossary

Acquired immunodeficiency syndrome (AIDS) An infectious disease syndrome caused by the human immunodeficiency retrovirus. The disease is characterized by the loss of a normal immune response, followed by increased susceptibility to opportunistic infections and an increased risk of some cancers.

Amino acid A small organic molecule that forms the unit of a polypeptide. A hydrogen atom, amino and carboxyl groups, and an R side group are covalently bonded to a central carbon atom.

Antibiotic A secondary metabolite, produced by some microorganisms, that kills or inhibits the growth of other microorganisms, often through transcriptional or translational regulation.

Antibody An antigen-binding immunoglobulin produced by the B cells of the immune system.

Anticodon The three-base-pair region of a tRNA molecule that base-pairs with the complementary codon of the mRNA.

Antigen A molecule, often a protein, which is recognized as foreign by the body and elicits an immune response.

Antisense A sequence of DNA or RNA that is complementary to a functional RNA, often mRNA; the DNA or gene that is transcribed.

Aquaculture The farming, often commercial, of a variety of shellfish, crustaceans and finfish. Aquaculture is a general term, but often refers to freshwater culture.

Assay The qualitative or quantitative determination of the components of a material.

Attenuation A weakening; a reduction in virulence.

Autoimmune disease An immunological disorder resulting from the attack on normal body cells and tissues by the body's lymphocytes.

Autoradiography A method by which a radiolabelled molecule such as DNA or RNA or a structure such as chromosome produces an image on photographic film. The image is called an autoradiogram or autoradiograph.

Bacteriophage A virus that infects bacteria; often called phage.

Baculovirus A virus that specifically infects arthropods, most often insects.

Batch culturing A method in which cells are cultured for a specific period of time after inoculation into the culture medium and then collected for processing.

Biolistics A method of delivering DNA into animal and plant cells by DNA-coated microprojectiles discharged under pressure at high velocity.

Biotechnology The commercial use of living organisms or their components to improve animal and human health, agriculture, and the environment.

Blunt end The DNA terminals produced, without overhanging 5′ or 3′ ends, after restriction endo-nuclease digestion.

B lymphocyte A type of lymphocyte or white blood cell, originating in the bone marrow, that synthesizes antibodies and secretes them as part of the immune response to a foreign antigen.

Cancer A malignant tumour that expands locally by invasion of surrounding tissues, and systemically by metastasis.

cDNA Complementary double-stranded DNA synthesized *in vitro* by reverse transcriptase and DNA polymerase with mRNA as the template.

Centromere A region of DNA on eukaryotic chromosomes to which the mitotic spindle fibres attach during mitosis and meiosis; the region where two chromosomes are attached after replication.

Chimera A recombinant DNA molecule or organisms that contain sequences from more than one organism.

Chromosomes Structures composed of DNA, RNA, and protein; mechanism by which hereditary information is transmitted from generation to generation. In eukaryotic cells, the chromosomes reside in the nucleus.

Clone Refers to an individual or cell from a genetically identical population derived by asexual division. Also a cell harbouring a recombinant DNA; a foreign DNA fragment contained within a recombinant clone.

Codon The triplet bases of an mRNA that translates into an amino acid or stop signal; the tRNA anticodon base-pairs with the codon during translation.

Cohesive ends The cohesive DNA termini produced, with overhanging 5′ or 3′ ends, after restriction endonuclease digestion; sometimes called sticky ends.

Cohesive termini *(cos* sites) The 12-base single-stranded complementary ends of bacteriophage λ DNA that is formed during packaging of concatemeric DNA into viral head particles.

Colonies Clumps of bacterial cells on solid medium; each colony is composed of genetically identical cells that arose by cell division from a single cell; each colony is called a clone.

Competent Describes cells, most often bacterial, that have the ability to take up DNA.

Complementary Describes a pair of nucleotides that can base-pair through hydrogen bonding. An example is cytosine (C) and guanine (G). Strands of DNA also can base-pair along their nucleotides.

Complementation A method by which two DNA fragments in a cell encode non-functional polypeptides that when synthesized together in the cell, become a functional protein.

Cosmid A hybrid vector comprising plasmid sequences and the *cos* sites of lambda bacteriophage, enabling the vector to be packaged *in vitro* with large DNA inserts.

Crossing over A process that generates new genetic combinations; homologous chromosomes line up and exchange segments of corresponding DNA during meiosis.

Cytokines Extracellular signalling proteins or peptides produced by the immune system and other tissues that are involved in cell–cell communication; includes interferons, colony-stimulating factors, interleukins and tumour necrosis factors.

Deoxyribonucleic acid (DNA) The nucleic acid that encodes genetic information that is passed to progeny; most frequently a double-stranded helical molecule composed of four nucleotides. The nucleotide unit of DNA has one or more nitrogenous organic bases— adenine, guanine, cytosine and thymine—attached to a deoxyribose sugar with a phosphate group (called a deoxyribonucleotide).

Diagnostics Sensitive assays used to determine the presence of a disease, DNA mutation, or pathogen.

Diploid Having two sets of chromosomes (2n); chromosomes of each pair are referred to as homologous.

DNA helicase Protein that hydrolyses ATP while moving along the DNA molecule and melting the double strand.

DNA library A collection of cloned DNA fragments often from specific genome.

DNA ligase An enzyme that reforms the phosphodiester bond between two nucleotides of DNA (the 5′ phosphate of one nucleotide and the 3′-OH group of the other).

DNA linker A synthetic oligonucleotide containing a restriction site ligated to the end of DNA fragments to create a restriction site for cloning.

DNA polymerase An enzyme that synthesizes a copy of a DNA template. DNA is synthesized in the 5′ to 3′ direction by adding nucleotides to the 3′ hydroxyl group of the newly synthesized strand.

DNA topoisomerase An enzyme that, during DNA replication, removes twists to relieve tension in the double-stranded molecule by breaking phosphodiester bonds and then reseals the breaks.

Electroporation A method for introducing DNA into cells; exposure to a brief electric pulse is thought to open transient pores through which DNA enters.

Embryo A multicellular structure (such as a blastula and gastrula in animals, sporophyte in plants) formed by subsequent cell divisions after fertilization.

Embryonic stem cells (ES) Cells from an early embryo that are not terminally differentiated but can divide to give rise to differentiated cells including germ-line cells.

Enhancer A regulatory DNA sequence element that enhances transcription of a eukaryotic gene; functional from thousands of base pairs away and in either orientation.

Enzyme A protein that catalyses a specific chemical reaction.

Epitope The smallest part of an antigen which is recognized by and combines with an antibody molecule. A single antigen may exhibit several different epitopes.

Eukaryotic cell A cell that is compartmentalized with a membrane-bound nucleus, other membrane-bound organelles, and a distinct cytoplasm; eukaryotic organisms are composed of one or more cells.

Fermentation The breakdown of organic compounds by cells or organisms in the absence of oxygen to generate ATP. In industrial applications, cells or microorganisms are cultured in fermenters (i.e., bioreactors) with or without oxygen to produce desirable products or to increase biomass.

Fermenter A growth chamber used for cultivating cells and microorganisms and used for the production of important compounds; also called a bioreactor.

Fingerprinting A method used to identify individual DNA banding patterns derived from hyper-variable regions of DNA; used in forensics, to establish paternity, and in conservation biology.

Gel electrophoresis A method used to separate DNA or RNA fragments by length; proteins can be separated by size and/or charge.

Gene A region of DNA that encodes information for a discrete product that can be protein or RNA; includes coding sequences, introns, and noncoding regulatory sequences.

Gene bank A facility where plant material is stored usually as seeds, tubers, or cultured tissue. (*See* germplasm.)

Gene cloning The insertion of a DNA into a vector and the transfer of the recombinant molecule into a host for propagation; also called recombinant DNA technology or gene splicing.

Gene therapy The use of genes to correct a genetic or acquired disorder.

Genetic code The 64 triplet codons that encode the 20 amino acids and three stop codons used in protein synthesis.

Genome The entire content of genetic material within an organism (including bacteria and viruses); sometimes, the amount of DNA in a haploid set of chromosomes in a eukaryotic organism.

Genotype The genetic composition of an organism; can include any number of gene pairs from one up to all the genes.

Germ cell An animal sperm or egg cell (i.e., gamete) or a precursor cell that gives rise to gametes.

Germplasm The genetic material within an organism; usually meaning plant genetic material.

Haemoglobin The constituent of red blood cells that gives them their colour and carries oxygen.

Haploid Having only one set of chromosomes (n); found in gametes, some fungi and protists, and plant gametophytes.

Hapten A small molecule which can act as an epitope but is incapable by itself of eliciting an antibody response.

Helper virus A virus that provides functional properties to another virus in the same cell that lacks specific functional proteins.

Heterozygous Having different forms or alleles of a gene at a specific locus on homologous chromosomes.

Hybridization Hydrogen bonding between two complementary single-stranded DNA sequences or between DNA and RNA sequences.

Hybridoma An immortal cell line produced from the fusion of antibody-secreting B lymphocytes to lymphocyte tumour cells and used in the production of monoclonal antibodies.

Immunization Any process that develops resistance (immunity) to a specific disease in a host.

Immunoassay A method that uses the specificity of antibodies to detect the presence of a specific antigen (usually protein or glycoprotein) in a biological sample, such as blood.

Immunoglobulin (*See* antibody.)

In situ (from Latin "in place") In the location in which the entity is normally found. In its usual place.

Insulin A hormone, secreted by β cells of the pancreas, that reduces the glucose levels in the blood by regulating glucose metabolism in animals.

Intellectual property Once a traditional area of law that encompassed patents, trade secrets, trademarks, copyrights, and plant variety protection; intellectual property now can be living organisms.

Interferon A group of protein molecules (cytokines), which are capable of conferring non-specific immunity to viral infection.

Interleukins (*See* cytokines.)

In vitro (from Latin "in glass") Outside the living host organism.

In vivo (from Latin "in the living being") Within the living host organism.

Isoenzyme Any one of the group of enzymes of different structural forms that possess identical (or nearly identical) catalytic properties. Also called isozymes.

kDa Kilodaltons. A measure of mass of a molecule.

Liposome transfer The transfer into eukaryotic cells of DNA, RNA, or therapeutic agents encapsulated in an artificial lipid bilayer vesicle formed from an aqueous suspension of phospholipid molecules.

Lysogens Bacteria that are carrying a viral prophage and have the potential of producing bacteriophages under the proper conditions.

Major histocompatibility complex (MHC) A complex genetic locus that specifies the cell surface glycoproteins of lymphocytes that are involved in antigen recognition.

Messenger RNA (mRNA) An RNA molecule that encodes the amino acid sequence of a polypeptide; translated into protein during the process of translation.

Microinjection Direct injection of DNA or a therapeutic agent into a cell with a microcapillary.

Microorganism A prokaryotic organism lacking membrane-bound organelles and a true nucleus.

Mollusc A member of the phylum Mollusca, which comprises seven classes and includes mussels, clams, snails, squids, and octopuses. Members of the phylum have a tissue fold called a mantle around a soft, fleshy body.

Monoclonal antibody (mAB) A single type of antibody specific for one antigenic determinant (portion of an antigen that binds to an antibody) that is secreted by a hybridoma clone derived from a single B cell.

Mutation A change in the nucleotide sequence of a chromosome that is inherited.

Myeloma Cancer of haematopoietic cells.

Neoplasm A synonym for cancerous disease.

Nucleic acid A polymer of nucleotides joined by phosphodiester bonds; a general name given to DNA or RNA.

Nucleotide The basic unit of DNA or RNA; a pentose sugar with a bound nitrogenous base and phosphate group.

Oligonucleotide A short, single-stranded DNA that usually is synthesized *in vivo* and often used as a probe in hybridizations or as primers for the polymerase chain reaction.

Ovary The reproductive organ in female animals where the eggs are made; where seeds are produced after fertilization in plants, a mature ovary comprises all or a portion of the fruit.

Patent A legal document that grants exclusive rights to sell or use an invention for a defined period of time (17 years in the United States). *(See* intellectual property.)

Pathogen An organism or virus that causes disease.

Peptide A short sequence of amino acids.

Peptide bond A covalent bond between the carboxyl group of one amino acid and the amino group of an adjacent one, formed during protein synthesis.

Phenotypic trait An observable characteristic of an organism or cell that is determined by the genotype or genetic composition, often in conjunction with the environment.

Plasmid An extrachromosomal circular double-stranded DNA molecule that replicates independently from the chromosome; frequently modified for use as a cloning vector.

Pluripotent Describes the ability of a cell to develop into a differentiated cell type; progenitor of other cells; often used in conjunction with stem cells (bone marrow, embryonic) in animals.

Polymer A macromolecule composed of repeating units. Examples include DNA, RNA, starch.

Polymerase chain reaction (PCR) A method for selectively amplifying regions of DNA by *in vitro* replication involving repeated denaturation and renaturation of the DNA template.

Promoter A region of DNA where RNA polymerase and other proteins involved in the regulation of transcription bind.

Pronuclear microinjection A method by which one of the nuclei in a fertilized egg is injected with DNA prior to fusion of the gamete nuclei. (*See* microinjection.)

Protease An enzyme that degrades proteins by hydrolysing the peptide bonds; an example is trypsin.

Protein A linear macromolecule that is composed of amino acids that are connected by peptide bonds; sometimes used interchangeably with polypeptide, although a protein can consist of more than one polypeptide (e.g. dimeric protein is composed of two polypeptides).

Protoplasts Plant, bacterial or yeast cells that have no cell wall; cell walls usually are removed enzymatically.

Protozoan A single-celled, free-living or parasitic, nonphotosynthetic and motile eukaryotic organism; examples are amoebae and paramecia; a type of protist.

Purine One of the two classes of nitrogen-containing ring structures in DNA and RNA. Adenine and guanine have double-ring structures.

Pyrimidine   One of two classes of nitrogen-containing ring structures in DNA and RNA. Cytosine, thymine (in DNA), and uracil (in RNA) have single-ring structures.

Receptor   A protein, located on the cell surface or intracellularly, that binds to a specific type of molecule (ligand) and elicits a cellular reaction or response.

Recombinant DNA   A sequence comprising DNA from different sources that have been joined together.

Reporter   A gene that encodes a product that can be assayed to determine whether a specific DNA construct is present and functional in a host organism or cell.

Repressor protein   A protein that binds to a gene promoter or operator site and blocks the binding of RNA polymerase, thereby preventing transcription.

Restriction endonuclease   A member of several classes of enzymes that cut DNA internally; each recognizes a specific short double-stranded DNA sequence and hydrolyses phosphodiester bonds on both strands (type II enzymes are used in recombinant DNA technology).

Restriction fragment length polymorphism (RFLP)   Region of DNA with nucleotide sequence variations that generates fragment length differences according to the presence or absence of type II restriction endonuclease recognition site(s).

Retrovirus   A virus with an RNA genome that replicates in a eukaryotic host cell by synthesizing a double-stranded DNA inter-mediate. The double-stranded form can integrate into the host chromo-some.

Reverse transcriptase   An RNA-dependent DNA polymerase that uses an RNA template for the synthesis of a complementary DNA (cDNA); present in viruses with RNA genomes.

Ribonuclease (RNase)   An enzyme that hydrolyses RNA by hydrolysing the phosphodiester bonds between the nucleotides.

RNA polymerase   The enzyme that catalyses the synthesis of RNA on a DNA template.

Southern blotting   A method of transferring denatured DNA from an agarose gel after gel electrophoresis to a membrane for hybridization to detect specific sequences.

Stem cell   A precursor cell that divides, and whose progeny give rise to differentiated cells.

Therapeutic Any agent (DNA, protein, synthetic compound, etc.) that is used in the treatment of disease.

Tissue culture *In vitro* propagation of plant and animals cells or tissues; used to regenerate whole plants from cultured cells and tissues.

T lymphocytes A type of cell that is responsible for cell-mediated immunity. Examples are helper T cells and cytotoxic cells.

Totipotency Describes plant cells or tissues able to develop into any differentiated cell or tissue type.

Transcription The synthesis of RNA that is catalysed by RNA polymerase and requires a DNA template.

Transfection The transfer of DNA into a eukaryotic cell.

Transformation The uptake of DNA into a cell, usually a bacterium or yeast, and its expression. It also refers to the conversion of a normal cell to a cancerous one.

Transgene A gene that has been introduced into the genome of a eukaryotic cell(s) or organism that originated from another source.

Transgenic organism A living organism that harbours an inserted functional gene in its germ line.

Triploid Three copies of homologous chromosomes (3n).

Vaccine An agent that is used to promote an antibody response that protects the organism against infection; confers immunity to disease. Examples include protein, DNA, living or nonliving virus or microorganism.

Vector A DNA molecule used to transfer foreign DNA into a host cell or organism. Examples include plasmids, bacteriophage, and cosmids.

Western blotting A method to transfer electrophoretically separated proteins to a membrane for antibody-binding to detect specific proteins.

Yeast artificial chromosome A linear recombinant DNA molecule that replicates in a yeast host cell; important components include a large fragment of foreign DNA (up to 1 million base pairs), a centromere, a yeast origin of replication, and yeast telomeres at the ends of the chromosome.

Zygote A diploid cell that is produced by the fusion of nuclei from a male and female gamete (i.e., sperm and egg); also called a fertilized egg.

References

Abbas, A.K., Lichman, A.H. and Prober, J.S. *Cellular and Molecular Immunology.* W.B. Saunders Publications, London.

Agellon, L.B., Emery, C.J., Jones, J.M., Davies, S.L., Dingle, A.D. and Chem, T.T. 1988. "Growth hormone enhancement by genetically engineered rainbow trout growth hormone." *Can. J. Fish. Aqua. Sci.* 45: 146–151.

Anderson, W.F. 1984. Prospects for human gene therapy. *Science.* 226: 401–409.

Anderson, W.F. 1992. "Human gene therapy." *Science.* 256:808–813.

Arakauva,T. *et al.* 1999. Transgenic Res. 6:403–413.

Ashok Pandey. *Advances in Biotechnology.* Educational Publishers and Distributors, New Delhi.

Brinster, R.L. 1993. "Stem cells and transgenic mice in the study of development." *Intl. J. Dev. Bio.* 37:89–99.

Brunt, J.V. 1988. "Molecular farming: transgenic animals as bioreactors." *Biotechnology.* 6: 1149–1154.

Bu'lock, J.D. 1987. "Introduction to basic biotechnology". In J.D. Bulock and B. Krishtiansen (eds.). *Basic Biotechnology.* Academic Press, London. pp.3–10.

Bu'lock, J.D. and Krishtianusen, B. 1987. *Basic Biotechnology.* Academic Press, London.

Campbell, K.H. 1999. "Nuclear transfer in farm animal species." *Cell Dev. Biol* . 10:245–252.

Campbell. K.H.S., McWhie, J., Ritchic, W.A. and Wilmut, I. 1996. "Sheep cloned by nuclear transfer from a cultured cell line." *Nature.* 380: 64–66.

Cappecchi, M.R. 1994. "Targeted gene replacement." *Scientific American.* 270: 34–41.

Chen, T.T. and Davis, P.L. 1990. "Transgenic fish." *Trends Biotechnol.* 8: 209–215.

Chen, T.T. and Powers, D.A. 1990. "Transgenic fish." *TIBTECH.* 8: 209–215.

Chen, T.T. and Powers, D.A. 1990. "Transgenic fish." *Trends in Biotechnology.* 8(8): 209–215.

Cloud, J.G. 1990. Strategies for introducing foreign DNA into the germ line of fish *J. Reprod. Fertil. Suppl.* 41: 107–118.

Dale, J.W. 1991. *Genetic Manipulations: Techniques and Applications.* Blackwell Scientific Publishers.

Danziger, D.J. and Dean, P.M. 1989. "Automated site-directed drug design." *Proc. Roy. Soc. Lond.* 26:101–162.

Darling, D.C. and Morgan, S.J. 1994. *Animal cell culture and media.* BIOS Scientific Publishers.

Di marchi, R., Brooker, G., Gali, C., Cracknell, V., Doel, T. and Mowat, N. 1986. "Protection of cattle against FMD by a synthetic peptide." *Science.* 232: 639–641.

Felgner, P.L. 1997. "Nonviral strategies for gene therapy." *Sci. Am.* 276: 102–106.

Fincham, J.R.G. and Ravter, T.R. 1991. *Genetically Engineered Organisms: Risks and Benefits.* University Press, Buckingham.

Fletcher, G.L. and Davies, P.L. 1991. "Transgenic fish for aquaculture." *Genetic Engineering, Principles and Methods.* J.Setlow (ed.). Plenum Press, New York. pp. 331–370.

Freshney, R.I. 1992. *Animal Cell Culture: A Practical Approach.* 2nd edn. Oxford Univ. Press, Oxford.

Friedmann, T. 1989. "Progress toward human gene therapy." *Science.* 244:1275–1280.

Frishchauf, A.M. 1991. "Construction of Genetic Libraries in Lambda and Cosmid Vectors." In *Essential Molecular Biology: A Practical Approach,* Vol.II. IRL Press, Oxford.

Gerald C. 1996. *Cell and Molecular Biology: Concept and Experiment.* John Wiley and Sons, Inc., U.S.A.

Glover, D.M. 1984. *Gene cloning — The mechanism of DNA manipulation.* Chapman and Hall, London. p. 222.

Golding, J.W. 1983. *Monoclonal Antibodies: Principles and Practice.* Academic Press, New York.

Golding, J.W. 1993. *Monoclonal Antibody:Principles and Practice,* 3rd edn. Academic Press, New York.

Gordon, J.W. 1989. "Transgenic animals." *Int. Rev. Cytol.* 155: 171–229.

Gupta, P.K. 1994. *Elements of Biotechnology.* Rastogi and Co., Meerut, India.

Hammer, R.E. *et al.* 1992. "Production of transgenic rabbits, sheep and pigs by microinjection," *Nature.* 315: 680–683.

Hammer, R.E., Pursel, V.G., Rexroad, C.E., Wall, R.J., Bolt, D.J., Ebert, K.M., Palmiter, R.D. and Brinster, R.L. 1985. "Production of transgenic rabbits, sheep and pigs by microinjection." *Nature.* 315: 680–683.

Hardy, K.G. 1987. *Plasmid.* IRL Press, Oxford.

Hasler, J.F. 2000. *"In vitro* production of cattle embryos; problems with pregnancies and parturition." *Human Reproduction.* 5: 47–58.

Higgis, I.J., Best, D.J. and Jones, J. 1985. *Biotechnology: Principles and Applications.* Blackwell Scientific Publishers, Oxford.

Higgis, I.J., Best, D.J. and Jones, J. 1985. *Biotechnology: Principles and Applications.* Blackwell Scientific Publishers, Oxford.

John, E.S. 1996. *Biotechnology,* 3rd edn. Cambridge University Press, Cambridge.

John, E.S. 1996. *Biotechnology,* 3rd edn. Cambridge University Press, Cambridge.

Khoo, H.W. 1995. "Transgenesis and its applications in aquaculture." *Asian Fish. Sci.* 8:1–25.

Kuby, J. *Immunology.* W.H. Freeman and Company, New York.

Miller, A.D. 1992. "Human gene therapy comes of age." *Nature.* 35: 455–460.

Milstein, C. 1980. "Monoclonal antibodies." *Sci. Amer.* 243: 66–74.

Mueller, M. and Brem, G.1994. "Transgenic strategies to increased disease resistance in livestock." *Reprod. Fertil. Dev.* 6: 605–613.

Mulligan, R.C. 1993. "The basic science of gene therapy." *Science.* 260: 926–932.

Old, R.W. and Primrose, S.B. 1990. "Principles of Gene manipulation." *An Introduction to Genetic Engineering.* Blackwell Scientific Publishers, London.

Palmiter, R.D. and Brinster, R.L. 1985. "Transgenic mice." *Cell.* 41: 343–345.

Paul, J. 1975. Cell and Tissue Culture, 5th edn. Livingstone, Edinburgh.

Pinket, C.A. 1994. *Transgenic Animal Technology: A Laboratory Handbook.* Academic Press, San Diego, California.

Powers, D.A. Chen, T.T and Dunham, R.A. 1992. "Transgenic fish." In J.A.H. Murray, (ed.) *Transgenesis.* John Wiley and Sons Ltd., Chichester. pp. 233–249.

Rose, N.R., Milgrom, F. and Carel, J.V.O. 1979. *Principles of Immunology,* 2nd edn. Macmillan Publishing Co., Inc., New York.

Stanbury, P.F. and Whitaker, A., 1984, *Principles of Fermentation Technology,* Pergamon Press, London.

Straughan, R. 1989. "The Genetic Manipulation of Plants, Animals and Microbes. The Social and Ethical Issues for consumers: A Discussion Paper." National Consumer Council, London.

Thompson, P.B. 1993. "Genetically modified animals: Ethical issues." *J. Anim.Sci.* 71 (suppl 3): 51–56.

Tizard, T.R. *Immunology: An Introduction.* Saunder's College Publishers, Philadelphia.

Trounson, A. and Conti, A. 1982. Research in human *in vitro* fertilization and embryo transfer. *Brit. Med. J.* 284:244–248.

Willadsen, S.M. 1989. "Cloning of sheep and cow embryos." *Genome.* 31: 956–962.

Windle, J.J., Albert, D.M., O'Brien, J.M., Marcus, D.M., Disteche, C.M., Bernards, R. and Mellon, P.L. 1990. "Retinoblastoma in transgenic mice." *Nature.* 343: 665–669.

Wiseman, A. 1983. *Principles of Biotechnology.* Chapman & Hall, New York.

Zanwar, S.G. 1989. "Embryo transfer in cattle." *Indian Dairyman.* 41. April issue. pp. 194–196.

Index